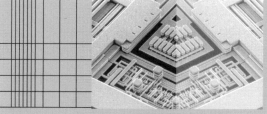

SEM image of copper wiring
in IBM's CMOS 7S technology
[Edelstein et al., 1997].
(Tom Way, IBM)

Atomic Force
Microscopy images
[Binnig et al., 1986]
of CD pits
(Brian Hubert, MIT)

Magnetic tracks
at 5, 10 and 20 kilobits/mm
[Gruetter et al., 1995]
(IBM)

Terabit/in^2
information storage
in titanium oxide
[Cooper et al., 1999]
(Scott Manalis, MIT).

Electron wave function density in a quantum
corral of iron atoms on a copper substrate
[Crommie et al., 1993]. (Don Eigler, IBM)

QAM CDMA vector
modulation analysis.
(Agilent) [Wright, 1998]

Ramsey cavity from a GPS
satellite cesium beam atomic
clock [Ramsey, 1980]. (Donald
Emmons and Michael Garvey,
Datum; Felice Frankel, MIT)

Microencapsulated
eletrophoretic electronic
ink display [Comiskey
et al., 1998]. (E Ink)

Synthetic aperture radar image
of Cambridge, MA, from the
Space Shuttle SIR-C/X-SAR
mission [Stofan et al., 1995].
(NASA/JPL).

The Physics of Information Technology

The familiar devices that we use to collect, transform, transmit, and interact with electronic information operate surprisingly close to very many fundamental physical limits. A handheld GPS receiver requires special and general relativistic corrections to the time reported by the system's atomic clocks; the typical distance between air molecules in a hard disk drive is larger than the height that the head flies above the platter; the linewidth in a VLSI circuit is approaching the size of a single atom; the performance of satellite receivers is limited by the echo of the Big Bang.

Given the economic and intellectual importance of these scaling limits, surprisingly few people are equipped to address them. Understanding how such devices work, and how they can (and cannot) be improved, requires deep insight into the character of physical law as well as engineering practice. *The Physics of Information Technology* provides this needed connection by introducing underlying governing equations and then deriving operational device principles. This self-contained volume will help both physical scientists and computer scientists see beyond the conventional division between hardware and software to understand the implications of physical theory for information manipulation. It is at this interface that many of the most dramatic advances in both domains are occurring.

The book starts with an introduction to units, forces, and the probabilistic foundations of noise and signalling, then progresses through the electromagnetics of wired and wireless communications, and the quantum mechanics of electronic, optical, and magnetic materials, to discussions of mechanisms for computation, storage, sensing, and display. Attention is drawn throughout to the remarkable opportunities associated with more closely integrating the physical and logical descriptions of classical and quantum information.

This textbook will be useful for advanced undergraduates and graduate students in physics, computer science, and electrical engineering, but its unique scope will also make it a handy reference and guide for working scientists, engineers, and technical leaders.

Cambridge Series on Information and the Natural Sciences

Academic and industrial research has separated the description of the information in a system from its physical properties, but many of today's most compelling opportunities and obstacles lie right at this interface. From controlling the coherent dynamics of atomic nuclei to compute beyond the scaling limits of integrated circuits, to programming the expression of genetic sequences to fabricate nanostructures, evolutionary technological progress has brought us to a revolutionary integration of the most profound physical theories with their practical application in systems that detect, transform, and deliver information. This series bridges the historical gulf between the fundamental enabling research and the domain-specific description of its engineering application through studies by leading researchers of both the theory and practice in these emerging fields.

The Physics of Information Technology

Neil Gershenfeld

CAMBRIDGE
UNIVERSITY PRESS

PUBLISHED BY THE PRESS SYNDICATE OF THE UNIVERSITY OF CAMBRIDGE
The Pitt Building, Trumpington Street, Cambridge, United Kingdom

CAMBRIDGE UNIVERSITY PRESS
The Edinburgh Building, Cambridge CB2 2RU, UK
40 West 20th Street, New York, NY 10011–4211, USA
477 Williamstown Road, Port Melbourne, VIC 3207, Australia
Ruiz de Alarcón 13, 28014 Madrid, Spain
Dock House, The Waterfront, Cape Town 8001, South Africa

http://www.cambridge.org

First published 2000
Reprinted 2002

Printed in the United Kingdom at the University Press, Cambridge

Typeset in Monotype Ehrhardt $10\frac{1}{2}$/13pt, in LaTeX 2_ε [EPC]

A catalogue record for this book is available from the British Library

Library of Congress Cataloguing in Publication data

ISBN 0 521 58044 7 hardback

for JOEL, who showed me the value of creating things
and for ALAN, who showed me the value of fixing them

Contents

Preface

How does the bandwidth of a telephone line relate to the bit rate that can be sent through it? Modems keep getting faster; how quickly can they operate? These sensible questions have unexpectedly profound answers. At MIT, I've been asked them by people ranging from undergrads to faculty. A good engineer might know about coding theory and the concept of channel capacity, but not understand the origin of the noise that limits the capacity. Conversely, a physicist might use the fluctuation–dissipation theorem to explain why resistors are necessarily noisy, but know nothing of information theory. And the computer scientist sending data over the phone line might not understand either side. The most interesting aspects of this problem can easily be missed among these poles. I've found this pattern to recur over and over: people may not appreciate the useful applications of fundamental results in the devices they use, or the deep implications of their practical knowledge, and may not have a good sense of how their formal academic training can relate to their personal passions.

The familiar computing and communications devices that we use to manipulate information operate near many remarkable physical limits. A handheld GPS receiver applies both special- and general-relativistic corrections to its timing measurements of signals from atomic clocks in satellites in order to maintain the system's global 1 ns accuracy. The head in a high-capacity disk drive flies within a single mean-free-path of an air molecule above the platter, and so the aerodynamic design problem can no longer be solved by modeling the airflow with continuum partial differential equations. This kind of tremendous ingenuity has gone into finding practical solutions to what had appeared to be impossible technological problems. However, the exponential improvements that we've come to rely on, such as processor speeds doubling every few years, must stop when current scaling trends run into basic physical limits. Circuits cannot have wires smaller than atoms, signals faster than light, or charge carriers less than an electron. Given such constraints, a CMOS chip that can perform 10^9 floating-point operations per second (a gigaflop) is feasible, but 10^{12} (a teraflop) is unlikely. Understanding these kinds of systems requires equal familiarity with fundamental physics and with very practical engineering. Because this kind of background is hard to develop given the traditional split between basic and applied science, it's easy for students (and practitioners) to run into either the Scylla of uncritically accepting the received wisdom of past practice, or the Charybdis of enthusiastically pursuing impossible alternatives.

This book grew out of lecture notes for a course that I've developed at MIT's Media Lab. The goal is to review basic physical governing equations in a number of areas relevant to information technology, and then work up through device mechanisms to a quantitative examination of practical implementations. There is a companion course called *The Nature of Mathematical Modeling* that studies the possible levels of description for mathematical modeling; these can loosely be summarized as covering the rules for the logical worlds inside computers and the physical world outside them. The breadth of research at the Media Lab, coupled with the strong scientific background and personal interests of many of the students, has provided a fertile environment for me to attempt this kind of synthesis. Because there was no suitable text for such a course I started by writing lecture notes for myself to teach from, then began handing these out in response to student requests, then put them on the computer when the students couldn't read them, and eventually they grew into this book. I hope that my presumption in covering so many important areas in such a limited space is justified by the value of covering so many important areas in such a limited space.

I've taken the liberty to broadly define "physics" to mean a number of branches of physical science and the supporting mathematics, and "information technology" to cover many kinds of systems that detect, transmit, transform, store, and deliver information. The breadth of subjects covered necessitates sacrificing some detail. And although the text is self-contained, the quick review of basic governing equations such as Maxwell's equations implicitly assumes some previous familiarity. I've tried to balance these demands by including enough background to introduce the key ideas in each section, and then providing pointers to equip the interested reader to access the specialized literature as needed. Particularly useful references are given at the end of each chapter. Wherever possible, I've tried to gather together the kind of information that I routinely use but regularly need to look up.

Any mathematical techniques that might not be familiar are covered in [Gershenfeld, 1999a], and the implications and applications of these technologies are explored in [Gershenfeld, 1999b]. Of these books, this one was started first and finished last because of the challenge of including enough depth to explain any one area and enough breadth to follow rapidly advancing developments. Throughout, I've tried to abstract the timeless principles that are as applicable to string telephones as to quantum teleporters.

I am deeply grateful to the Media Lab for the opportunity to prepare and teach this material, to the great students who suffered through drafts that have ranged from incomplete to incorrect, to the valuable feedback from my many readers, to Susan Murphy-Bottari for the cheerful efficiency with which she handles the unreasonable demands of overly ambitious projects, and to the many Media Lab sponsors for their generous support of, and participation in, the research that lies behind and ahead of this book.

Cambridge, MA   Neil Gershenfeld
March, 2000

1 Introduction

Why does computation require energy?

Because there must be some irreversibility to ensure that calculations go forward (from inputs to outputs) and not in reverse, and because logical erasure necessarily implies dissipation because of the compression of phase-space.

What is a quantum computer?

One that operates on quantum bits that can be in a superposition of many different states simultaneously and that maintain a connection (called entanglement) following an interaction. These properties change the computational order of many important problems, such as reducing factoring from requiring a time that is exponential to polynomial in the number of bits.

What limits the bit density for semiconductor memory?

Lithography (constrained by the wavelength used to pattern a memory cell, and the resulting yield), electromigration (when too few atoms are used in a wire they move in response to currents), and capacitance (when too few electrons are used, the fluctuation in their number becomes significant).

What limits the bit density in a typical hard disk?

Magnetic domain wall energies, and the head height.

What limits the bit density for optical storage?

The diffraction limit for focusing light, which is proportional to the wavelength.

Why are twisted pairs twisted, and coaxial cables coaxial?

To reduce the generation of unwanted radiation and the sensitivity to interference, and to effectively guide the signal. Twisted pairs are best at low frequencies, and coaxial cables at high frequencies.

Where does electronic noise come from, and how does it limit data rates?

Thermodynamic fluctuations, defect scattering, and finite-size statistics. The capacity of a communications channel grows as the logarithm of the ratio of the energy in the signal and the noise.

What is a liquid crystal, and how does it modulate light?

It is a material that maintains long-range orientational ordering without translational ordering. Under an applied field it is able to rotate the direction of polarization of light, thereby modulating the intensity of the light if the material is enclosed between polarizers.

These questions are examples of the many ways in which familiar devices that detect, transmit, process, store, and deliver information operate surprisingly near fundamental physical limits. The goal of this book is to explore how such devices function, how they can be used, what the limits on their performance are, and how they might be improved. This will require developing familiarity with the physical governing equations for a range of types of behavior, and with the mathematical tools necessary to manipulate these equations. One important aim is to equip the reader to work out quantitative answers to questions such as these.

A note about pedagogy: reading about physics is as satisfying as reading about food or exercise. It can be useful, but there is no substitute for experience solving problems. Each chapter has problems that apply and develop the preceeding ideas, ranging from trivial calculations to open research questions. Since another goal of this book is to help develop problem-solving skills, consulting the supplied answers before a problem is attempted is entirely counter-productive because the real problems that will come after this book don't come with such handy answers.

And a note about epistemology: it is important to keep in mind the distinction between truth and models. I will be describing models for a variety of types of behavior; these are the product of both experimental observations and theoretical inferences. A good model should compactly explain what you already know and allow you to predict new things that you did not know, but it does not necessarily contain any guide to an underlying "truth." Some physicists believe that there is an ultimate "correct" answer that these models are approaching, and some violently disagree, yet all agree on the usefulness of the current set of models and on how to manipulate them. *Truth* and *Meaning* are concepts that one may choose to associate with these models, but their presence or absence does not affect the models' use. At most, they do guide what you choose to think about. This distinction is very important because, when faced with unexpected claims or results, there is a recurring danger of seeing particular models as privileged correct answers rather than being open-minded about judging evidence on its merits. The history of science is littered with conflicts arising from prior beliefs that were stronger than experimental observations.

2 Interactions, Units, and Magnitudes

Modern information technology operates over a spectacular range of scales; bits from a memory cell with a size of 10^{-7} meters might be sent 10^7 meters to a geosynchronous satellite. It is important to be comfortable with the orders of magnitudes and associated interaction mechanisms that are useful in practice. Our first task will be to review the definitions of important units, then survey the types of forces, and finally look at typical numbers in various regimes.

2.1 UNITS

Many powers of ten have been named because it is much easier to say something like "a femtosecond optical pulse" than "a 0.000 000 000 000 001 second optical pulse" when referring to typical phenomena at that scale (a cycle of light takes on the order of a femtosecond). The dizzying growth of our ability to work with large and small systems pushes the bounds of this nomenclature; data from terabyte storage systems is read out into femtofarad memory cells. It is well worth memorizing the prefixes in Table 2.1.

Physical quantities must of course be measured in a system of units; there are many alternatives that are matched to different regimes and applications. Because of their inter-relationships it is necessary only to define a small number of fundamental quantities to be able to derive all of the other ones. The choice of which fundamental definitions to use changes over time to reflect technological progress; once atomic clocks made it possible to measure time with great *precision* (small variance) and *accuracy* (small bias), it became more reliable to define the meter in terms of time and the speed of light rather than a reference bar kept at the Bureau International des Poids et Mesures (BIPM, http://www.bipm.fr) in Sevres, France. The kilogram is still defined in terms of a platinum–iridium cylinder held at BIPM instead of a fundamental physical process, a source of great frustration in the metrology community. Aside from the difficulty in duplicating it, the accumulation of contaminants on the surface increases the mass by about 1 part in 10^9 per year, requiring that it be measured only after a special cleaning procedure [Girard, 1994].

The most common set of base defined quantities in use is the *Système International d'Unités* (*SI*) [BIPM, 1998]:

length: *meter* (m)
> The meter is the length of path traveled by light in vacuum during a time interval of 1/299 792 458 of a second.

Table 2.1. *Orders of magnitude.*

Magnitude	Prefix	Symbol	Magnitude	Prefix	Symbol
10^{-24}	yocto	y	10^{24}	yotto	Y
10^{-21}	zepto	z	10^{21}	zetta	Z
10^{-18}	atto	a	10^{18}	exa	E
10^{-15}	femto	f	10^{15}	peta	P
10^{-12}	pico	p	10^{12}	tera	T
10^{-9}	nano	n	10^{9}	giga	G
10^{-6}	micro	μ	10^{6}	mega	M
10^{-3}	milli	m	10^{3}	kilo	k
10^{-2}	centi	c	10^{2}	hecto	h
10^{-1}	deci	d	10^{1}	deka	da

mass: *kilogram* (kg)

The kilogram is the unit of mass; it is equal to the mass of the international prototype of the kilogram.

time: *second* (s)

The second is the duration of 9 192 631 770 periods of the radiation corresponding to the transition between the two hyperfine levels of the ground state of the cesium-133 atom.

current: *ampere* (A)

The ampere is that constant current which, if maintained in two straight parallel conductors of infinite length, of negligible circular cross-section, and placed 1 meter apart in vacuum, would produce between these conductors a force equal to 2×10^{-7} newtons per meter of length. (See Problem 5.4.)

temperature: *kelvin* (K)

The kelvin, the unit of thermodynamic temperature, is the fraction of $1/273.16$ of the thermodynamic temperature of the triple point of water. (Temperatures in degrees Celsius are equal to temperatures in kelvin + 273.15. The triple point is the temperature and pressure at which the liquid, solid, and gas phases of water co-exist. It is fixed at 0.01 °C, and provides a more reliable reference than the original centigrade definition of 0 °C as the freezing point of water at atmospheric pressure.)

quantity: *mole* (mol)

The mole is the amount of substance of a system which contains as many elementary entities as there are atoms in 0.012 kg of carbon 12 (i.e., Avogadro's constant $6.022 \ldots \times 10^{23}$).

intensity: *candela* (cd)

The candela is the luminous intensity, in a given direction, of a source that emits monochromatic radiation of frequency 540×10^{12} hertz and that has a radiant intensity in the direction of $1/683$ watts per steradian. (The frequency corresponds to the wavelength of 555 nm where the eye is most sensitive, the factor of 683 comes from matching an earlier definition based on the emission from solidifying platinum, and a steradian is the solid angle subtended by a unit area on the surface of a sphere with unit radius; see Chapter 11 for more on luminosity.)

From these seven fundamental units many other ones are derived in terms of them, including:

capacitance: *farad* F ($m^{-2} \cdot kg^{-1} \cdot s^4 \cdot A^2$)
> The farad is the capacitance of a capacitor between the plates of which there appears a difference of potential of 1 volt when it is charged by a quantity of electricity equal to 1 coulomb.

charge: *coulomb* C ($A \cdot s$)
> The coulomb is the quantity of electricity transported in 1 second by a current of 1 ampere.

energy: *joule* J ($m^2 \cdot kg \cdot s^{-2}$)
> The joule is the work done when the point of application of a force of 1 newton is displaced a distance of 1 meter in the direction of the force. (Remember that energy equals force times distance.)

force: *newton* N ($m \cdot kg \cdot s^{-2}$)
> The newton is that force which, when applied to a body having a mass of 1 kilogram, gives it an acceleration of 1 meter per second squared. (Remember that force equals mass time acceleration.)

illuminance: *lux* lx ($cd \cdot m^{-2}$)
> The lux is equal to an illuminance of 1 lumen per square meter.

inductance: *henry* H ($m^2 \cdot kg \cdot s^{-2} \cdot A^{-2}$)
> The henry is the inductance of a closed circuit in which an electromotive force of 1 volt is produced when the electric current in the circuit varies uniformly at a rate of 1 ampere per second.

luminous flux: *lumen* lm (cd)
> The lumen is the luminous flux emitted within a unit solid angle of 1 steradian by a point source having a uniform intensity of 1 candela.

magnetic flux: *weber* Wb ($m^2 \cdot kg \cdot s^{-2} \cdot A^{-1}$)
> The weber is the magnetic flux which, linking a circuit of 1 turn, produces in it an electromotive force of 1 volt as it is reduced to zero at a uniform rate in 1 second.

magnetic flux density: *tesla* T ($kg \cdot s^{-2} \cdot A^{-1}$)
> The tesla is the magnetic flux density given by a magnetic flux of 1 weber per square meter.

power: *watt* W ($m^2 \cdot kg \cdot s^{-3}$)
> The watt is the power which gives rise to the production of energy at the rate of 1 joule per second.

pressure: *pascal* Pa ($m^{-1} \cdot kg \cdot s^{-2}$)
> The pascal is the pressure of 1 newton per square meter.

potential: *volt* V ($m^2 \cdot kg \cdot s^{-3} \cdot A^{-1}$)
> The volt is the difference of electric potential between two points of a conductor carrying a constant current of 1 ampere, when the power dissipated between these points is equal to 1 watt.

Table 2.2. *Selected conversion factors.*

1 dyne $(\text{gm} \cdot \text{cm} \cdot \text{s}^{-2})$	=	1×10^{-5} N
1 erg $(\text{gm} \cdot \text{cm}^2 \cdot \text{s}^{-2})$	=	1×10^{-7} J
1 horsepower (hp)	=	745.7 W
1 atmosphere (atm)	=	101325 Pa
1 ton (short)	=	2000 pounds
	=	907.18474 kg
1 electron volt (eV)	=	$1.602176462 \times 10^{-19}$ J
1 amu	=	$1.66053873 \times 10^{-27}$ kg
1 ångstrom (Å)	=	1×10^{-10} m
1 fermi (fm)	=	1×10^{-15} m
1 parsec (pc)	=	3.085678×10^{16} m
1 mile (mi)	=	1609.344 m
1 foot (ft)	=	0.3048 m
1 inch (in)	=	0.0254 m
1 liter (L)	=	0.001 m^3
1 pound (lb)	=	0.45359237 kg
1 pound-force (lbf)	=	4.44822 N

resistance: *ohm* Ω $(\text{m}^2 \cdot \text{kg} \cdot \text{s}^{-3} \cdot \text{A}^{-2})$

The ohm is the electric resistance between two points of a conductor when a constant difference of potential of 1 volt, applied between these two points, produces in this conductor a current of 1 ampere. (These derivative definitions of the volt and ohm have more recently been replaced by fundamental ones fixing them in terms of the voltage across a *Josephson junction* and the resistance steps in the *quantum Hall effect* [Zimmerman, 1998], and capacitance may be defined by counting electrons on a *Single-Electron Tunneling* (*SET*) device [Keller *et al.*, 1999].)

It is important to pay attention to the units in these definitions. Many errors in calculations can be caught by making sure that the final units are correct, and it can be possible to make a rough estimate of an answer to a problem simply by collecting relevant terms with the right units (this is the subject of *dimensional analysis*). Electromagnetic units are particularly confusing; we will consider them in more detail in Chapter 5. The SI system is also called *MKS* because it bases its units on the meter, the kilogram, and the second. For some problems it will be more convenient to use *CGS* units (based on the centimeter, the gram, and the second); MKS is more common in engineering and CGS in physics. A number of other units have been defined by characteristic features or by historical practice; some that will be useful later are given in Table 2.2.

It's often more relevant to know the value on one quantity relative to another one, rather than the value itself. The ratio of two values X_1 and X_2, measured in *decibels* (*dB*), is defined to be

$$\text{dB} = 20 \log_{10} \frac{X_1}{X_2} \quad . \tag{2.1}$$

If the *power* (energy per time) in two signals is P_1 and P_2, then

$$\text{dB} = 10 \log_{10} \frac{P_1}{P_2} \quad . \tag{2.2}$$

This is because the power is the mean square amplitude (Chapter 3), and so to be consistent with equation (2.1) a factor of 2 is brought in to account for the exponent. An increase of 10 db therefore represents a increase by a factor of 10 in the relative power of two signals, or a factor of 3.2 in their values. A change of 3 dB in power is a change by a factor of 2.

The name decibel comes from Bell Labs. Engineers there working on the telephone system found it convenient to measure the gain or loss of devices on a logarithmic scale. Because the log of a product of two numbers is equal to the sum of their logs, this let them find the overall gain of a system by adding the logs of the components, and using logarithms also made it more convenient to express large numbers. They called the base-10 logarithm the *bel* in honor of Alexander Graham Bell; multiplying by 10 to bring up one more significant digit gave them a tenth-bel, or a decibel.

Some decibel reference levels occur so commonly that they are given names; popular ones include:

- *dBV* measures an electrical signal relative to 1 volt
- *dBm* measures relative to a 1 mW signal. The power will depend on the (usually unspecified) load, which traditionally is 50 Ω for radiofrequency signals and 600 Ω for audio ones (loads will be covered in Chapter 6). In audio recording, this is also called the *Volume Unit* or *VU*.
- *dBspl*, for *Sound Pressure Level* (or just *SPL*), measures sound pressure relative to a reference of 2×10^{-5} Pa, the softest sound that the ear can perceive. The sound of a jet taking off is about 140 dBspl.

Finally, there are a number of fundamental observed constants in nature that we will use, shown in Table 2.3. In this list the digits in parentheses are the standard deviation uncertainty (see Chapter 3) in the corresponding digits, so that for example the error in the value for G is 0.010×10^{-11} (which, compared to the other constants, is an embarrasingly large uncertainty [Gundlach *et al.*, 1996]).

The speed of light no longer really belongs here, because its value has been defined exactly as part of the SI system. All the others are determined by exquisite metrology experiments. Each fundamental constant can appear in many different types of measurements, and these are done by many different groups, leading to multiple values that unfortunately don't always agree to within their careful error estimates. For this reason, the International Council of Scientific Unions in 1966 formed the *Committee on Data for Science and Technology* (*CODATA*) to do global optimizations over all these data to come up with an internally-consistent set of values. The most recent adjustment was done in 1998, and is available at `http://physics.nist.gov`.

2.2 PARTICLES AND FORCES

The world is built out of elementary particles and their interactions. There are a number of natural divisions in organization, energy, and length that occur between the structure of the nucleus of an atom and the structure of the universe; it will be useful to briefly survey this range in order to understand the relevant regimes for present and prospective information technologies.

Table 2.3. *Selected fundamental constants.*

gravitational constant (G)	=	$6.673(10) \times 10^{-11}$ m$^3 \cdot$ kg$^{-1} \cdot$ s^{-2}
speed of light (c)	=	2.99792458×10^8 m/s
elementary charge (e)	=	$1.602176462(63) \times 10^{-19}$ C
Boltzmann constant (k)	=	$1.3806503(24) \times 10^{-23}$ J/K
Planck constant (h)	=	$6.62606876(52) \times 10^{-34}$ J$\cdot$s
$\hbar = h/2\pi$	=	$1.054571596(82) \times 10^{-34}$ J$\cdot$s
Avogadro constant (N_A)	=	$6.02214199(47) \times 10^{23}$ mol^{-1}
electron mass (m_e)	=	$9.10938188(72) \times 10^{-31}$ kg
proton mass (m_p)	=	$1.67262158(13) \times 10^{-27}$ kg
gas constant (R)	=	$8.314472(15)$ J$\cdot$mol$^{-1} \cdot$K^{-1}
vacuum permittivity (ϵ_0)	=	$10^7/(4\pi c^2) = 8.854188\ldots \times 10^{-12}$ F/m
vacuum permeability (μ_0)	=	$4\pi \times 10^{-7}$ H/m

This story starts with quantum mechanics, the laws that govern things that are very small. Around 1900 Max Planck was led by his inability to explain the spectrum of light from a hot oven to propose that the energy of light is quantized in units of $E = h\nu = hc/\lambda$, where ν is the frequency and λ is the wavelength; $h = 6.626\ldots \times 10^{-34}$ J$\cdot$s is now called *Planck's constant*. From there, in 1905 Einstein introduced the notion of massless photons as the discrete constituents of light, and in 1924 de Broglie suggested that the wavelength relationship applies to massive as well as massless particles by $\lambda = h/p$; λ is the *de Broglie wavelength*, and is a consequence of the *wave–particle duality*: all quantum particles behave as both waves and particles. An electron, or a photon, can diffract like a wave from a periodic grating, but a detector will register the arrival of individual particles. Quantum effects usually become significant when the de Broglie wavelength becomes comparable to the size of an object.

Quantum mechanical particles can be either *fermions* (such as an electron) or *bosons* (such as a photon). Fermions and bosons are as unlike as anything can be in our universe. We will later see that bosons are particles that exist in states that are symmetric under the interchange of particles, they have an integer spin quantum number, and multiple bosons can be in the same quantum state. Fermions have half-integer spin, exist in states that are antisymmetric under particle interchange, and only one fermion can be in a particular quantum state. Spin is an abstract property of a quantum particle, but it behaves just like an angular momentum (as if the particle is spinning).

Particles can interact through four possible forces: *gravitational, electromagnetic, weak,* and *strong*. The first two are familiar because they have infinite range; the latter two operate on short ranges and are associated with nuclear and subnuclear processes (the characteristic lengths are approximately 10^{-15} m for the strong force and 10^{-18} m for the weak force). The electromagnetic force is so significant because of its strength: if a quantum atom was held together by gravitational forces alone (like a miniature solar system) its size would be on the order of 10^{23} m instead of 10^{-10} m. The macroscopic forces that we feel, such as the hardness of a wall, are transmitted to us by the electromagnetic force through the electrons in our atoms interacting with electrons in the adjoining atoms in the surface, but can be much more simply described in terms of fictitious effective forces ("the wall is hard").

All forces were originally thought to be transmitted by an intervening medium, the long-sought *ether* for electromagnetic forces. We now understand that forces operate by the exchange of spin-1 gauge bosons – the photon for the electromagnetic interaction (electric and magnetic fields), the $W^{\pm}$ and Z^0 bosons for the weak interaction, and eight gluons for the strong interaction (there is not yet a successful quantum theory of gravity). *Quantum ElectroDynamics (QED)* is the theory of the quantum electromagnetic interaction, and *Quantum ChromoDynamics (QCD)* the theory of the strong interaction. The weak and electromagnetic interactions are united in the *electroweak theory*, which, along with QCD is the basis for the *Standard Model*, the current summary of our understanding of particle physics. This amalgam of experimental observations and theoretical inferences successfully predicts most observed behavior extremely accurately, with two important catches: the theory has 20 or so adjustable parameters that must be determined from experiments, and it cannot explain gravitation. *String theory* [Giveon & Kutasov, 1999], a reformulation of particle theory that starts from loops rather than points as the primitive mathematical entity, appears to address both these limitations, and so is of intense interest in the theoretical physics community even though it is still far from being able to make experimentally testable predictions.

The most fundamental massive particles that we are aware of are the *quarks* and *leptons*. There's no reason to assume that there's nothing below them (i.e., turtles all the way down); there's just not a compelling reason right now to believe that there is. Quarks and leptons appear in the scattering experiments used to study particle physics to be point-particles without internal structure, and are spin-1/2 fermions. The leptons interact through the electromagnetic and weak interactions, and come in pairs: the *electron* and the electron *neutrino* (e^-, ν_e), the *muon* and its neutrino (μ^-, ν_μ), and the *tau lepton* and its neutrino (τ^-, ν_τ). Muons and tau leptons are unstable, and therefore are seen only in accelerators, particle decay products, and cosmic rays. Because neutrinos interact only through the weak force, they can pass unhindered though a light-year of lead. But they are profoundly important for the energy balance of the universe, and if they have mass [Fukuda, 1998] it will have enormous implications for the fate of the universe. Quarks interact through the strong as well as weak and electromagnetic interactions, and they come in pairs: *up* and *down*, *charm* and *strange*, and *top* and *bottom*. These fanciful names are just labels for the underlying abstract states. The first member of each pair has charge $+2/3$, the second member has charge $-1/3$, and each charge flavor comes in three colors (once again, flavor and color are just descriptive names for quantum numbers).

Quarks combine to form *hadrons*; the best-known of which are the two *nucleons*. A proton comprises two ups and a down, and the neutron an up and two downs. The nucleons, along with their excited states, are called *baryons* and are fermions. Transitions between baryon states can absorb or emit spin-1 boson hadrons, called *mesons*. The size of hadrons is on the order of 10^{-15} m, and the energy difference between excited states is on the order of 10^9 electron volts (1 GeV).

The nucleus of an atom is made up of some number of protons and neutrons, bound into ground and excited states by the strong interaction. Typical nuclear sizes are on the order of 10^{-14} m, and energies for nuclear excitations are on the order of 10^6 eV (1 MeV). Atoms consist of a nucleus and electrons bound by the electromagnetic interaction; typical sizes are on the order of 1 ångstrom (Å, 10^{-10} m) and the energy difference between states is on the order of 1 eV. Notice the large difference in size between the atom and the nucleus: atoms are mostly empty space. Atoms can exist in different *isotopes* that have the same number of protons but differing numbers of neutrons, and *ions* are atoms that have had electrons removed or added.

Atoms can bond to form molecules; bond energies are on the order of 1 eV and bond lengths are on the order of 1 Å. Molecular sizes range from simple diatomic molecules up to enormous biological molecules with 10^6–10^9 atoms. Large molecules fold into complex shapes; this is called their *tertiary structure*. These shapes are responsible for the geometrical constraints in molecular interactions that govern many biochemical pathways. Predicting tertiary structure is one of the most difficult challenges in chemistry.

Macroscopic materials are described by the arrangement of their constituent atoms, and include crystals (which have complete long-range ordering), liquids and glasses (which have short-range order but little long-range order), and gases (which have little short-range order). There are also very interesting intermediate cases, such as quasiperiodic alloys called *quasicrystals* that have deterministic translational order without translational periodicity [DiVincenzo & Steinhardt, 1991], and *liquid crystals* that maintain orientational but not translational ordering [Chandrasekhar, 1992]. Most solids do not contain just a single phase; there are usually defects and boundaries between different kinds of domains.

The atomic weight of an element is equal to the number of grams equal to one mole ($N_A \approx 10^{23}$) of atoms. It is approximately equal to the number of protons and neutrons in an atom, but differs because of the mix of naturally occuring isotopes. 22.4 liters of an ideal gas at a pressure of 1 atmosphere and at room temperature will also contain a mole of atoms.

The structure of a material at more fundamental levels will be invisible and can be ignored unless energies are larger than its characteristic excitations. Although we will rarely need to descend below atomic structure, there are a number of important applications of nuclear transitions, such as nuclear power and the use of nuclear probes to characterize materials.

2.3 ORDERS OF MAGNITUDE

Understanding what is possible and what is preposterous requires being familiar with the range of meaningful numbers for each unit; the following lists include some significant ones:

Time

10^{-43} s: the Planck time (Problem 2.7)

10^{-15} s: this is the period of visible light, and a typical time scale for chemical reactions

10^{-9} s: atomic excitations and molecular rotations typically have lifetimes on the order of nanoseconds, and this is the clock cycle for the fastest computers

10^{-3} s: the shortest time difference that is consciously perceptible by people

10^{17} s: the approximate age of the observable universe

Power and Energy

1 eV: atomic excitations
10^6 eV: nuclear excitations
10^9 eV: subnuclear excitations
10^{28} eV: the Planck energy
10 W: laptop computer
100 W: workstation; human
10^4 W: car
10^5 W: supercomputer; heating and lighting a building
10^{26} W: luminosity of the sun
10^{-12} W/m^2: softest sound that can be heard
1 W/m^2: loudest sound that can be tolerated
10^7 J/kg: energy density of food
10^9 J: energy in a ton of TNT
10^{20} J: energy consumption in the US per year

Temperature

10^{-7} K: lowest temperatures obtained in solids in the laboratory
2.75 K: microwave background radiation from the Big Bang
77 K: temperature of liquid nitrogen
6000 K: temperature of the surface of the sun

Mass

10^{-27} kg: proton mass
10^{-12} kg: typical cell
10^{-5} kg: small insect
10^{16} kg: Earth's biomass
5.98×10^{24} kg: the mass of the Earth
10^{42} kg: approximate mass of the Milky Way

Length

10^{-35} m: the Planck distance
10^{-15} m: size of a proton
10^{-10} m: size of an atom
4×10^5 m: height of a Low Earth Orbit satellite above the surface
6.378×10^6 m: radius of the Earth
4×10^7 m: height of a geosynchronous satellite above the equator
10^{11} m: distance from the Earth to the Sun
10^{20} m: Milky Way radius
10^{26} m: size of the observable universe

Electromagnetic spectrum

< 0.1 Å: gamma rays
0.1–100 Å: X-rays

100–4000 Å: UV (atomic ionization energy)

4000–7000 Å: visible (this coincides with a transmission band through the atmosphere, and corresponds to 10^{14}–10^{15} Hz)

0.7–100 μm: IR (thermal radiation)

0.01–10 cm: microwave (GHz)

0.1–10^3 m: radio (MHz–kHz)

Magnetic and Electric Fields

10^{-12} tesla: magnetic field needed for radio reception

10^{-6} tesla: magnetic field generated by a cordless phone

3×10^{-5} tesla: magnetic field at the Earth's surface

20 tesla: large superconducting/normal hybrid magnet

10^4 A: lightning bolt current

10^8 V: potential across a lightning bolt

3×10^6 V/m: breakdown voltage in air

Number

10^5: number of DNA bases in a bacteriophage

4×10^6: bytes in the Bible

10^9: number of DNA bases in a mammal

10^{13}: number of synapses in the human cortex

10^{14}: bytes passing through the Internet backbone during December, 1994

10^{70}: number of atoms in the universe

2.4 SELECTED REFERENCES

[Anderson, 1989] Anderson, H. (ed). (1989). *A Physicist's Desk Reference*. New York: Institute of Physics.

This is a handy summary of units, conversion factors, and governing equations.

[Lerner & Trigg, 1991] Lerner, Rita G., & Trigg, George L. (1991). *Encyclopedia of Physics*. 2nd edn. New York: VCH.

A Who's Who of interesting physical phenomena.

[Morrison & Morrison, 1982] Morrison, Philip, & Morrison, Phylis. (1982). *Powers Of Ten: A Book About the Relative Size of Things*. Redding: Scientific American Library.

The Morrisons provide a marvelous tour through the characteristic phenomena at many length scales.

[Nachtmann, 1990] Nachtmann, Otto. (1990). *Elementary Particle Physics: Concepts and Phenomena*. New York: Springer-Verlag. Translated by A. Lahee and W. Wetzel.

A nice introduction to particles and forces.

2.5 PROBLEMS

(2.1) (*a*) How many atoms are there in a yoctomole?

(*b*) How many seconds are there in a nanocentury? Is the value near that of any important constants?

(2.2) A large data storage system holds on the order of a terabyte. How tall would a 1 terabyte stack of floppy disks be? How does that compare to the height of a tall building?

(2.3) If all the atoms in our universe were used to write an enormous binary number, using one atom per bit, what would that number be (converted to base 10)?

(2.4) Compare the gravitational acceleration due to the mass of the Earth at its surface to that produced by a 1 kg mass at a distance of 1 m. Express their ratio in decibels.

(2.5) (*a*) Approximately estimate the chemical energy in a ton of TNT. You can assume that nitrogen is the primary component; think about what kind of energy is released in a chemical reaction, where it is stored, and how much there is.

(*b*) Estimate how much uranium would be needed to make a nuclear explosion equal to the energy in a chemical explosion in 10 000 tons of TNT (once again, think about where the energy is stored).

(*c*) Compare this to the *rest mass energy* $E = mc^2$ of that amount of material (Chapter 14), which gives the maximum amount of energy that could be liberated from it.

(2.6) (*a*) What is the approximate de Broglie wavelength of a thrown baseball?

(*b*) Of a molecule of nitrogen gas at room temperature and pressure? (This requires either the result of Section 3.4.2, or dimensional analysis.)

(*c*) What is the typical distance between the molecules in this gas?

(*d*) If the volume of the gas is kept constant as it is cooled, at what temperature does the wavelength become comparable to the distance between the molecules?

(2.7) (*a*) The potential energy of a mass m a distance r from a mass M is $-GMm/r$. What is the *escape velocity* required to climb out of that potential?

(*b*) Since nothing can travel faster than the speed of light (Chapter 14), what is the radius within which nothing can escape from the mass?

(*c*) If the rest energy of a mass M is converted into a photon, what is its wavelength?

(*d*) For what mass does its equivalent wavelength equal the size within which light cannot escape?

(*e*) What is the corresponding size?

(*f*) What is the energy?

(*g*) What is the period?

(2.8) Consider a pyramid of height H and a square base of side length L. A sphere is placed so that its center is at the center of the square at the base of the pyramid, and so that it is tangent to all of the edges of the pyramid (intersecting each edge at just one point).

(*a*) How high is the pyramid in terms of L?

(*b*) What is the volume of the space common to the sphere and the pyramid?

(This question comes from an entrance examination for humanities students at Tokyo University [*Economist*, 1993].)

3 Noise in Physical Systems

Understanding noise is central to understanding the design and performance of almost any device. Noise sets essential limits on how small a bit can reliably be stored and on how fast it can be sent; effective designs must recognize these limits in order to approach them. Our first step will be an introduction to random variables and some of their important probability distributions, then we will turn to noise generation mechanisms, and close with some more general thermodynamic insights into noise. Although the study of noise can be surprisingly interesting in its own right, this chapter primarily provides concepts that we will use throughout the book.

3.1 RANDOM VARIABLES

3.1.1 Expectation Values

Consider a fluctuating quantity $x(t)$, such as the output from a noisy amplifier. If x is a *random variable*, it is drawn from a probability distribution $p(x)$. This means that it is not possible to predict the value of x at any instant, but knowledge of the distribution does let precise statements be made about the average value of quantities that depend on x. The *expected value* of a function $f(x)$ can be defined by an integral either over time or over the distribution:

$$\langle f(x) \rangle \equiv \lim_{T \to \infty} \frac{1}{T} \int_{-T/2}^{T/2} f(x(t)) \, dt$$
$$= \int f(x) p(x) \, dx \tag{3.1}$$

(or a sum if the distribution is discrete). Taking $f(x) = 1$ shows that a probability distribution must be *normalized*:

$$\int_{-\infty}^{\infty} 1 \cdot p(x) \, dx = 1 \quad . \tag{3.2}$$

If $p(x)$ exists and is independent of time then the distribution is said to be *stationary*.

The *moments* of a distribution are the expectation values of powers of the observable $\langle x^n \rangle$. The first moment is the *average*

$$\langle x \rangle = \int x \, p(x) \, dx \quad , \tag{3.3}$$

and the mean square deviation from this is the *variance*:

$$\sigma^2 = \langle (x - \langle x \rangle)^2 \rangle$$
$$= \langle x^2 - 2x\langle x \rangle + \langle x \rangle^2 \rangle$$
$$= \langle x^2 \rangle - \langle x \rangle^2 \quad . \tag{3.4}$$

The square root of the variance is the *standard deviation* σ.

   The probability distribution contains no information about the temporal properties of the observed quantity; a useful probe of this is the *autocovariance function*:

$$\langle x(t)x(t - \tau) \rangle = \lim_{T \to \infty} \frac{1}{T} \int_{-T/2}^{T/2} x(t)x(t - \tau) \, dt \quad . \tag{3.5}$$

If the autocovariance is normalized by the variance then it is called the *autocorrelation function*, ranging from 1 for perfect correlation to 0 for no correlation to -1 for perfect anticorrelation. The rate at which it decays as a function of τ provides one way to determine how quickly a function is varying. In the next chapter we will introduce the *mutual information*, a much more general way to measure the relationships among variables.

3.1.2 Spectral Theorems

The *Fourier transform* of a fluctuating quantity is

$$X(f) = \lim_{T \to \infty} \int_{-T/2}^{T/2} e^{i2\pi ft} x(t) \, dt \tag{3.6}$$

and the inverse transform is

$$x(t) = \lim_{F \to \infty} \int_{-F/2}^{F/2} e^{-i2\pi ft} X(f) \, df \quad . \tag{3.7}$$

The Fourier transform is also a random variable. The *Power Spectral Density* (*PSD*) is defined in terms of the Fourier transform by taking the average value of the square magnitude of the transform

$$S(f) = \langle |X(f)|^2 \rangle = \langle X(f)X^*(f) \rangle$$
$$= \lim_{T \to \infty} \frac{1}{T} \int_{-T/2}^{T/2} e^{i2\pi ft} x(t) \, dt \int_{-T/2}^{T/2} e^{-i2\pi ft'} x(t') \, dt' \quad . \tag{3.8}$$

X^* is the complex conjugate of X, replacing i with $-i$, and we'll assume that x is real. The power spectrum might not have a well-defined limit for a non-stationary process; *wavelets* and *Wigner functions* are examples of *time–frequency transforms* that retain both temporal and spectral information for non-stationary signals [Gershenfeld, 1999a].

   The Fourier transform is defined for negative as well as positive frequencies. If the sign of the frequency is changed, the imaginary or sine component of the complex exponential changes sign while the real or cosine part does not. For a real-valued signal this means that the transform for negative frequencies is equal to the complex conjugate of the transform for positive frequencies. Since the power spectrum is used to measure energy as a function of frequency, it is usually reported as the *single-sided* power spectral density found by adding the square magnitudes of the negative- and positive-frequency components. For a real signal these are identical, and so the single-sided density differs from the *two-sided* density by an (occasionally omitted) factor of 2.

The Fourier transform can also be defined with the 2π in front,

$$X(\omega) = \lim_{T \to \infty} \int_{-T/2}^{T/2} e^{i\omega t} x(t) \, dt$$

$$x(t) = \lim_{\Omega \to \infty} \frac{1}{2\pi} \int_{-\Omega/2}^{\Omega/2} e^{-i\omega t} X(\omega) \, d\omega \quad . \tag{3.9}$$

ν measures the frequency in cycles per second; ω measures the frequency in *radians* per second (2π radians = 1 cycle). Defining the transform in terms of ν eliminates the errors that arise from forgetting to include the 2π in the inverse transform or in converting from radians to cycles per second, but it is less conventional in the literature. We will use whichever is more convenient for a problem.

The power spectrum is simply related to the autocorrelation function by the *Wiener–Khinchin Theorem*, found by taking the inverse transform of the power spectrum:

$$\int_{-\infty}^{\infty} S(f) e^{-i2\pi f \tau} \, df$$

$$= \int_{-\infty}^{\infty} \langle X(f) X^*(f) \rangle e^{-i2\pi f \tau} \, df$$

$$= \lim_{T \to \infty} \frac{1}{T} \int_{-\infty}^{\infty} \int_{-T/2}^{T/2} e^{i2\pi f t} x(t) \, dt \int_{-T/2}^{T/2} e^{-i2\pi f t'} x(t') \, dt' \, e^{-i2\pi f \tau} \, df$$

$$= \lim_{T \to \infty} \frac{1}{T} \int_{-\infty}^{\infty} \int_{-T/2}^{T/2} \int_{-T/2}^{T/2} e^{i2\pi f (t - t' - \tau)} \, df \, x(t) x(t') \, dt \, dt'$$

$$= \lim_{T \to \infty} \frac{1}{T} \int_{-T/2}^{T/2} \int_{-T/2}^{T/2} \delta(t - t' - \tau) x(t) x(t') \, dt \, dt'$$

$$= \lim_{T \to \infty} \frac{1}{T} \int_{-T/2}^{T/2} x(t) x(t - \tau) \, dt$$

$$= \langle x(t) x(t - \tau) \rangle \quad , \tag{3.10}$$

using the Fourier transform of a *delta function*

$$\int_{-\infty}^{\infty} e^{i2\pi xy} \, dx = \delta(y)$$

$$\int_{-\infty}^{\infty} f(x) \delta(x - x_0) \, dx = f(x_0) \tag{3.11}$$

(one way to derive these relations is by taking the delta function to be the limit of a Gaussian with unit norm as its variance goes to zero).

The Wiener–Khinchin Theorem shows that the Fourier transform of the autocovariance function gives the power spectrum; knowledge of one is equivalent to the other. An important example of this is white noise: a memoryless process with a delta function autocorrelation will have a flat power spectrum, regardless of the probability distribution

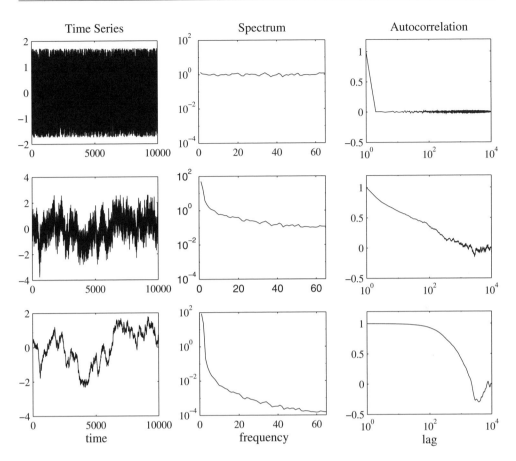

Figure 3.1. Illustration of the Wiener–Khinchin Theorem: as the power spectrum decays more quickly, the autocorrelation function decays more slowly.

for the signal. As the autocorrelation function decays more slowly, the power spectrum will decay more quickly (Figure 3.1).

Taking $\tau = 0$ in the Wiener–Khinchin Theorem yields *Parseval's Theorem*:

$$\langle x(t)x(t-\tau)\rangle = \int_{-\infty}^{\infty} S(f)e^{-i2\pi f\tau}\,df = \int_{-\infty}^{\infty} \langle |X(f)|^2\rangle e^{-i2\pi f\tau}\,df$$

$$\Rightarrow \langle |x|^2(t)\rangle = \int_{-\infty}^{\infty} \langle |X(f)|^2\rangle\,df \quad . \tag{3.12}$$

The average value of the square of the signal (which is equal to the variance if the signal has zero mean) is equal to the integral of the power spectral density. This means that true white noise has an infinite variance in the time domain, although the finite bandwidth of any real system will roll off the frequency response, and hence determine the variance of the measured signal. If the division by T is left off in the limiting process defining the averages on both sides of Parseval's Theorem, then it reads that the total energy in the signal equals the total energy in the spectrum (the integral of the square of the magnitude).

3.2 PROBABILITY DISTRIBUTIONS

So far we have taken the probability distribution $p(x)$ to be arbitrary. In practice, three probability distributions recur so frequently that they receive most attention: *binomial*, *Poisson* and *Gaussian*. Their popularity is due in equal parts to the common conditions that give rise to them and to the convenience of working with them. The latter reason sometimes outweighs the former, leading these distributions to be used far from where they apply. For example, many physical system have *long-tailed distributions* that fall off much more slowly than these ones do [Crisanti *et al.*, 1993; Boguna & Corral, 1997].

3.2.1 Binomial

Consider many trials of an event that can have one outcome with probability p (such as flipping a coin and seeing a head), and an alternative with probability $1 - p$ (such as seeing a tail). In n trials, the probability $p_n(x)$ to see x heads and $n - x$ tails, independent of the particular order in which they were seen, is found by adding up the probability for each outcome times the number of equivalent arrangements:

$$p_n(x) = \binom{n}{x} p^x (1 - p)^{n-x} \quad , \tag{3.13}$$

where

$$\binom{n}{x} = \frac{n!}{(n - x)!\, x!} \tag{3.14}$$

(read "n choose x"). This is the *binomial distribution*. The second line follows by dividing the total number of distinct arrangements of n objects ($n!$) by the number of equivalent distinct arrangements of heads $x!$ and tails $(n-x)!$. The easiest way to convince yourself that this is correct is to exhaustively count the possibilites for a small case.

3.2.2 Poisson

Now consider events such as radioactive decays that occur randomly in time. Divide time into n very small intervals so that there are either no decays or one decay in any one interval, and let p be the probability of seeing a decay in an interval. If the total number of events that occur in a given time is recorded, and this is repeated many times to form an ensemble of measurements, then the distribution of the total number of events recorded will be given by the binomial distribution. If the number of intervals n is large, and the probability p is small, the binomial distribution can be approximated by using $\ln(1 + x) \approx x$ for small x and *Stirling's approximation* for large n:

$$n! \approx \sqrt{2\pi}\, n^{n+\frac{1}{2}} e^{-n}$$
$$\ln n! \approx n \ln n - n \quad , \tag{3.15}$$

to find the *Poisson distribution* (Problem 3.1):

$$p(x) = \frac{e^{-N} N^x}{x!} \quad , \tag{3.16}$$

where $N = np$ is the average number of events. This distribution is very common for measurements that require counting independent occurrences of an event. Naturally, it is normalized:

$$\sum_{x=0}^{\infty} \frac{e^{-N}N^x}{x!} = e^{-N} \underbrace{\sum_{x=0}^{\infty} \frac{N^x}{x!}}_{e^N} = 1 \quad . \tag{3.17}$$

If x is drawn from a Poisson distribution then its *factorial moments*, defined by the following equation, have a simple form (Problem 3.1):

$$\langle x(x-1)\cdots(x-m+1)\rangle = N^m \quad . \tag{3.18}$$

This relationship is one of the benefits of using a Poisson approximation. With it, it is easy to show that $\langle x \rangle = N$ and $\sigma = \sqrt{N}$, which in turn implies that the relative standard deviation in a Poisson random variable is

$$\frac{\sigma}{\langle x \rangle} = \frac{1}{\sqrt{N}} \quad . \tag{3.19}$$

The fractional error in an estimate of the average value will decrease as the square root of the number of samples. This important result provides a good way to make a quick estimate of the expected error in a counting measurement.

3.2.3 Gaussian

The *Gaussian* or *normal* distribution

$$p(x) = \frac{1}{\sqrt{2\pi\sigma^2}} \, e^{-(x-\mu)^2/2\sigma^2} \tag{3.20}$$

has mean μ, a standard deviation σ, and the integral from $-\infty$ to ∞ is 1. The partial integral of a Gaussian is an *error function*:

$$\frac{1}{\sqrt{2\pi\sigma^2}} \int_0^y e^{-x^2/2\sigma^2} \, dx = \frac{1}{2}\text{erf}\left(\frac{y}{\sqrt{2\sigma^2}}\right) \quad . \tag{3.21}$$

Since the Gaussian is normalized, $\text{erf}(\infty) = 1$.

The Gaussian distribution is common for many reasons. One way to derive it is from an expansion around the peak of the binomial distribution for large n [Feller, 1968]:

$$p(x) = \frac{n!}{(n-x)! \, x!} \, p^x(1-p)^{n-x}$$
$$\ln p(x) = \ln n! - \ln(n-x)! - \ln x! + x \ln p + (n-x)\ln(1-p) \quad . \tag{3.22}$$

Finding the peak by treating these large integers as continuous variables and setting the first derivative to zero shows that this has a maximum at $x \approx np$, and then expanding in a power series around the maximum gives the coefficient of the quadratic term to be $-1/(2np(1-p))$. Because the lowest non-zero term will dominate the higher orders for large n, this is therefore approximately a Gaussian with mean np and variance $np(1-p)$. In the next section we will also see that the Gaussian distribution emerges via the *Central Limit Theorem* as the limiting form for an ensemble of variables with almost

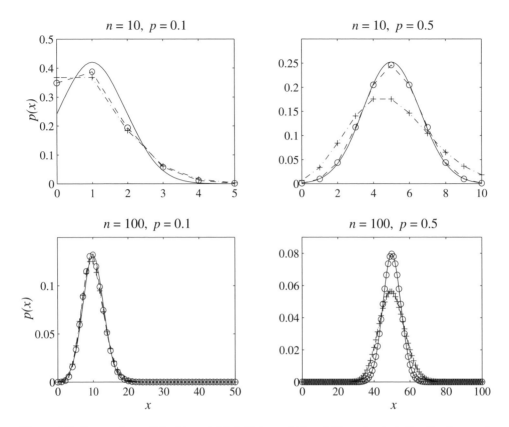

Figure 3.2. Comparison of the binomial ($\circ$), Poisson ($+$) and Gaussian ($-$) distributions: n is the number of trials, and p is the probability of seeing an event. By definition, the binomial distribution is correct. For a small probability of seeing an event, the Poisson distribution is a better approximation (although the difference is small for a large number of events), while for a large probability of seeing an event the Gaussian distribution is closer.

any distribution. For these reasons, it is often safe (and certainly common) to assume that an unknown distribution is Gaussian.

The Fourier transform of a Gaussian has a particularly simple form, namely a Gaussian with the inverse of the variance

$$\frac{1}{\sqrt{2\pi\sigma^2}} \int_{-\infty}^{\infty} e^{-x^2/2\sigma^2} e^{ikx} \, dx = e^{-k^2\sigma^2/2} \quad . \tag{3.23}$$

Remember this: you should never need to look up the transform of a Gaussian, just invert the variance. Because of this relationship, the product of the variance of a Gaussian and the variance of its Fourier transform will be a constant; this is the origin of many classical and quantum uncertainty relationships.

Figure 3.2 compares the binomial, Poisson, and Gaussian distributions for $n = 10$ and 100, and for $p = 0.1$ and 0.5, showing where they are and are not good approximations.

3.2.4 Central Limit Theorem

What is the probability distribution for the noise in a room full of people talking? This may sound like a nonsensical question, because the answer will depend on how many people there are, and on what is being said in what language. The remarkable result from the *Central Limit Theorem* is that if there is a large number of people in the room, then the distribution will approximately be Gaussian, independent of the details of what they say.

If two random variables x_1 and x_2 are added (perhaps the sounds from two random people), the probability distribution for their sum $y = x_1 + x_2$ is found by counting all of the outcomes that give the same final result, weighted by the joint probability for that event:

$$p(y) = \int_{-\infty}^{\infty} p_1(x)p_2(y-x)\, dx$$
$$\equiv p_1(x) * p_2(x) \quad . \tag{3.24}$$

The distribution for the sum is the *convolution* of the individual distributions. Now consider the average of N variables

$$y = \frac{x_1 + x_2 + \cdots + x_N}{N} \tag{3.25}$$

that are independent and identically distributed (abbreviated as *iid*). The distribution of y is equal to the distribution for x convolved with itself N times, and since taking a Fourier transform turns convolution into multiplication, the Fourier transform of the distribution of y is equal to the product of the transforms of the distribution of x. It is convenient to take the transform of a probability distribution by using the *characteristic function*, which is the expectation value of a complex exponential

$$\langle e^{iky} \rangle = \int_{-\infty}^{\infty} e^{iky} p(y)\, dy \quad . \tag{3.26}$$

The characteristic function is equal to the Fourier transform of the probability distribution, and when evaluated with time-dependent quantities it plays an interesting role in studying the dynamics of a system [Gershenfeld, 1999a]. Now let's look at the characteristic function for the deviation of y from the average value $\langle x \rangle$:

$$
\begin{aligned}
\langle e^{ik(y-\langle x \rangle)} \rangle &= \langle e^{ik(x_1+x_2+\cdots+x_N - N\langle x \rangle)/N} \rangle \\
&= \langle e^{ik[(x_1-\langle x \rangle)+\cdots+(x_N-\langle x \rangle)]/N} \rangle \\
&= \langle e^{ik(x-\langle x \rangle)/N} \rangle^N \\
&= \left\langle 1 + \frac{ik}{N}(x-\langle x \rangle) - \frac{k^2}{2N^2}(x-\langle x \rangle)^2 + \mathcal{O}\left(\frac{k^3}{N^3}\right) \right\rangle^N \\
&= \left[1 + 0 - \frac{k^2\sigma^2}{2N^2} + \mathcal{O}\left(\frac{k^3}{N^3}\right) \right]^N \\
&\approx e^{-k^2\sigma^2/2N} \quad .
\end{aligned}
\tag{3.27}
$$

This derivation assumes that the variance $\sigma^2 = \langle (x-\langle x \rangle)^2 \rangle$ exists, and drops terms of third order and higher in the Taylor series expansion of the exponential because they will become vanishing small compared to the lower-order terms in the limit $N \rightarrow \infty$. The last line follows because an exponential can be written as

$$\lim_{N \rightarrow \infty} \left(1 + \frac{x}{N} \right)^N = e^x \quad , \tag{3.28}$$

which can be verified by comparing the Taylor series of both sides. To find the probability distribution for y we now take the inverse transform

$$p(y - \langle x \rangle) = \frac{1}{2\pi} \int_{-\infty}^{\infty} e^{-k^2 \sigma^2 / 2N} e^{-ik(y - \langle x \rangle)} \, dk$$

$$= \sqrt{\frac{N}{2\pi\sigma^2}} e^{-N(y - \langle x \rangle)^2 / 2\sigma^2} \tag{3.29}$$

(remember that the Fourier transform of a Gaussian is also a Gaussian). This proves the Central Limit Theorem [Feller, 1974]. The average of N iid variables has a Gaussian distribution, with a standard deviation $\sigma / \sqrt{N}$ reduced by the square root of the number of variables just as with Poisson statistics. It can be a surprisingly good approximation even with just tens of samples. The Central Limit Theorem also contains the *Law of Large Numbers*: in the limit $N \rightarrow \infty$, the average of N random variables approaches the mean of their distribution. Although this might appear to be a trivial insight, lurking behind it is the compressibility of data that is so important to digital coding (Section 4.1).

3.3 NOISE MECHANISMS

Now that we've seen something about how to describe random systems we will turn to a quantitative discussion of some of the most important fundamental noise mechanisms: *shot noise*, *Johnson noise*, and *1/f noise*. Chapter 13 will consider other practical sources of noise, such as interference from unwanted signals.

3.3.1 Shot Noise

A current, such as electrons in a wire or rain on a roof, is made up of the discrete arrival of many carriers. If their interactions can be ignored so that they arrive independently, this is an example of a Poisson process. For an electrical signal, the average current is $\langle I \rangle = qN/T$ for N electrons with charge q arriving in a time T. If the electrons arrive far enough apart so that the duration during which they arrive is small compared to the time between the arrival of successive electrons, then the current can be approximated as a sum of delta functions.

$$I(t) = q \sum_{n=1}^{N} \delta(t - t_n) \quad , \tag{3.30}$$

where t_n is the arrival time for the nth electron. The Fourier transform of this impulse train is.

$$I(f) = \lim_{T \rightarrow \infty} \int_{-T/2}^{T/2} e^{i2\pi ft} q \sum_{n=1}^{N} \delta(t - t_n) \, dt$$

$$= q \sum_{n=1}^{N} e^{i2\pi ft_n} \quad . \tag{3.31}$$

Therefore, the power spectrum is

$$S_I(f) = \langle I(f)I^*(f)\rangle$$

$$= \lim_{T\to\infty} \frac{q^2}{T}\left(\sum_{n=1}^{N} e^{i2\pi f t_n} \sum_{m=1}^{N} e^{-i2\pi f t_m}\right)$$

$$= \lim_{T\to\infty} \frac{q^2 N}{T}$$

$$= q\langle I\rangle \tag{3.32}$$

(the cross terms $n \neq m$ vanish in the expectation because their times are independent). We see that the power spectrum of carrier arrivals is white (flat) and that the magnitude is linearly proportional to the current. This is called *shot noise* or *Schottky noise*. If the carriers do not really arrive as delta functions then the broadening of the impulses will roll the spectrum off for high frequencies, so the flat power spectrum is a good approximation up to the inverse of the characteristic times in the system.

To find the fluctuations associated with shot noise, we can use Parseval's Theorem to relate the average total energy in the spectrum to the average variance. If the bandwidth of the system is infinite this variance will be infinite, because for ideal shot noise there is equal power at all frequencies. Any real measurement system will have a finite bandwidth, and this determines the amplitude of the noise. Multiplying the power spectrum by $2\Delta f$, where Δf is the bandwidth in hertz and the factor of 2 comes from including both positive and negative frequencies,

$$\langle I_{\text{noise}}^2\rangle = 2q\langle I\rangle\Delta f \quad . \tag{3.33}$$

Shot noise will be important only if the number of carriers is small enough for the rate of arrival to be discernible; Problem 3.2 looks at this limit for detecting light.

3.3.2 Johnson Noise

Johnson (or *Nyquist*) noise is the noise associated with the relaxation of thermal fluctuations in a resistor. Small voltage fluctuations are caused by the thermal motion of the electrons, which then relax back through the resistance. We will calculate this in Section 3.4.3, but the result is simple:

$$\langle V_{\text{noise}}^2\rangle = 4kTR\Delta f \tag{3.34}$$

(where R is resistance, Δf is the bandwidth of the measuring system, T is the temperature, and k is Boltzmann's constant). Once again, this is white noise, but unlike shot noise it is independent of the current. The resistor is acting almost like a battery, driven by thermodynamic fluctuations. The voltage produced by these fluctuations is very real and very important: it sets a basic limit on the performance of many kinds of electronics. Unfortunately, it is not possible to take advantage of Johnson noise by rectifying the fluctuating voltage across a diode to use a resistor as a power source (hint: what temperature is the diode?).

Johnson noise is an example of a *fluctuation–dissipation* relationship (Section 3.4.3) – the size of a system's thermodynamic fluctuations is closely related to the rate at which the system relaxes to equilibrium from a perturbation. A system that is more strongly damped has smaller fluctuations, but it dissipates more energy.

3.3.3 $1/f$ Noise and Switching Noise

In a wide range of transport processes, from electrons in resistors, to cars on the highway, to notes in music, the power spectrum diverges at low frequencies inversely proportionally to frequency: $S(f) \propto f^{-1}$. Because such *1/f noise* is scale-invariant (the spectrum looks the same at all time scales [Mandelbrot, 1983]) and is so ubiquituous, many people have been lured to search for profound general explanations for the many particular examples. While this has led to some rather bizarre ideas, there is a reasonable theory for the important case of electrical $1/f$ noise.

In a conductor there are usually many types of defects, such as lattice vacancies or dopant atoms. Typically, the defects can be in a few different inequivalent types of sites in the material, which have different energies. This means that there is a probability for a defect to be thermally excited into a higher-energy state, and then relax down to the lower-energy state. Because the different sites can have different scattering cross-sections for the electron current, this results in a fluctuation in the conductivity of the material. A process that is thermally activated between two states, with a characteristic time τ to relax from the excited state, has a Lorentzian power spectrum of the form

$$S(f) = \frac{2\tau}{1 + (2\pi f \tau)^2} \tag{3.35}$$

(we will derive this in Problem 3.4). If there is a distribution of activation times $p(\tau)$ instead of a single activation time in the material, and if the activated scatterers don't interact with each other, then the spectrum will be an integral over this:

$$S(f) = \int_0^\infty \frac{2\tau}{1 + (2\pi f \tau)^2}\, p(\tau)\, d\tau \quad . \tag{3.36}$$

If the probability of the defect having an energy equal to a barrier height E goes as $e^{-E/kT}$ (Section 3.4), then the characteristic time τ to be excited over the barrier will be inversely proportional to probability

$$\tau = \tau_0 e^{E/kT} \quad . \tag{3.37}$$

This is called a *thermally activated* process. If the distribution of barrier heights $p(E)$ is flat then $p(\tau) \propto 1/\tau$, and putting this into equation (3.36) shows that $S(f) \propto 1/f$ (Problem 3.4) [Dutta & Horn, 1981].

This is the origin of $1/f$ noise: scatterers with a roughly flat distribution of activation energies. Cooling a sample to a low enough temperature can turn off the higher-energy scatterers and reveal the individual Lorentzian components in the spectrum [Rogers & Buhrman, 1984]. In this regime, the noise signal in time is made up of jumps between discrete values, called *switching noise*. This can be seen unexpectedly and intermittently at room temperature, for example if a device has a very bad wire-bond so that the current passes through a narrow constriction.

Unlike Johnson noise, $1/f$ noise is proportional to the current in the material because it is a conductivity rather than a voltage fluctuation, and it increases as the cross-sectional

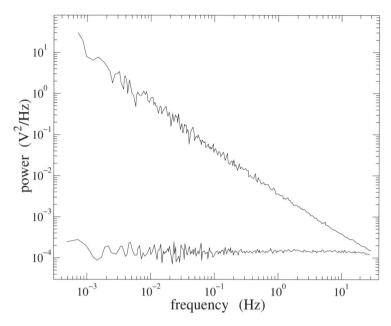

Figure 3.3. Noise in a 50 Ω resistor with and without a current.

area of the material is decreased because the relative influence of a single defect is greater. That is why $1/f$ noise is greater in carbon resistors, which have many small contacts between grains, than in metal film resistors. Low-noise switches have large contact areas, and wiping connections that slide against each other as the switch is closed, to make sure that the conduction is not constrained to small channels.

The power spectrum of the noise from a resistor will be flat because of Johnson noise if there is no current flowing; as the current is increased the $1/f$ noise will appear, and the frequency below which it is larger than the Johnson noise will depend on the applied current as well as on the details of the material. $1/f$ noise is not an intrinsic property: the magnitude is a function of how a particular sample is prepared. Figure 3.3 shows the Johnson and $1/f$ noise for a carbon resistor. Because $1/f$ noise diverges at low frequencies, it sets a time limit below which measurements cannot be made; a common technique to avoid $1/f$ noise is to modulate the signal up to a higher frequency (we will discuss this in Chapter 13).

3.3.4 Amplifier Noise

Any device that detects a signal must contend with these noise mechanisms in its workings. Johnson noise leads to the generation of voltage noise by an amplifier. Since the power spectral density is flat, the mean square noise magnitude will be proportional to the bandwidth, or the *Root Mean Square* (*RMS*) magnitude will increase as the square root of the bandwidth. The latter quantity is what is conventionally used to characterize an amplifier; for a low-noise device it can be on the order of $1 \text{ nV}/\sqrt{\text{Hz}}$. Likewise, shot noise is responsible for the generaton of current noise at an amplifier's output; this is also flat and for a low-noise amplifier can be on the order of $1 \text{ pA}/\sqrt{\text{Hz}}$.

Given the practical significance of detecting signals at (and beyond) these limits, it can be more relevant to relate the noise an amplifier introduces to the noise that is input to it. Signals and noise are usually compared on a logarithmic scale to cover a large dynamic range; the *Signal-to-Noise Ratio (SNR)*, measured in decibels, is

$$
\begin{aligned}
\text{SNR} &= 10 \log_{10} \left(\frac{\langle V_{\text{signal}}^2 \rangle}{\langle V_{\text{noise}}^2 \rangle} \right) \\
&= 20 \log_{10} \left(\frac{\langle V_{\text{signal}}^2 \rangle^{1/2}}{\langle V_{\text{noise}}^2 \rangle^{1/2}} \right) \\
&= 20 \log_{10} \left(\frac{V_{\text{RMS signal}}}{V_{\text{RMS noise}}} \right) \quad .
\end{aligned}
\tag{3.38}
$$

It can be defined in terms of the mean square values of the signal and noise (equal to the variances if the signals have zero mean), or the RMS values by bringing out a factor of 2.

One way to describe the performance of an amplifier is to ask how much more noise appears at its output than was present at its input, assuming that the input is responsible for Johnson noise due to its source impedance R_{source} (Chapter 6). This ratio, measured in decibels, is called the *noise figure*:

$$
\begin{aligned}
\text{NF} &= 10 \log_{10} \left(\frac{\text{output noise power}}{\text{input noise power}} \right) \\
&= 10 \log_{10} \left(\frac{4kTR_{\text{source}}\Delta f + \langle V_{\text{noise}}^2 \rangle}{4kTR_{\text{source}}\Delta f} \right) \\
&= 10 \log_{10} \left(1 + \frac{\langle V_{\text{noise}}^2 \rangle}{4kTR_{\text{source}}\Delta f} \right) \quad .
\end{aligned}
\tag{3.39}
$$

V_{noise} is the noise added by the amplifier to the source; it is what would be measured if the input impedance was cooled to absolute zero. The noise figure is often plotted as noise contours as a function of the input impedance and frequency (Figure 3.4). There is a "sweet spot" in the middle: it gets worse at low source impedances because the source thermal noise is small compared to the amplifier thermal noise; it gets worse at high source impedances and high frequencies because of capacitive coupling; and it degrades at low frequencies because of $1/f$ noise.

Amplifier noise can also be measured by the *noise temperature*, defined to be the temperature T_{noise} to which the input impedance must be raised from its actual temperature T_{source} for its thermal noise to match the noise added by the amplifier:

$$
\begin{aligned}
\text{NF} &= 10 \log_{10} \left(1 + \frac{\langle V_{\text{noise}}^2 \rangle}{4kT_{\text{source}}R\Delta f} \right) \\
&= 10 \log_{10} \left(1 + \frac{4kT_{\text{noise}}R}{4kT_{\text{source}}R\Delta f} \right) \\
&= 10 \log_{10} \left(1 + \frac{T_{\text{noise}}}{T_{\text{source}}} \right) \quad .
\end{aligned}
\tag{3.40}
$$

In a GaAs *HEMT (High-Electron-Mobility Transistor*, Chapter 10) most of the electron scattering mechanisms have been eliminated and so the mean-free-path can be as

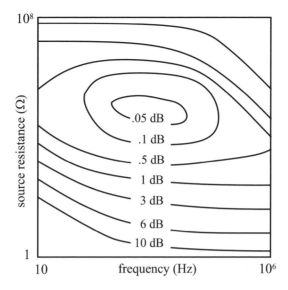

Figure 3.4. Noise contours for a low-noise amplifier.

large as the device. Since inelastic scattering is the origin of resistance and hence of the thermodynamic coupling of the conduction electrons to the material, this means that the noise temperature can be much lower than room temperature. In the best devices it gets down to just a few kelvins. One of the places where this sensitivity is particularly important is for detecting the weak signals from space for satellite communications and radio astronomy.

3.4 THERMODYNAMICS AND NOISE

Thermal fluctuations and noise are intimately related. This section turns to a more general discussion of this connection, starting with a brief review of macroscopic thermodynamics and its origin in microscopic statistical mechanics, and then looking at the Equipartition Theorem (which relates temperature to the average energy stored in a system's degrees of freedom) and the Fluctuation–Dissipation Theorem (which relates fluctuations to the dissipation in a system).

3.4.1 Thermodynamics and Statistical Mechanics

A thermodynamic system can be described by a *temperature* T, an *internal energy* E, and an *entropy* S. The internal energy is the sum off all of the energy stored in all of the degrees of freedom of the system. The entropy provides a relationship between heat and temperature: if the system is kept at a constant temperature, and a heat current δQ flows into or out of the system, the change in entropy is

$$\delta Q = T \, dS \quad . \tag{3.41}$$

This is written as δQ rather than dQ because energy that flows in and increases the entropy of a system cannot be reversibly recovered to do work. In any spontaneous process the entropy cannot decrease:

$$dS \geq 0 \quad . \tag{3.42}$$

This is the *Second Law of Thermodynamics*, with equality holding for a *reversible* process (the first law is conservation of energy). Because of the second law, all of the internal energy in a system is not available to do work on another system, only that part that was not associated with an entropy increase. Integrating both sides of equation (3.41) shows that the total heat energy in a system is $Q = TS$. Therefore the *free energy*, A, defined to be

$$A = E - TS \quad , \tag{3.43}$$

is the difference between the internal energy and the heat energy. It measures the energy in the system that is available to do work. We will see that a system seeks to minimize its free energy, by both reducing the internal energy and increasing the entropy.

Entropy was originally introduced in the early 1800s as a phenomenological quantity to help explain the efficiency of heat engines; one of the greatest problems in modern science has been explaining its microscopic origin. Although this was essentially solved by Maxwell, Boltzmann, and Gibbs around 1870, many subtle questions remained. This quest helped lead to many other areas of inquiry, including the development of information theory that we will see in the next chapter [Leff & Rex, 1990].

All systems are quantized so that they are restriced to a discrete set of possbile states, even though the spacing between the states may be so small that macroscopically they appear to be continuous. Let i index the possible states of a system, p_i be the probability to be in the ith state, and let there be Ω total states. Then the microscopic definition of the entropy is

$$S = -k \sum_{i=1}^{\Omega} p_i \log p_i \quad , \tag{3.44}$$

where k is Boltzmann's constant, which has units of energy over temperature. If all of the states are equally likely, then

$$p_i = \frac{1}{\Omega} \implies S = k \log \Omega \quad . \tag{3.45}$$

This equation was so important to Boltzmann that it appears on his grave!

According to the postulates of statistical mechanics, if there are Ω microscopic configurations of a system compatible with a given macroscopic state, then the probability p of seeing that macroscopic state is proportional to the number of states,

$$p \propto \Omega = e^{S/k} \quad . \tag{3.46}$$

The system is equally likely to be in any of the available microscopic states. If the total energy is fixed, then the probability to be in any state with that energy is the same (equation 3.45). This is called a *microcanonical ensemble*. In the real world it is much more common to be able to determine the average energy (which we'll see is closely related to the temperature) rather than the exact energy. That case is called a *canonical ensemble*. To work out its properties we need one more postulate from statistical mechanics: the system chooses the distribution of probabilities that maximizes its entropy, subject to the constraints that we impose. Justifying this essentially experimental fact is the subject of endless mathematical if not mystical discussion; Boltzmann's *H-Theorem* provides a derivation in the context of scattering in a dilute gas [Reichl, 1998].

For the canonical ensemble there are two constraints: the probability distribution must be normalized

$$\sum_{i=1}^{\Omega} p_i = 1 \quad , \tag{3.47}$$

and the average energy must be a constant E

$$\sum_{i=1}^{\Omega} E_i p_i = E \quad . \tag{3.48}$$

To do a constrained maximization we will use the method of *Lagrange multipliers*. Define a quantity I to be the entropy plus Lagrange multipliers times the constraint equations

$$I = -k \sum_{i=1}^{\Omega} p_i \log p_i + \lambda_1 \sum_{i=1}^{\Omega} p_i + \lambda_2 \sum_{i=1}^{\Omega} E_i p_i \quad . \tag{3.49}$$

We want to find the values for the p_i's that make this extremal:

$$\frac{\partial I}{\partial p_i} = 0 \quad . \tag{3.50}$$

We can do this because the two terms that we've added are just constants, Equations (3.47) and (3.48); we just need to choose the values of the Lagrange multipliers to make sure that they have the right values. Solving,

$$\frac{\partial I}{\partial p_i} = 0 = -k \log p_i - k + \lambda_1 + \lambda_2 E_i \tag{3.51}$$

$$\Rightarrow p_i = e^{(\lambda_1/k)+(\lambda_2 E_i/k)-1} \quad . \tag{3.52}$$

If we sum this over i,

$$\sum_{i=1}^{\Omega} p_i = 1 = e^{\lambda_1/k-1} \sum_{i=1}^{\Omega} e^{\lambda_2 E_i/k} \quad . \tag{3.53}$$

This can be rearranged to define the *partition function* $\mathcal{Z}$

$$\mathcal{Z} \equiv e^{1-\lambda_1/k} = \sum_{i=1}^{\Omega} e^{\lambda_2 E_i/k} \quad . \tag{3.54}$$

Another equation follows from multiplying equation (3.51) by p_i and summing:

$$\sum_{i=1}^{\Omega} p_i \frac{\partial I}{\partial p_i} = S - k + \lambda_1 + \lambda_2 E = 0 \quad . \tag{3.55}$$

Since

$$\mathcal{Z} = e^{1-\lambda_1/k} \quad , \tag{3.56}$$

$$k \log \mathcal{Z} = k - \lambda_1 \quad , \tag{3.57}$$

and so equation (3.55) can be written as

$$S - k \log \mathcal{Z} + \lambda_2 E = 0 \quad . \tag{3.58}$$

Comparing this to the definition of the free energy $A = E - TS$, we see that

$$S - \underbrace{k \log \mathcal{Z}}_{-A/T} + \underbrace{\lambda_2}_{-1/T} E = 0 \quad . \tag{3.59}$$

This provides a connection between the macroscopic thermodynamic quantities and the microscopic statistical mechanical ones.

Putting the value of λ_2 into equation (3.54) shows that the partition function is given by

$$\mathcal{Z} = \sum_{i=1}^{\Omega} e^{-E_i/kT} \equiv \sum_{i=1}^{\Omega} e^{-\beta E_i} \quad . \tag{3.60}$$

Returning to equation (3.52) we see that

$$p_i = e^{\lambda_1/k-1} e^{-E_i/kT} = \frac{e^{-E_i/kT}}{\mathcal{Z}} \quad . \tag{3.61}$$

In terms of this, the expected value of a function f_i that depends on the state of the system is

$$\langle f \rangle = \sum_{i=1}^{\Omega} f_i p_i = \frac{\sum_{i=1}^{\Omega} f_i e^{-E_i/kT}}{\mathcal{Z}} \tag{3.62}$$

3.4.2 Equipartition Theorem

The *Equipartition Theorem* is a simple, broadly applicable result that can give the magnitude of the thermal fluctuations associated with energy storage in independent degrees of freedom of a system. Assume that the state of a system is specified by variables $x_0, \ldots, x_n$, and that the internal energy of the system is given in terms of them by

$$E = E(x_0, \ldots, x_n) \quad . \tag{3.63}$$

Now consider the case where one of the degrees of freedom splits off additively in the energy:

$$E = E_0(x_0) + E_1(x_1, \ldots, x_n) \quad . \tag{3.64}$$

E might be the energy in a circuit, and $E_0 = CV_0^2/2$ the energy in a particular capacitor in terms of the voltage V_0 across it, or $E_0 = mv_0^2/2$ the kinetic energy of one particle in terms of its velocity v_0.

If we now assume that the overall system is in equilibrium at a temperature T, the expectation value for E_0 is given by the canonical statistical mechanical distribution (here taken as an integral instead of a discrete sum for a continuous system)

$$\langle E_0 \rangle = \frac{\int_{-\infty}^{\infty} e^{-\beta E(x_0,\ldots,x_n)} E_0(x_0)\, dx_0 \cdots dx_n}{\int_{-\infty}^{\infty} e^{-\beta E(x_0,\ldots,x_n)}\, dx_0 \cdots dx_n} \quad (\beta \equiv kT)$$

$$= \frac{\int_{-\infty}^{\infty} e^{-\beta[E_0(x_0)+E_1(x_1,\ldots,x_n)]} E_0(x_0)\, dx_0 \cdots dx_n}{\int_{-\infty}^{\infty} e^{-\beta[E_0(x_0)+E_1(x_1,\ldots,x_n)]}\, dx_0 \cdots dx_n}$$

$$= \frac{\int_{-\infty}^{\infty} e^{-\beta E_0(x_0)} E_0(x_0)\, dx_0 \int_{-\infty}^{\infty} e^{-\beta E_1(x_1,\ldots,x_n)}\, dx_1 \cdots dx_n}{\int_{-\infty}^{\infty} e^{-\beta E_0(x_0)}\, dx_0 \int_{-\infty}^{\infty} e^{-\beta E_1(x_1,\ldots,x_n)}\, dx_1 \cdots dx_n}$$

$$= \frac{\int_{-\infty}^{\infty} e^{-\beta E_0(x_0)} E_0(x_0)\, dx_0}{\int_{-\infty}^{\infty} e^{-\beta E_0(x_0)}\, dx_0}$$

$$= -\frac{\partial}{\partial \beta} \ln \int_{-\infty}^{\infty} e^{-\beta E_0(x_0)}\, dx_0 \quad . \tag{3.65}$$

If $E_0 = ax_0^2$ for some constant a, we can simplify the integral further:

$$\langle E_0 \rangle = -\frac{\partial}{\partial \beta} \ln \int_{-\infty}^{\infty} e^{-\beta E_0(x_0)}\, dx_0$$

$$= -\frac{\partial}{\partial \beta} \ln \int_{-\infty}^{\infty} e^{-\beta a x_0^2}\, dx_0$$

$$= -\frac{\partial}{\partial \beta} \ln \left[\frac{1}{\sqrt{\beta}} \int_{-\infty}^{\infty} e^{-ay^2}\, dy \right] \quad (y^2 \equiv \beta x_0^2)$$

$$= -\frac{\partial}{\partial \beta} \left[-\frac{1}{2} \ln \beta + \ln \int_{-\infty}^{\infty} e^{-ay^2}\, dy \right]$$

$$a\langle x_0^2 \rangle = \frac{1}{2} kT \quad . \tag{3.66}$$

Each independent thermalized quadratic degree of freedom has an average energy of $kT/2$ due to fluctuations.

3.4.3 Fluctuation–Dissipation Theorem

The Equipartition Theorem relates the size of thermal fluctuations to the energy stored in independent degrees of freedom of a system; the Fluctuation–Dissipation Theorem relates the thermal fluctuations to the amount of dissipation. We will start with a simple example and then discuss the more general theory. Consider an ideal inductor L connected in parallel with a resistor R. Because of thermal fluctuations there will be a voltage across the resistor; model that by a fluctuating voltage source V in series with a noiseless resistor (Figure 3.5).

In Chapter 6 we will show that the energy stored in an inductor is $LI^2/2$. Since the inductor is the only energy storage element, from the equipartition theorem we know what the current across it due to thermal fluctuations must be:

$$\left\langle \frac{1}{2} LI^2 \right\rangle = \frac{1}{2} kT \quad . \tag{3.67}$$

Ohm's Law (Section 6.1.3) still applies, so this current must also be equal to the fluctuating thermal voltage divided by the total impedance Z of the circuit. Written in terms

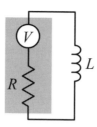

Figure 3.5. Resistor modeled as a fluctuating voltage source in series with a noiseless resistor, connected in parallel with an inductor.

of the frequency components,

$$I(\omega) = \frac{V(\omega)}{Z(\omega)} = \frac{V(\omega)}{R + i\omega L(\omega)} \tag{3.68}$$

(we will explain why the impedance of an inductor is $i\omega L$ when we derive the circuit equations from Maxwell's equations). Writing the equipartition result in terms of frequency components,

$$\begin{aligned}
\frac{1}{2}kT &= \left\langle \frac{1}{2}LI^2 \right\rangle = \frac{L}{2}\langle I^2 \rangle \\
&= \frac{L}{2}\int_{-\infty}^{\infty} \langle |I(\omega)|^2 \rangle \, d\omega \quad \text{(Parseval's Theorem)} \\
&= \frac{L}{2}\int_{-\infty}^{\infty} \left\langle \frac{|V(\omega)|^2}{|Z(\omega)|^2} \right\rangle \, d\omega \\
&= \frac{L}{2}\int_{-\infty}^{\infty} \frac{\langle |V(\omega)|^2 \rangle}{R^2 + \omega^2 L^2} \, d\omega \quad .
\end{aligned} \tag{3.69}$$

Since this is assumed to be an ideal resistor with no time constant from an inductive or capacitative component, it's a reasonable assumption to take the fluctuating voltage V to have a delta function autocorrelation (this can be justified by a microscopic derivation). And since that implies that the power spectrum of the fluctuations is flat, V does not depend on ω and can come out of the intergral:

$$\frac{1}{2}kT = \frac{L\langle V^2(\omega) \rangle}{2}\int_{-\infty}^{\infty} \frac{1}{R^2 + \omega^2 L^2} \, d\omega \quad . \tag{3.70}$$

This integration can then be done analytically,

$$\frac{1}{2}kT = \frac{\pi\langle V^2(\omega) \rangle}{2R} \quad . \tag{3.71}$$

Therefore,

$$\begin{aligned}
\frac{\pi\langle V^2(\omega) \rangle}{2R} &= \frac{1}{2}kT \\
\langle V^2(\omega) \rangle &= \frac{kTR}{\pi} \\
\langle V^2(f) \rangle &= 4kTR \quad .
\end{aligned} \tag{3.72}$$

In the last line there is a factor of 2π to convert from radian per second to cycles per second, and a factor of 2 to convert to the single-sided distribution with only positive frequencies used for power spectra. This is the familiar Johnson noise formula we saw earlier.

Johnson noise is a simple example of a much more general relationship between the fluctuations in a system and the dissipation in the system: more dissipation implies smaller fluctuations. Let's start by assuming that the macroscopic state of a system is indexed by a single degree of freedom x (say, the current in a circuit). This means that the entropy is a function of this variable, $S(x)$. In equilibrium, the system will be in the state that maximizes the entropy. Taking for convenience this to be at $x = 0$, the entropy can be approximated by a Taylor series around its maximum:

$$S = S_0 - \frac{1}{2}k\alpha x^2 \tag{3.73}$$

(remember that there is no linear term around a maximum). The constant α determines how sharply peaked the entropy is. The probability to see a state away from the maximum is then

$$p(x) \propto e^{S(x)/k} \propto e^{-\alpha x^2/2} \quad . \tag{3.74}$$

x is a random variable; we see that it has a Gaussian distribution with a variance

$$\sigma^2 = \langle x^2 \rangle = \int x^2 \, p(x) \, dx = \frac{1}{\alpha} \quad . \tag{3.75}$$

As α grows larger the entropy becomes more sharply peaked and the distribution becomes narrower. We now see why fluctuations such as Johnson noise so often have a Gaussian distribution (remember the Central Limit Theorem?).

If the state of the system is away from equilibrium there will be an effective restoring force that moves it back. This will depend on how the entropy varies with the state; the simplest assumption good for small deviations takes the time rate of change of the state to be proportional to the slope of the entropy

$$R\frac{dx}{dt} = \frac{dS}{dx} \quad . \tag{3.76}$$

There are many familiar examples of this kind of linear restoring force, such as Ohm's Law $IR = V$ for which the flux dx/dt is the current I, the slope of the entropy dS/dx is the driving force V, and the damping constant R is the resistance. These are the subject of linear non-equilibrium thermodynamics [de Groot & Mazur, 1984; Callen, 1985].

Equation (3.73) can be plugged into equation (3.76) to give a stochastic differential equation for the relaxation of x (called a *Langevin equation* [Gershenfeld, 1999a])

$$R\frac{dx}{dt} = -k\alpha x \quad . \tag{3.77}$$

Squaring both sides and averaging over time gives

$$R^2 \left\langle \left(\frac{dx}{dt}\right)^2 \right\rangle = k^2\alpha^2 \langle x^2 \rangle \quad . \tag{3.78}$$

The right hand side is just the variance, the size of the fluctuations. To understand the left hand side we need to return to equation (3.76). Remember that force times displacement gives energy, and that energy per time gives power. Therefore multiplying the driving force dS/dx by dx and dividing by dt gives the power P being dissipated,

$$P = \frac{dS}{dx}\frac{dx}{dt} = R\left(\frac{dx}{dt}\right)^2 \quad . \tag{3.79}$$

Therefore equation (3.77) shows that

$$P = k^2\frac{\alpha^2}{R}\langle x^2\rangle = k^2\frac{\alpha}{R} \quad . \tag{3.80}$$

If the entropy is sharply peaked (α large relative to R), then the fluctuations will be small but the dissipation will be large. If the entropy is flatter (α small), the fluctuations will be large but the dissipation will be small. A related equation is found by multiplying both sides of equation (3.77) by x and averaging:

$$R\ \underbrace{x\frac{dx}{dt}}_{\frac{1}{2}\frac{dx^2}{dt}} = -k\alpha x^2$$

$$\frac{d\langle x^2\rangle}{dt} = -2k\frac{\alpha}{R}\langle x^2\rangle \quad . \tag{3.81}$$

If the system is perturbed, the variance also relaxes at a rate proportional to α/R. It doesn't go to zero, of course, because we've left off the noise source term in the Langevin equation that drives the fluctuations.

Equation (3.80) is a simple example of the *Fluctuation–Dissipation Theorem*. The generalization is straightforward to systems with more degrees of freedom [Montroll & Lebowitz, 1987; Reichl, 1998] and to quantum systems [Balian, 1991]. In higher dimensions the relaxation constant R becomes a matrix, and if the system has time reversal invariance so that the governing equations are the same if $t \rightarrow -t$ then this matrix is symmetrical ($R_{ij} = R_{ji}$, called the *Onsager reciprocal relationship*).

The fluctuation dissipation theorem can be understood by remembering that a change in entropy is associated with a heat current $\delta Q = TdS$; if the entropy is sharply peaked then the fluctuations lead to larger changes in the entropy. This is an essential tradeoff in the design of any system: the faster and more accurately you want it to do something, the more power it will require. For example, one of the most important lessons in the design of low-power electronics is to make sure that the system does not produce results any faster than they are needed. This also shows why, without knowing anything else about electronics, low-noise amplifiers require more power than noisy ones.

3.5 SELECTED REFERENCES

[Feller, 1968] Feller, William. (1968). *An Introduction to Probability Theory and Its Applications*. 3rd edn. New York: Wiley.

[Feller, 1974] Feller, William. (1974). *An Introduction to Probability Theory and Its Applications*. 2nd edn. Vol. II. New York: Wiley.

A definitive probability reference.

[Balian, 1991] Balian, Roger. (1991). *From Microphysics to Macrophysics: Methods and Applications of Statistical Physics*. New York: Springer–Verlag. Translated by D. ter Haar and J.F. Gregg, 2 volumes.

A good introduction to all aspects of the statistical properties of physical systems.

[Horowitz & Hill, 1993] Horowitz, Paul, & Hill, Winfield. (1993). *The Art of Electronics*. 2nd edn. New York: Cambridge University Press.

The first place to start for any kind of practical electronics questions, including noise.

3.6 PROBLEMS

(3.1) (a) Derive equation (3.16) from the binomial distribution and Stirling's approximation,

(b) use it to derive equation (3.18), and

(c) use that to derive equation (3.19).

(3.2) Assume that photons are generated by a light source independently and randomly with an average rate N per second. How many must be counted by a photodetector per second to be able to determine the rate to within 1%? To within 1 part per million? How many watts do these cases correspond to for visible light?

(3.3) Consider an audio amplifier with a 20 kHz bandwidth.

(a) If it is driven by a voltage source at room temperature with a source impedance of 10 kΩ how large must the input voltage be for the SNR with respect to the source Johnson noise to be 20 dB?

(b) What size capacitor has voltage fluctuations that match the magnitude of this Johnson noise?

(c) If it is driven by a current source, how large must it be for the RMS shot noise to be equal to 1% of that current?

(3.4) This problem is much harder than the others. Consider a stochastic process $x(t)$ that randomly switches between $x = 0$ and $x = 1$. Let $\alpha\, dt$ be the probability that it makes a transition from 0 to 1 during the interval dt if it starts in $x = 0$, and let $\beta\, dt$ be the probability that it makes a transition from 1 to 0 during dt if it starts in $x = 1$.

(a) Write a matrix differential equation for the change in time of the probability $p_0(t)$ to be in the 0 state and the probability $p_1(t)$ to be in the 1 state.

(b) Solve this by diagonalizing the 2×2 matrix.

(c) Use this solution to find the autocorrelation function $\langle x(t)x(t + \tau)\rangle$.

(d) Use the autocorrelation function to show that the power spectrum is a Lorentzian.

(e) At what frequency is the magnitude of the Lorentzian reduced by half relative to its low-frequency value?

(f) For a thermally activated process, show that a flat distribution of barrier energies leads to a distribution of switching times $p(\tau) \propto 1/\tau$, and in turn to $S(f) \propto 1/\nu$.

4 Information in Physical Systems

What is information? A good answer is that information is what you don't already know. You do not learn much from being told that the sun will rise tomorrow morning; you learn a great deal if you are told that it will not. Information theory quantifies this intuitive notion of surprise. Its primary success is an explanation of how noise and energy limit the amount of information that can be represented in a physical system, which in turn provides insight into how to efficiently manipulate information in the system.

In the last chapter we met some of the many ways that devices can introduce noise into a signal, effectively adding unwanted information to it. This process can be abstracted into the concept of a *communications channel* that accepts an input and then generates an output. A telephone connection is a channel, as is the writing and subsequent reading of bits on a hard disk. In all cases there is assumed to be a set of known input symbols (such as 0 and 1), possibly a device that maps them into other symbols in order to satisfy constraints of the channel, the channel itself which has some probability for modifying the message due to noise or other errors, and possibly a decoder that turns the received symbols into an output set. We will assume that the types of messages and types of channel errors are sufficiently stationary to be able to define probability distributions $p(x)$ to see an input message x, and $p(y|x)$ for the channel to deliver a y if it is given an input x. This also assumes that the channel has no memory so that the probability distribution depends only on the current message. These are important assumptions: the results of this chapter will not apply to non-stationary systems.

4.1 INFORMATION

Let x be a random variable that takes on X possible values indexed by $i = 1, \ldots, X$, and let the probability of seeing the ith value be p_i. For example, x could be the letters of the alphabet, and p_i could be the probability to see letter i. How much information is there on average in a value of x drawn from this distribution? If there is only one possible value for x then we learn very little from successive observations because we already know everything; if all values are equally likely we learn as much as possible from each observation because we start out knowing nothing. An information functional $H(p)$ (a functional is a function of a function) that captures this intuitive notion should have the following reasonable properties:

- $H(p)$ is continuous in p. Small changes in the distribution should lead to small changes in the information.
- $H(p) \geq 0$, and $H(p) = 0$ if and only if just one p_i is non-zero. You always learn something unless you already know everything.
- $H(p) \leq C(X)$, where $C(X)$ is a constant that depends on the number of possible values X, with $H(p) = C(X)$ when all values are equally likely, and $X' > X \Rightarrow C(X') > C(X)$. The more options there are, the less you know about what will happen next.
- If x is drawn from a distribution p and y is independently drawn from a distribution q, then $H(p,q) = H(p) + H(q)$, where $H(p,q)$ is the information associated with seeing a pair (x,y). The information in independent events is the sum of the information in the events individually.

While it might appear that this list is not sufficient to define $H(p)$, it can be shown [Ash, 1990] that these desired properties are uniquely satisfied by the function

$$H(p) = -\sum_{i=1}^{X} p_i \log p_i \quad . \tag{4.1}$$

This is the definition of the *entropy* of a probability distribution, the same definition that was used in the last chapter in statistical mechanics. To make the dependence on x clear, we will usually write this as $H(x)$ instead of $H(p(x))$ or $H(p)$. The choice of the base of the logarithm is arbitrary; if the base is 2 then the entropy is measured in *bits*, and if it is base e then the entropy units are called *nats* for the *natural logarithm*. Note that to change an entropy formula from bits to nats you just change the logarithms from $\log_2$ to $\log_e$, and so unless otherwise noted the base of the logarithms in this chapter is arbitrary.

Now consider a string of N samples $(x_1, \ldots, x_N)$ drawn from p, and let N_i be the number of times that the ith value of x was actually seen. Because of the independence of the observations, the probability to see a particular string is the product of the individual probabilities

$$p(x_1, \ldots, x_N) = \prod_{n=1}^{N} p(x_n) \quad . \tag{4.2}$$

This product of terms can be regrouped in terms of the possible values of x,

$$p(x_1, \ldots, x_N) = \prod_{i=1}^{X} p_i^{N_i} \quad . \tag{4.3}$$

Taking the log and multiplying both sides by $-1/N$ then lets this be rewritten as

$$
\begin{aligned}
-\frac{1}{N} \log p(x_1, \ldots, x_N) &= -\frac{1}{N} \log \prod_{i=1}^{X} p_i^{N_i} \\
&= -\sum_{i=1}^{X} \frac{N_i}{N} \log p_i \\
&\approx -\sum_{i=1}^{X} p_i \log p_i \\
&= H(x) \quad . \tag{4.4}
\end{aligned}
$$

The third line follows from the Law of Large Numbers (Section 3.2.4): as $N \to \infty$, $N_i/N \to p_i$. Equation (4.4) can be inverted to show that

$$p(x_1, \ldots, x_N) \approx 2^{-NH(x)} \qquad (4.5)$$

(taking the entropy to be defined base 2). Something remarkable has happened: the probability of seeing a particular long string is independent of the elements of that string. This is called the *Asymptotic Equipartition Property* (*AEP*). Since the probability of occurrence for a string is a constant, its inverse $1/p = 2^{NH(x)}$ gives the effective number of strings of that length. However, the actual number of strings is larger, equal to

$$X^N = 2^{N \log_2 X} \qquad . \qquad (4.6)$$

The difference between these two values is what makes data compression possible. It has two very important implications [Blahut, 1988]:

- Since samples drawn from the distribution can on average be described by $H(x)$ bits rather than $\log_2 X$ bits, a coder can exploit the difference to store or transmit the string with $NH(x)$ bits. This is *Shannon's First Coding Theorem*, also called the *Source Coding Theorem* or the *Noiseless Coding Theorem*.

- The compressibility of a typical string is made possible by the vanishing probability to see rare strings, the ones that violate the Law of Large Numbers. In the unlikely event that such a string appears the coding will fail and a longer representation must be used. Because the Law of Large Numbers provides an increasingly tight bound on this occurrence as the number of samples increases, the failure probability can be made arbitrarily small by using a long enough string. This is the *Shannon–McMillan Theorem*.

Because the entropy is a maximum for a flat distribution, an efficient coder will represent information with this distribution. This is why modems "hiss": they make best use of the telephone channel if the information being sent appears to be as random as possible. The value of randomness in improving a system's performance will recur throughout this book, particularly in Chapter 13.

We see that the entropy (base 2) gives the average number of bits that are required to describe a sample drawn from the distribution. Since the entropy is equal to

$$-\sum_{i=1}^{X} p_i \log p_i = \langle -\log p_i \rangle \qquad (4.7)$$

it is natural to interpret $-\log p_i$ as the information in seeing event p_i, and the entropy as the expected value of that information.

Entropy can be applied to systems with more degrees of freedom. The joint entropy for two variables with a joint distribution $p(x, y)$ is

$$H(x, y) = -\sum_x \sum_y p(x, y) \log p(x, y) \qquad . \qquad (4.8)$$

This can be rewritten as

$$H(x, y) = -\sum_x \sum_y p(x, y) \log p(x, y)$$

$$= -\sum_x \sum_y p(x, y) \log[p(x|y)p(y)]$$

$$= -\sum_x \sum_y p(x, y) \log p(x|y) - \sum_x \sum_y p(x, y) \log p(y)$$

$$= -\sum_x \sum_y p(x, y) \log p(x|y) - \sum_y p(y) \log p(y)$$

$$= H(x|y) + H(y) \tag{4.9}$$

by using *Bayes' rule* $p(x, y) = p(x|y)p(y)$. The entropy in a conditional distribution $H(x|y)$ is the expected value of the information $\langle -\log p(x|y) \rangle$. The entropy of both variables equals the entropy of one of them plus the entropy of the other one given the observation of the first.

The *mutual information* between two variables is defined to be the information in them taken separately minus the information in them taken together

$$I(x, y) = H(x) + H(y) - H(x, y)$$

$$= H(y) - H(y|x)$$

$$= H(x) - H(x|y)$$

$$= \sum_x \sum_y p(x, y) \log \frac{p(x, y)}{p(x)p(y)} \tag{4.10}$$

(these different forms are shown to be equal in Problem 4.2). This measures how many bits on average one sample tells you about the other. It vanishes if the variables are independent, and it is equal to the information in one of them if they are completely dependent. The mutual information can be viewed as an information-theoretic analog of the cross-correlation function $\langle x(t)y(t) \rangle$, but the latter is useful only for measuring the overlap among signals from linear systems [Gershenfeld, 1993].

In a sequence of N values $(x_1, x_2, \ldots, x_N)$ the *joint* (or *block entropy*)

$$H_N(x) = -\sum_{x_1} \sum_{x_2} \cdots \sum_{x_N} p(x_1, x_2, \ldots, x_N) \log p(x_1, x_2, \ldots, x_N) \tag{4.11}$$

is the average number of bits needed to describe the string. The limiting rate at which this grows

$$h(x) = \lim_{N \to \infty} \frac{1}{N} H_N(x) = \lim_{N \to \infty} H_{N+1} - H_N \tag{4.12}$$

is called the *source entropy*. It is the rate at which the system generates new information.

So far we've been discussing random variables that can take on a discrete set of values; defining entropy for continuous variables requires some care. If x is a real number, then $p(x) \, dx$ is the probability to see a value between x and $x + dx$. The information in such an observation is given by its logarithm, $-\log[p(x) \, dx] = -\log p(x) - \log dx$. As $dx \to 0$ this will diverge! The divergence is in fact the correct answer, because a single real number can contain an infinite amount of information if it can be specified to any resolution. The *differential entropy* is the part of the entropy that does not diverge:

$$H(x) = - \int_{-\infty}^{\infty} p(x) \log p(x) \, dx \quad . \tag{4.13}$$

Unlike the discrete entropy this can be positive or negative. The particular value of the differential entropy is not meaningful, because we have ignored the diverging part due to the infinitesimal limit, but differences between differential entropies are meaningful, because the diverging parts would cancel.

To understand mutual information for the continuous case we first need *Jensen's Theorem* [Cover & Thomas, 1991]: for a convex function $f(x)$ (one that has a non-negative second derivative, such as $-\log$)

$$\langle f(x) \rangle \geq f(\langle x \rangle) \quad . \tag{4.14}$$

This implies that for two normalized distributions p and q

$$D(p, q) \equiv \int_{-\infty}^{\infty} p \log \frac{p}{q} \tag{4.15}$$

$$= - \int_{-\infty}^{\infty} p \log \frac{q}{p} \tag{4.16}$$

$$\geq - \log \int_{-\infty}^{\infty} p \frac{q}{p}$$

$$= - \log \int_{-\infty}^{\infty} q$$

$$= - \log 1$$

$$= 0 \quad .$$

$D(p, q)$ is non-negative, vanishing if $p = q$. It is called the *Kullback–Leibler distance* between two probability functions, and $D[p(x, y), p(x)p(y)]$ is the continuous analog of the mutual information. The Kullback–Leibler distance arises naturally as a measure of the distance between two distributions, but it is not a true distance function: it is not symmetric in f and g (it can change value if they are interchanged), and it does not satisfy the triangle inequality ($D(p, q) + D(q, r)$ is not necessarily greater than or equal to $D(p, r)$).

4.2 CHANNEL CAPACITY

Claude Shannon is best known for finding a surprisingly simple solution to what had been thought to be a hard problem. The use of telephones grew faster than the available capacity of the phone system and so it became increasingly important to make good use of that capacity, raising an essential question: how many phone calls can be sent through a phone line? This is not easy to answer because a phone line is an analog channel with limited SNR and bandwidth. Clever modulation schemes can let more messages share the same cable; is there any limit to how much of an improvement is possible? Shannon's answer was a simple quantitative yes.

Consider a long string of N symbols $(x_1, x_2, \ldots, x_N)$ drawn independently from $p(x)$

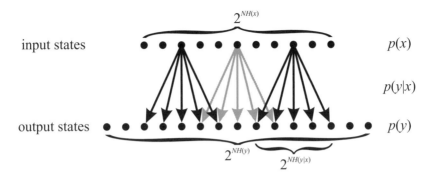

Figure 4.1. Effective number of states input to, added by, and output from a channel.

that are input to a channel specified by $p(y|x)$. On average each sample contains $H(x)$ bits of information, so this input string of N symbols can represent roughly $2^{NH(x)}$ different states. After being sent through the channel an output string $(y_1, y_2, \ldots, y_N)$ can represent $2^{NH(y)}$ states. However, it is possible that because of noise in the channel different input states can produce the same output state and hence garble the message; $2^{NH(y|x)}$ is the average number of different output states that are produced by an input state, the extra information in y given knowledge of x. In order to make sure that each input state typically leads to only one output state it is necessary to reduce the number of allowable output states by the excess information generated by the channel (Figure 4.1)

$$\frac{2^{NH(y)}}{2^{NH(y|x)}} = 2^{N[H(y)-H(y|x)]} = 2^{NI(x,y)} \quad . \tag{4.17}$$

We see that the probability distribution that maximizes the mutual information between the input and the output leads to the maximum number of distinct messages that can reliably be sent through the channel. The *channel capacity* is this maximum bit rate:

$$C = \max_{p(x)} \ I(x,y) \quad . \tag{4.18}$$

Applying the Shannon–McMillan Theorem to the input and output of the channel taken together shows that, if the data rate is below the channel capacity and the block length is long enough, then messages can be decoded with an arbitrarily small error. On the other hand, it is impossible to send data error-free through the channel at a rate greater than the capacity. This is *Shannon's Second Coding Theorem* (also called the *Channel Coding Theorem* or the *Noisy Coding Theorem*). If you're sending information at a rate below the channel capacity you are wasting part of the channel and should seek a better code (Chapter 13 will look at how to do this); if you're sending information near the capacity you are doing as well as possible and there is no point in trying to improve the code; and there is no hope of reliably sending messages much above the capacity.

A few points about channel coding:

- As the transmission rate increases it might be expected that the best-case error rate will also increase; it is surprising that the error rate can remain zero until the capacity is reached (Figure 4.2).
- This proves the existence of zero-error codes but it doesn't help find them, and

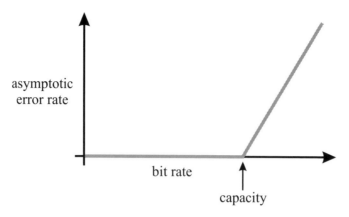

Figure 4.2. Onset of errors at the channel capacity in transmitting a long string.

once they are found they may not be useful. In particular, the coding/decoding effort or latency may become enormous as the rate approaches the capacity. For example, the length of the required code word may become prohibitively long.

- This is not a fundamental limit like the speed of light. The channel capacity holds for long strings of symbols independently drawn from a stationary probability distribution; it does not apply to short strings, non-stationary systems, or correlated variables. These approximations may not be justified, but can nevertheless be useful to make a rough estimate of the properties of a system. High-speed modems, for example, can exceed the theoretical capacity of a phone line (Problem 4.4) by adaptively modeling and coding for the channel errors.

- In many domains, such as broadcasting video, error-free transmission is irrelevant. All that matters is that the errors are not apparent; this is the subject of *lossy compression*. By taking advantage of what is known about human perception much higher bit rates are possible. Although you wouldn't want a money machine to do lossy compression on your bank balance when it communicates with the bank, your ear doesn't respond to a soft sound with a frequency immediately adjacent to that of a louder sound, and your eye cannot recognize the details of families of image textures beyond their statistical properties. These kinds of insights are used in the standards developed by the *Moving Pictures Experts Group* (*MPEG*) for variable lossy compression of video and audio. The MPEG-4 standard goes further to abandon a description based on arbitrary bit patterns and instead decomposes sights and sounds into high-level descriptions of their constituent elements [Koenen, 1999].

4.3 THE GAUSSIAN CHANNEL

In the last chapter we saw that the Central Limit Theorem explains why Gaussian noise is so common. It is therefore natural to consider a channel that adds Gaussian noise: $y_i = x_i + \eta_i$, where η_i is drawn from a Gaussian distribution. This might represent the Johnson noise in the input stage of a telephone amplifier, along with the accumulated effect of many small types of interference. Gaussian distributions are particularly important in information theory because, for a given mean and variance, they maximize the differential entropy. This makes it easy to calculate the maximum in equation (4.18). To see this, let $\mathcal{N}(x)$ be a Gaussian distribution

$$\mathcal{N}(x) = \frac{1}{\sqrt{2\pi\sigma_\mathcal{N}^2}}\, e^{-(x-\mu_\mathcal{N})^2/2\sigma_\mathcal{N}^2} \quad , \tag{4.19}$$

and let $p(x)$ be an arbitrary distribution with mean μ_p and variance σ_p^2. Then

$$-\int_{-\infty}^{\infty} p(x)\ln\mathcal{N}(x)\,dx$$

$$= -\int_{-\infty}^{\infty} p(x)\left[-\ln\sqrt{2\pi\sigma_\mathcal{N}^2} - \frac{(x-\mu_\mathcal{N})^2}{2\sigma_\mathcal{N}^2}\right]\,dx$$

$$= \ln\sqrt{2\pi\sigma_\mathcal{N}^2} + \frac{\sigma_p^2 + \mu_p^2 - 2\mu_p\mu_\mathcal{N} + \mu_\mathcal{N}^2}{2\sigma_\mathcal{N}^2} \quad . \tag{4.20}$$

This depends only on the mean and variance of $p(x)$ and so if $q(x)$ has the same mean and variance then

$$-\int_{-\infty}^{\infty} p(x)\ln\mathcal{N}(x)\,dx = -\int_{-\infty}^{\infty} q(x)\ln\mathcal{N}(x)\,dx \quad . \tag{4.21}$$

Now consider the difference in the entropy between a Gaussian distribution $\mathcal{N}$ and another one p with the same mean and variance:

$$H(\mathcal{N}) - H(p) = -\int_{-\infty}^{\infty} \mathcal{N}(x)\ln\mathcal{N}(x)\,dx + \int_{-\infty}^{\infty} p(x)\ln p(x)\,dx$$

$$= -\int_{-\infty}^{\infty} p(x)\ln\mathcal{N}(x)\,dx + \int_{-\infty}^{\infty} p(x)\ln p(x)\,dx$$

$$= \int_{-\infty}^{\infty} p(x)\ln\frac{p(x)}{\mathcal{N}(x)}\,dx$$

$$= D(p,\mathcal{N}) \geq 0 \quad . \tag{4.22}$$

The differential entropy in any other distribution will be less than that of a Gaussian with the same mean and variance. This differs from the discrete case, where the maximum entropy distribution was a constant, or an exponential if the energy is fixed.

Now return to our Gaussian channel $y = x + \eta$. Typically the input signal will be constrained to have some maximum power $S = \langle x^2 \rangle$. The capacity must be found by maximizing with respect to this constraint:

$$C = \max_{p(x):\langle x^2 \rangle \leq S} I(x,y) \quad . \tag{4.23}$$

The mutual information is

$$\begin{aligned} I(x,y) &= H(y) - H(y|x) \\ &= H(y) - H(x+\eta|x) \\ &= H(y) - H(\eta|x) \\ &= H(y) - H(\eta) \quad , \end{aligned} \tag{4.24}$$

where the last line follows because the noise is independent of the signal. The differential entropy of a Gaussian process is straightforward to calculate (Problem 4.3):

$$H(\mathcal{N}) = \frac{1}{2}\log(2\pi e N) \tag{4.25}$$

(where $N = \sigma_{\mathcal{N}}^2$ is the noise power). The mean square channel output is

$$\begin{aligned}
\langle y^2 \rangle &= \langle (x+\eta)^2 \rangle \\
&= \langle x^2 \rangle + 2\langle x \rangle \underbrace{\langle \eta \rangle}_{0} + \langle \eta^2 \rangle \\
&= S + N \quad .
\end{aligned} \tag{4.26}$$

Since the differential entropy of x must be bounded by that of a Gaussian process with the same variance, the mutual information will be a maximum for

$$\begin{aligned}
I(x,y) &= H(y) - H(\eta) \\
&\leq \frac{1}{2}\log[2\pi e(S+N)] - \frac{1}{2}\log(2\pi e N) \\
&= \frac{1}{2}\log\left(1 + \frac{S}{N}\right) \quad .
\end{aligned} \tag{4.27}$$

The capacity of a Gaussian channel grows as the logarithm of the ratio of the signal power to the channel noise power.

Real channels necessarily have finite bandwidth. If a signal is sampled with a period of $1/2\Delta f$ then by the *Nyquist Theorem* the bandwidth will be Δf. If the (one–sided, white) noise power spectral density is N_0, the total energy in a time T is $N_0 \Delta f T$, and the noise energy per sample is $(N_0 \Delta f T)/(2\Delta f T) = N_0/2$. Similarly, if the signal power is S, the signal energy per sample is $S/2\Delta f$. This means that the capacity per sample is

$$\begin{aligned}
C &= \frac{1}{2}\log\left(1 + \frac{S}{N}\right) \\
&= \frac{1}{2}\log\left(1 + \frac{S}{2\Delta f}\frac{2}{N_0}\right) \\
&= \frac{1}{2}\log_2\left(1 + \frac{S}{N_0\Delta f}\right) \quad \frac{\text{bits}}{\text{sample}} \quad .
\end{aligned} \tag{4.28}$$

If the signal power equals the noise power, then each samples carries $1/2$ bit of information.

Since there are $2\Delta f$ samples per second the information rate is

$$\begin{aligned}
C &= \Delta f \log\left(1 + \frac{S}{N}\right) \\
&= \Delta f \log_2\left(1 + \frac{S}{N_0\Delta f}\right) \quad \frac{\text{bits}}{\text{second}} \quad .
\end{aligned} \tag{4.29}$$

This is the most important result in this chapter: the capacity of a band-limited Gaussian channel. It increases as the bandwidth and input power increase, and decreases as the noise power increases.

4.4 FISHER INFORMATION

There is a natural connection between the information in a measurement and the accuracy with which the measurement can be made, and so not surprisingly entropy shows up here also. Let $p_\alpha(x)$ be a probability distribution that depends on a parameter α, and let $f(x_1, x_2, ..., x_N)$ be an estimator for the value of α given N measurements of x drawn from $p_\alpha(x)$. The function f is a *biased* estimator if $\langle f(x_1, x_2, \ldots, x_N) \rangle \neq \alpha$, and it is *consistent* if in the limit $N \to \infty$ the probability to see $|f(x_1, x_2, \ldots, x_N) - \alpha| > \epsilon$ goes to 0 for any ϵ. An estimator f_1 *dominates* f_2 if $\langle (f_1(x_1, x_2, \ldots, x_N) - \alpha)^2 \rangle \leq \langle (f_2(x_1, x_2, \ldots, x_N) - \alpha)^2 \rangle$. This raises the question of what is the minimum variance possible for an unbiased estimator of α? The answer is given by the *Cramér–Rao bound*.

Start by defining the *score*:

$$V = \frac{\partial}{\partial \alpha} \log p_\alpha(x) = \frac{1}{p_\alpha(x)} \frac{\partial p_\alpha(x)}{\partial \alpha} \quad . \tag{4.30}$$

The mean value of the score is

$$\begin{aligned}
\langle V \rangle &= \int_{-\infty}^{\infty} p_\alpha(x) \frac{1}{p_\alpha(x)} \frac{\partial p_\alpha(x)}{\partial \alpha} \, dx \\
&= \int_{-\infty}^{\infty} \frac{\partial p_\alpha(x)}{\partial \alpha} \, dx \\
&= \frac{\partial}{\partial \alpha} \int_{-\infty}^{\infty} p_\alpha(x) \, dx \\
&= \frac{\partial}{\partial \alpha} 1 \\
&= 0 \quad . \tag{4.31}
\end{aligned}$$

Therefore the variance of the score is just the mean of its square, $\sigma^2(V) = \langle V^2 \rangle$. The variance of the score is called the *Fisher information*:

$$\begin{aligned}
J(\alpha) &= \langle V^2 \rangle \\
&= \left\langle \left[\frac{\partial \log p_\alpha(x)}{\partial \alpha} \right]^2 \right\rangle \\
&= \left\langle \left[\frac{1}{p_\alpha(x)} \frac{\partial p_\alpha(x)}{\partial \alpha} \right]^2 \right\rangle \\
&= \int_{-\infty}^{\infty} \frac{1}{p_\alpha(x)} \left[\frac{\partial p_\alpha(x)}{\partial \alpha} \right]^2 \, dx \quad . \tag{4.32}
\end{aligned}$$

The score for a set of independent, identically distributed variables is the sum of the individual scores:

$$\begin{aligned}
V(x_1, x_2, \ldots, x_N) &= \frac{\partial}{\partial \alpha} \log p_\alpha(x_1, x_2, \ldots, x_N) \\
&= \frac{\partial}{\partial \alpha} \log \prod_{i=1}^{N} p_\alpha(x_i)
\end{aligned}$$

$$= \sum_{i=1}^{N} \frac{\partial \log p_\alpha(x_i)}{\partial \alpha}$$

$$= \sum_{i=1}^{N} V(x_i) \tag{4.33}$$

and so the Fisher information for N measurements is

$$J_N(\alpha) = \left\langle \left(\frac{\partial}{\partial \alpha} \log p_\alpha(x_1, x_2, \ldots, x_N) \right)^2 \right\rangle$$

$$= \langle V^2(x_1, x_2, \ldots, x_N) \rangle$$

$$= \left\langle \left(\sum_{i=1}^{N} V(x_i) \right)^2 \right\rangle$$

$$= \sum_{i=1}^{N} \langle V^2(x_i) \rangle$$

$$= N J(\alpha) \quad . \tag{4.34}$$

The sum can be taken out of the expectation because the variables are uncorrelated.

The *Cramér–Rao inequality* states that the mean square error of an unbiased estimator f of α is lower bounded by the reciprocal of the Fisher information:

$$\sigma^2(f) \geq \frac{1}{J(\alpha)} \quad . \tag{4.35}$$

To prove this, start with the *Cauchy–Schwarz inequality*

$$\langle (V - \langle V \rangle)(f - \langle f \rangle) \rangle^2 \leq \langle (V - \langle V \rangle)^2 \rangle \langle (f - \langle f \rangle)^2 \rangle$$

$$\langle Vf - \langle V \rangle f - \langle f \rangle V + \langle V \rangle \langle f \rangle \rangle^2 \leq \langle V^2 \rangle \langle (f - \langle f \rangle)^2 \rangle$$

$$\langle Vf \rangle^2 \leq J(\alpha) \sigma^2(f) \quad . \tag{4.36}$$

The expectation of the left hand side equals one:

$$\langle Vf \rangle = \int_{-\infty}^{\infty} p_\alpha(x) \frac{1}{p_\alpha(x)} \frac{\partial p_\alpha(x)}{\partial \alpha} f(x) \, dx$$

$$= \int_{-\infty}^{\infty} \frac{\partial p_\alpha(x)}{\partial \alpha} f(x) \, dx$$

$$= \frac{\partial}{\partial \alpha} \int_{-\infty}^{\infty} p_\alpha(x) f(x) \, dx$$

$$= \frac{\partial}{\partial \alpha} \langle f(x) \rangle$$

$$= \frac{\partial \alpha}{\partial \alpha}$$

$$= 1 \quad , \tag{4.37}$$

thus proving the Cramér–Rao inequality. Just like the channel capacity, the Cramér–Rao bound sets a lower limit on what is possible but does not provide any guidance in actually finding the minimum variance estimator. In fact, in practice a biased estimator might be preferable because it could be easier to calculate, or might converge more quickly.

The Cramér–Rao inequality measures how tightly a distribution constrains a parameter. To relate it to the differential entropy $H(x)$ of a distribution $p(x)$, consider what happens when a random Gaussian variable η is added to x. The new probability distribution is found by convolution:

$$p(\underbrace{x + \eta}_{\equiv\, y}) = \int_{-\infty}^{\infty} p(x) \frac{1}{\sqrt{2\pi\sigma^2}} e^{-(y-x)^2/2\sigma^2}\, dx \quad . \tag{4.38}$$

Differentiating,

$$\frac{\partial p}{\partial \sigma^2} = \int_{-\infty}^{\infty} p(x) \frac{1}{\sqrt{2\pi\sigma^2}} \left[\frac{(y-x)^2 - \sigma^2}{2\sigma^4} \right] e^{-(y-x)^2/2\sigma^2}\, dx$$

$$\frac{\partial^2 p}{\partial y^2} = \int_{-\infty}^{\infty} p(x) \frac{1}{\sqrt{2\pi\sigma^2}} \left[\frac{(y-x)^2 - \sigma^2}{\sigma^4} \right] e^{-(y-x)^2/2\sigma^2}\, dx$$

$$\Rightarrow \quad \frac{\partial p}{\partial \sigma^2} = \frac{1}{2} \frac{\partial^2 p}{\partial y^2} \quad . \tag{4.39}$$

This has the form of a diffusion equation: the added noise smooths out the distribution. Now taking the gradient of the differential entropy with respect to the noise variance, we see that

$$\begin{aligned}
\frac{\partial H}{\partial \sigma^2} &= -\frac{\partial}{\partial \sigma^2} \int_{-\infty}^{\infty} p(y) \log p(y)\, dy \\
&= -\frac{\partial}{\partial \sigma^2} \underbrace{\int_{-\infty}^{\infty} p(y)\, dy}_{0} - \int_{-\infty}^{\infty} \frac{\partial p}{\partial \sigma^2} \log p(y)\, dy \\
&= -\frac{1}{2} \int_{-\infty}^{\infty} \frac{\partial^2 p}{\partial y^2} \log p(y)\, dy \quad \left(\int_{-\infty}^{\infty} u\, dv = uv \big|_{-\infty}^{\infty} - \int_{-\infty}^{\infty} v\, du \right) \\
&= -\frac{1}{2} \frac{\partial p(y)}{\partial y} \log p(y) \bigg|_{-\infty}^{\infty} + \frac{1}{2} \int_{-\infty}^{\infty} \frac{1}{p(y)} \left(\frac{\partial p(y)}{\partial y} \right)^2 dy \\
&= 0 + \frac{1}{2} J(y) \quad . \tag{4.40}
\end{aligned}$$

The first term on the left vanishes because although the logarithm diverges as $p \to 0$ when $y \to \infty$, the slope $\partial p/\partial y$ must be vanishing faster than logarithmically for the probability distribution to be normalized. Taking the limit $\sigma \to 0$,

$$\frac{\partial H}{\partial \sigma^2} \bigg|_{\sigma^2 = 0} = \frac{1}{2} J(x) \quad . \tag{4.41}$$

The growth rate of the differential entropy with respect to the variance of an added Gaussian variable is equal to the Fisher information of the distribution. This is *de Bruijn's identity*. It can be interpreted as saying that the entropy measures the information in the volume of a distribution, and the Fisher information measures the information associated with its surface (as probed by smoothing it with the noise).

4.5 INFORMATION AND THERMODYNAMICS

We introduced entropy through statistical mechanics in Section 3.4, and in this chapter developed it as a powerful tool for analyzing probability distributions. The connection between thermodynamics and information theory is much deeper, providing a great example of how hard it can be to draw a clear boundary between basic and applied research in the evolution of significant ideas.

The important concept of the maximum efficiency of a heat engine was introduced by Sadi Carnot in 1824, motivated by the practical problem of understanding the limits on the performance of steam engines. This led to the macroscopic definition of entropy $\delta Q = T dS$ by Lord Kelvin (then William Thomson) and Rudolf Clausius around 1850–1860. Clausius named entropy for the Greek word for continuous transformation.

Statistical mechanics then grew out of the search for a microscopic explanation for macroscopic thermodynamics. This started with Maxwell's kinetic model of a gas, and the crucial connection $S = k \log \Omega$ was made by Boltzmann in 1877. Boltzmann, through his *H-Theorem*, provided a microscopic explanation for the macroscopic observation that a system moves to the available state with the maximum entropy. One of the (many) paradoxes in statistical mechanics was introduced by Maxwell in 1867: a microscopic creature (later called a *Maxwell Demon*) could open and close a door between two containers, separating fast from slow gas molecules without doing any work on them. This appears to violate the Second Law of Thermodynamics, because the hot and cold gases could be used to run a heat engine, making a perpetual motion machine. Leo Szilard studied this problem in 1929, reducing it to a single molecule that can be on either side of a partition, arguably the first introduction of the notion of a bit of information. While Szilard did not explain the paradox of the Demon, Shannon was inspired by this analysis to use entropy as a measure of information to build information theory, which later helped create the very important and practical modern theory of coding [Slepian, 1974].

The real resolution of Maxwell's Demon did not come until 1961, when Rolf Landauer showed that the irreversibility in the Demon arises when it forgets what it has done; any computer that erases information necessarily dissipates energy [Landauer, 1961]. A stored bit can be in one of two states; if you initially have no idea what was stored then the minimum entropy associated with this bit is

$$S = k \log \Omega = k \log 2 \quad . \tag{4.42}$$

A real bit may represent much more entropy than this because many electrons (for example) are used to store it, but this is the minimum possible. Erasing the bit reduces the number of possible microscopic states down to one. It compresses the phase space of the computer, and so the dissipation associated with this erasure is

$$\delta Q = T\, dS = kT \log 2 - kT \log 1 = kT \log 2 \quad . \tag{4.43}$$

No matter how a computer is built, erasing a bit costs a minimum energy on the order of $kT \log 2$.

This result contains a strong assumption that the bit is near enough to thermal equilibrium for statistical mechanics to apply and for temperature to be a meaningful concept. Also, kT at room temperature is on the order of 0.02 eV, far below the energy of bits stored in common computers. Nevertheless, Landauer's result is very important: whatever sets the energy scale of a stored bit (this might be the size of thermal fluctuations, or quantization in a small system), there is an energy penalty for erasing information. This has significant immediate implications for the design of low-power computers and algorithms [Gershenfeld, 1996].

Charles Bennett later showed in 1973 that it is possible to compute with *reversible computers* that never erase information and so can use arbitrarily little energy, depending on how long you are willing to wait for a sufficiently correct answer [Bennett, 1973]. We will return to this possibility in Section 10.5 when we look at the limits on computer performance.

4.6 SELECTED REFERENCES

[Cover & Thomas, 1991] Cover, Thomas M., & Thomas, Joy A. (1991). *Elements of Information Theory*. New York, NY: Wiley.

  A clear modern treatment of information theory.

[Balian, 1991] Balian, Roger. (1991). *From Microphysics to Macrophysics : Methods and Applications of Statistical Physics*. New York: Springer–Verlag. Translated by D. ter Haar and J.F. Gregg, 2 volumes.

  Statistical mechanics beautifully introduced from an information-theoretical point of view.

[Slepian, 1974] Slepian, David (ed.). (1974). *Key Papers in the Development of Information Theory*. New York: IEEE Press.

  The seminal papers in the development of information theory.

[Brush, 1976] Brush, Stephen G. (1976). *The Kind of Motion We Call Heat: A History of the Kinetic Theory of Gases in the 19th Century*. New York: North-Holland. 2 volumes.

  The history of statistical mechanics.

[Leff & Rex, 1990] Leff, Harvey S. & Rex, Andrew F. (eds.). (1990). *Maxwell's Demon: Entropy, Information, Computing*. Princeton: Princeton University Press.

  Key papers relating entropy and computing.

4.7 PROBLEMS

(4.1) Verify that the entropy function satisfies the required properties of continuity, non-negativity, monotonicity, and independence.

(4.2) Prove the relationships in equation (4.10).

(4.3) Calculate the differential entropy of a Gaussian process.

(4.4) A standard telephone line is specified to have a bandwidth of 3300 Hz and an SNR of 20 dB.

(*a*) What is the capacity?

(*b*) What SNR would be necessary for the capacity to be 1 Gbit/s?

(4.5) Let $(x_1, x_2, \ldots, x_n)$ be drawn from a Gaussian distribution with variance σ^2 and unknown mean value x_0. Show that $f(x_1, \ldots x_n) = n^{-1} \sum_{i=1}^{n} x_i$ is an estimator for x_0 that is unbiased and achieves the Cramér–Rao lower bound.

5 Electromagnetic Fields and Waves

James Clerk Maxwell's unification of electromagnetic phenomena, published in 1865, is perhaps the best example of a successful modern scientific theory [Maxwell, 1998]. In just a few simple equations he was able to show that the apparently distinct phenomena of electricity and magnetism were actually intimately related through a common theoretical framework that contained unexpected predictions, such as electromagnetic waves, and led to significant succeeding discoveries including the theory of special relativity. This chapter studies these *Maxwell's equations* for static and time-varying electric and magnetic fields. This theory is called *electrodynamics* because it describes the time variation of electromagnetic phenomena, and it will be the foundation of much the rest of the book.

5.1 VECTOR CALCULUS

Working with Maxwell's equations will require differentiating and integrating field vectors, and so our first step will be a review of the necessary vector calculus.

5.1.1 Vectors

Let $\vec{x} \equiv (x, y, z) \equiv (x_1, x_2, x_3)$ be the coordinates of a point expressed as a vector in rectangular coordinates (Figure 5.1). The *magnitude* of a vector is

$$|\vec{x}| = \sqrt{x_1^2 + x_2^2 + x_3^2} \quad . \tag{5.1}$$

For two vectors $\vec{A}$ and $\vec{B}$ with an angle θ between them, the *dot product* measures their overlap:

$$\begin{aligned}
\vec{A} \cdot \vec{B} &= |A||B|\cos(\theta) \\
&= A_1 B_1 + A_2 B_2 + A_3 B_3 \\
&= \sum_{i=1}^{3} A_i B_i \\
&\equiv A_i B_i \quad .
\end{aligned} \tag{5.2}$$

Because such sums recur frequently in manipulating vectors, the last line introduces the *Einstein summation convention* of summing over repeated indices.

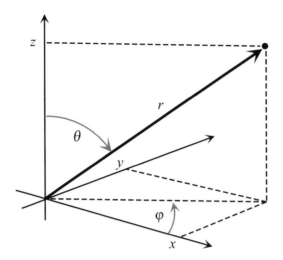

Figure 5.1. Rectangular, cylindrical, and spherical coordinate systems.

The *cross product* of these vectors is

$$\vec{A} \times \vec{B} = \begin{vmatrix} \hat{x}_1 & \hat{x}_2 & \hat{x}_3 \\ A_1 & A_2 & A_3 \\ B_1 & B_2 & B_3 \end{vmatrix}$$
$$= (A_2 B_3 - A_3 B_2)\hat{x}_1 + (A_3 B_1 - A_1 B_3)\hat{x}_2 + (A_1 B_2 - A_2 B_1)\hat{x}_3 \quad , \quad (5.3)$$

where $\hat{x}_1$ (pronounced "x hat") is a unit vector in the x_1 direction. The magnitude of the cross product is equal to the product of the lengths of the vectors times the sine of the angle between them, and its direction is perpendicular to the plane containing the vectors. The orientation can be remembered by the *right hand rule*: if the fingers of your right hand curl from $\vec{A}$ towards $\vec{B}$, your thumb points in the direction of their cross product. The *i*th term of the cross product can be written in terms of the summation convention as

$$(\vec{A} \times \vec{B})_i = \epsilon_{ijk} A_j B_k \qquad (5.4)$$

by using the *antisymmetric tensor*

$$\epsilon_{ijk} = \begin{cases} 1 & \text{if } (ijk) = (123),\ (231),\ \text{or } (312)\ \text{(cyclic permutation)} \\ -1 & \text{if } (ijk) = (132),\ (321),\ \text{or } (213)\ \text{(anticyclic permutation)} \\ 0 & \text{otherwise} \end{cases} \quad . \quad (5.5)$$

Interchanging indices shows that the cross product is *anticommutative*:

$$\vec{A} \times \vec{B} = -\vec{B} \times \vec{A} \quad . \qquad (5.6)$$

A useful expansion for the product of antisymmetric tensors is

$$\epsilon_{ijk}\epsilon_{klm} = \delta_{il}\delta_{jm} - \delta_{im}\delta_{jl} \quad , \qquad (5.7)$$

where δ_{ij} is the *Kroenecker delta*

$$\delta_{ij} = \begin{cases} 1 & \text{if } i = j \\ 0 & \text{otherwise} \end{cases} \quad . \qquad (5.8)$$

5.1.2 Differential Operators

Now let $\varphi(\vec{x})$ be a scalar function of $\vec{x}$. The *gradient* of φ is defined to be

$$\nabla\varphi(\vec{x}) = \frac{\partial\varphi(\vec{x})}{\partial x_1}\,\hat{x}_1 + \frac{\partial\varphi(\vec{x})}{\partial x_2}\,\hat{x}_2 + \frac{\partial\varphi(\vec{x})}{\partial x_3}\,\hat{x}_3 \quad . \tag{5.9}$$

The gradient is a vector that points in the direction of the fastest change of φ, and its magnitude is equal to the rate of change in that direction. If $\varphi(\vec{x})$ is the height of a hill, then a ball released at $\vec{x}$ would roll down the hill in the direction $-\nabla\varphi(\vec{x})$. The vector operator ∇ is called "del."

For a vector-valued function $\vec{A}(\vec{x}) = (A_1(\vec{x}), A_2(\vec{x}), A_3(\vec{x}))$, the *divergence* is

$$\begin{aligned}
\nabla \cdot \vec{A}(\vec{x}) &= \frac{\partial A_1}{\partial x_1} + \frac{\partial A_2}{\partial x_2} + \frac{\partial A_3}{\partial x_3} \\
&= \sum_i \frac{\partial A_i}{\partial x_i} \\
&\equiv \sum_i \partial_i A_i \\
&= \partial_i A_i \quad .
\end{aligned} \tag{5.10}$$

The divergence is a number that measures the rate at which the vector field is locally expanding or contracting.

The *curl* of a vector field in three dimensions is defined to be

$$\nabla \times \vec{A} = (\partial_2 A_3 - \partial_3 A_2)\hat{x}_1 + (\partial_3 A_1 - \partial_1 A_3)\hat{x}_2 + (\partial_1 A_2 - \partial_2 A_1)\hat{x}_3 \quad . \tag{5.11}$$

The curl points in the direction of circulation of the vector field. Written in the summation convention,

$$(\nabla \times \vec{A})_i = \epsilon_{ijk}\partial_j A_k \quad . \tag{5.12}$$

Plugging in the definitions shows that the curl of a gradient vanishes,

$$(\nabla \times \nabla\varphi)_i = \epsilon_{ijk}\partial_j\partial_k\varphi = 0 \quad , \tag{5.13}$$

as does the divergence of a curl,

$$\nabla \cdot \nabla \times \vec{A} = \epsilon_{ijk}\partial_i\partial_j A_k = 0 \quad . \tag{5.14}$$

The *Laplacian* of a scalar quantity is

$$\begin{aligned}
\nabla^2\varphi &= \nabla \cdot \nabla\varphi \\
&= \frac{\partial^2\varphi}{\partial x_1^2} + \frac{\partial^2\varphi}{\partial x_2^2} + \frac{\partial^2\varphi}{\partial x_3^2} \\
&= \partial_i\partial_i\varphi \quad .
\end{aligned} \tag{5.15}$$

It measures the curvature at a point. The Laplacian of a vector is a vector-valued quantity defined to be the Laplacian of each of the components of the vector

$$(\nabla^2\vec{A})_j = \partial_i\partial_i A_j \quad . \tag{5.16}$$

In addition to rectangular coordinates, there are two other common coordinate systems, *cylindrical* (r, φ, z) and *spherical* (r, θ, φ), also shown in Figure 5.1. When the coordinate system reflects the symmetries of a problem the math is much simpler. Inserting the trigonometric relationships among the varaibles into the definitions of the differential operators in rectangular coordinates, and taking the appropriate partial derivatives, shows that in cylindrical and spherical coordinates a differential volume element is

$$
\begin{aligned}
dV &= dx\, dy\, dz \\
&= r\, dr\, d\theta\, dz \\
&= r^2 \sin\theta\, dr\, d\theta\, d\varphi \quad,
\end{aligned}
\tag{5.17}
$$

the gradient is

$$
\begin{aligned}
\nabla\Phi &= \frac{\partial\Phi}{\partial x}\hat{x} + \frac{\partial\Phi}{\partial y}\hat{y} + \frac{\partial\Phi}{\partial z}\hat{z} \\
&= \frac{\partial\Phi}{\partial r}\hat{r} + \frac{1}{r}\frac{\partial\Phi}{\partial\varphi}\hat{\varphi} + \frac{\partial\Phi}{\partial z}\hat{z} \\
&= \frac{\partial\Phi}{\partial r}\hat{r} + \frac{1}{r}\frac{\partial\Phi}{\partial\theta}\hat{\theta} + \frac{1}{r\sin\theta}\frac{\partial\Phi}{\partial\varphi}\hat{\varphi} \quad,
\end{aligned}
\tag{5.18}
$$

the divergence is

$$
\begin{aligned}
\nabla\cdot\vec{A} &= \frac{\partial A_x}{\partial x} + \frac{\partial A_y}{\partial y} + \frac{\partial A_z}{\partial z} \\
&= \frac{1}{r}\frac{\partial}{\partial r}(rA_r) + \frac{1}{r}\frac{\partial A_\varphi}{\partial\varphi} + \frac{\partial A_z}{\partial z} \\
&= \frac{1}{r^2}\frac{\partial}{\partial r}(r^2 A_r) + \frac{1}{r\sin\theta}\frac{\partial}{\partial\theta}(A_\theta\sin\theta) + \frac{1}{r\sin\theta}\frac{\partial A_\varphi}{\partial\varphi} \quad,
\end{aligned}
\tag{5.19}
$$

the Laplacian is

$$
\begin{aligned}
\nabla^2\Phi &= \frac{\partial^2\Phi}{\partial x^2} + \frac{\partial^2\Phi}{\partial y^2} + \frac{\partial^2\Phi}{\partial z^2} \\
&= \frac{1}{r}\frac{\partial}{\partial r}\left(r\frac{\partial\Phi}{\partial r}\right) + \frac{1}{r^2}\frac{\partial^2\Phi}{\partial\varphi^2} + \frac{\partial^2\Phi}{\partial z^2} \\
&= \underbrace{\frac{1}{r^2}\frac{\partial}{\partial r}\left(r^2\frac{\partial\Phi}{\partial r}\right)}_{\frac{1}{r}\frac{\partial^2}{\partial r^2}(r\Phi)} + \frac{1}{r^2\sin\theta}\frac{\partial}{\partial\theta}\left(\sin\theta\frac{\partial\Phi}{\partial\theta}\right) + \frac{1}{r^2\sin^2\theta}\frac{\partial^2\Phi}{\partial\varphi^2} \quad,
\end{aligned}
\tag{5.20}
$$

and the curl is

$$
\begin{aligned}
\nabla\times\vec{A} &= \left(\frac{\partial A_z}{\partial y} - \frac{\partial A_y}{\partial z}\right)\hat{x} + \left(\frac{\partial A_x}{\partial z} - \frac{\partial A_z}{\partial x}\right)\hat{y} + \left(\frac{\partial A_y}{\partial x} - \frac{\partial A_x}{\partial y}\right)\hat{z} \\
&= \left(\frac{1}{r}\frac{\partial A_z}{\partial\varphi} - \frac{\partial A_\varphi}{\partial z}\right)\hat{r} + \left(\frac{\partial A_r}{\partial z} - \frac{\partial A_z}{\partial r}\right)\hat{\varphi} + \frac{1}{r}\left(\frac{\partial(rA_\varphi)}{\partial r} - \frac{\partial A_r}{\partial\varphi}\right)\hat{z} \\
&= \frac{1}{r\sin\theta}\left(\frac{\partial(A_\varphi\sin\theta)}{\partial\theta} - \frac{\partial A_\theta}{\partial\varphi}\right)\hat{r} + \frac{1}{r}\left(\frac{1}{\sin\theta}\frac{\partial A_r}{\partial\varphi} - \frac{\partial(rA_\varphi)}{\partial r}\right)\hat{\theta} \\
&\quad + \frac{1}{r}\left(\frac{\partial(rA_\theta)}{\partial r} - \frac{\partial A_r}{\partial\theta}\right)\hat{\varphi} \quad.
\end{aligned}
\tag{5.21}
$$

Table 5.1. *Vector and trigonometric identities.*

$$\nabla \times \nabla\varphi = 0$$
$$\nabla \cdot (\nabla \times \vec{A}) = 0$$
$$\nabla \cdot (\varphi\vec{A}) = \vec{A} \cdot \nabla\varphi + \varphi\nabla \cdot \vec{A}$$
$$\nabla \times (\varphi\vec{A}) = \nabla\varphi \times \vec{A} + \varphi\nabla \times \vec{A}$$
$$\vec{A} \times (\vec{B} \times \vec{C}) = \vec{B}(\vec{A} \cdot \vec{C}) - \vec{C}(\vec{A} \cdot \vec{B})$$
$$\nabla \cdot (\vec{A} \times \vec{B}) = \vec{B} \cdot (\nabla \times \vec{A}) - \vec{A} \cdot (\nabla \times \vec{B})$$
$$\cos(A \pm B) = \cos A \cos B \mp \sin A \sin B$$
$$\sin(A \pm B) = \sin A \cos B \pm \cos A \sin B$$
$$\cos(2A) = \cos^2 A - \sin^2 A$$
$$\sin(2A) = 2 \sin A \cos A$$
$$\sin^2 A + \cos^2 A = 1$$

Finally, Table 5.1 lists some vector and trigonometric relationships that will be needed later.

5.1.3 Integral Relationships

The differential operators introduced in the last section measure the local properties of scalar and vector fields. Not surprisingly, there are intimate relationships between these local properties and the global properties of the fields. These will be very useful for relating fields to their sources. Without proof, here are two important cases of general theorems relating local and global properties:

- *Divergence Theorem* (or *Gauss' Theorem*)

$$\int_V \nabla \cdot \vec{E} \, dV = \int_S \vec{E} \cdot d\vec{A} \quad . \tag{5.22}$$

  The volume integral of the divergence is equal to the surface integral of the normal component of the field. V is an arbitrary volume, S is its surface, dV is a volume element, and $d\vec{A}$ is a surface area element. $d\vec{A}$ is the same as $\hat{n} \, dA$, where $\hat{n}$ is an outward-pointing unit vector that is perpendicular to the patch dA. Adding up the net flux into or out of the volume is equivalent to adding up all the local sources and sinks.

- *Stokes' Theorem*

$$\int_S \nabla \times \vec{E} \cdot d\vec{A} = \oint_l \vec{E} \cdot d\vec{l} \quad . \tag{5.23}$$

  The line integral around a closed path is equal to the surface integral of the curl over any arbitrary surface bounded by the path (Figure 5.2). If your fingers curl in the direction of the line integral, your thumb points in the direction of the surface normal. The total circulation around the path is equal to the sum of all the local circulations.

Figure 5.2. Definition of Stokes' Theorem.

5.2 STATICS

This section reviews the governing equations for time-independent electromagnetic phenomena; the following one turns on time dependence to arrive at Maxwell's equations. Together these will serve as the foundation for much of the rest of the book. As surprising as the remarkable phenomenology contained in these apparently simple relationships is the sophistication of the techniques needed to reveal it.

5.2.1 Electrostatics

The force in a vacuum between two charges is given by *Coulomb's Law*:

$$\vec{F} = \frac{1}{4\pi\epsilon_0}\frac{q_1 q_2}{r^2}\hat{r} \quad \text{(N)} \quad . \tag{5.24}$$

$\epsilon_0 = 10^7/(4\pi c^2) = 8.854\times 10^{-12}$ F/m is a constant called the *permittivity of free space*, q_1 and q_2 are the size of the charges in coulombs, r^2 is the distance between them in meters, and $\hat{r}$ is a unit vector pointing between them. This relationship was determined experimentally by Charles Augustin Coulomb in 1785.

The force on one charge due to an applied electric field is

$$\vec{F} = q\vec{E} \tag{5.25}$$

and so the electric field due to a single charge is

$$\vec{E} = \frac{q}{4\pi\epsilon_0 r^2}\hat{r} \quad \left(\frac{\text{V}}{\text{m}}\right) \quad . \tag{5.26}$$

With this definition the electric field points from positive charge towards negative charge. The field diverges as you approach the charge; very close to a charge the expression is no longer valid and quantum electrodynamics is needed to describe the field.

We will shortly see that the curl of the electric field vanishes if there are no time-varying magnetic fields, which according to equation (5.13) means that the electric field can be written as the gradient of a *potential* Φ

$$\vec{E} = -\nabla\Phi \quad . \tag{5.27}$$

Given this definition, the potential from a point charge is

$$\Phi = \frac{q}{4\pi\epsilon_0 r} \quad \text{(V)} \quad . \tag{5.28}$$

Since the electric field is the gradient of the potential, the potential difference between two points $\vec{x}$ and $\vec{y}$ can be found by integrating the electric field along an arbitrary path between the points

$$V(\vec{x}, \vec{y}) = \Phi(\vec{y}) - \Phi(\vec{x})$$

$$= -\int_{\vec{x}}^{\vec{y}} \vec{E} \cdot d\vec{l} \quad (\mathrm{V}) \quad . \tag{5.29}$$

It is convenient to view the electric field in terms of fictitious *field lines*, which are perpendicular to the lines of constant potential and have an areal density proportional to the field strength.

An electric field inside a material is modified by the response of the material to the field. In a *dielectric*, the charge is bound so that it is not free to move, but an applied electric field will polarize the bound charge. Because of this polarization, the strength of the field generated by a free charge in the material will be reduced by a factor called the *permittivity* ϵ

$$\vec{E} = \frac{q}{4\pi\epsilon r^2}\hat{r} \quad . \tag{5.30}$$

The permittivity ϵ equals the *relative permittivity* of the material ϵ_r times the permittivity of free space

$$\epsilon = \epsilon_0\epsilon_r \quad . \tag{5.31}$$

ϵ_r, also called the *dielectric constant*, is 1 in a vacuum, between 2 and 5 for typical plastics, and can be over 100 in a material such as $SrTiO_3$. Depending on the symmetry of the material the permittivity can be a tensor that depends on direction, and for strong fields (such as those generated by a laser, Problem 5.5) it will depend nonlinearly on the field. This latter property is very useful for mixing and generating harmonics of incident beams of light.

Let's take the divergence of the field due to a charge and integrate over an infinitesimal spherical volume of radius r around the charge. According to Gauss' Theorem,

$$\int_V \nabla \cdot \vec{E} \, dV = \int_S \vec{E} \cdot d\vec{A}$$

$$= \int_S \frac{q}{4\pi\epsilon r^2}\hat{r} \cdot \hat{r} \, dA$$

$$= \frac{q}{4\pi\epsilon r^2}4\pi r^2$$

$$= \frac{q}{\epsilon}$$

$$= \int_V \frac{\rho}{\epsilon} \, dV \quad . \tag{5.32}$$

In the last line we've introduced the *charge density* ρ, which for this point charge is just a delta function. Since the left side must equal the right side independent of the volume, the integrands must be equal, and we see that

$$\nabla \cdot \vec{E} = \frac{\rho}{\epsilon} = \frac{\rho}{\epsilon_0\epsilon_r} \quad . \tag{5.33}$$

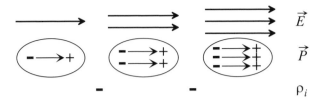

Figure 5.3. Relationship between an electric field with a gradient, the resulting spatial variation of the polarization, and the net charge induced by the local charge imbalance.

If we define $\vec{D} = \epsilon \vec{E}$, and if ϵ is a constant, this reads

$$\nabla \cdot \vec{D} = \rho \quad . \tag{5.34}$$

$\vec{D}$ is called the *displacement* field, and this is the differential form of *Gauss' Law*. $\vec{E}$ is the real physical field that exerts a force on charges; $\vec{D}$ is the effective field that results from source charges.

If the $\vec{E}$ field in the material is uniform, the induced polarization will be constant. However, if the $\vec{E}$ field varies in space (Figure 5.3) then there will be a spatially varying induced polarization, leading to an average induced charge density. To understand this, let's return to equation (5.33). The electric field in the material can be viewed as being the sum of the field due to any free charge ρ_{free} that would be there if the material was not present ($\epsilon_r = 1$), and the field due to the induced charge ρ_{induced}. The induced charge is conventionally defined in terms of a *polarization vector* $\vec{P}$

$$\rho_{\text{induced}} \equiv -\nabla \cdot \vec{P} \quad , \tag{5.35}$$

and so

$$\nabla \cdot \epsilon_0 \vec{E} = \rho_{\text{free}} + \rho_{\text{induced}} \tag{5.36}$$
$$= \rho_{\text{free}} - \nabla \cdot \vec{P}$$

or

$$\nabla \cdot \underbrace{(\epsilon_0 \vec{E} + \vec{P})}_{\equiv \vec{D}} = \rho_{\text{free}} \quad . \tag{5.37}$$

If the field is not too strong then $\vec{P}$ will be linearly related to $\vec{E}$, and this relationship defines the *electric susceptibility* χ_e

$$\vec{D} \equiv \epsilon_0 \vec{E} + \vec{P} = \epsilon_0 (1 + \chi_e) \vec{E} = \epsilon_0 \epsilon_r \vec{E} = \epsilon \vec{E} \quad . \tag{5.38}$$

The *dipole moment* of a charge distribution is defined as the integral of the charge times the position $\int \vec{x} \rho(\vec{x}) \, dV$. To relate this to $\vec{P}$, first note that by differentiating $x\vec{P}$ and writing out the terms,

$$\nabla \cdot (x\vec{P}) = x\nabla \cdot \vec{P} + P_x \quad . \tag{5.39}$$

Therefore,

$$\int x \rho_{\text{induced}} \, dV = -\int x \nabla \cdot \vec{P} \, dV$$

$$= \int P_x \, dV - \int \nabla \cdot (x\vec{P}) \, dV$$

$$= \int P_x \, dV - \int x\vec{P} \cdot d\vec{A} \quad . \tag{5.40}$$

In the limit that the volume goes to zero, $\vec{P}$ will be uniform and so the second term will vanish. Dropping it and repeating the calculation for the y and z components gives

$$\int \vec{x} \rho_{\text{induced}} \, dV = \int \vec{P} \, dV \quad . \tag{5.41}$$

Since this must be true for any volume, we see that the polarization vector is equal to the local density of the dipole moment. Note that unlike the $\vec{E}$ field, the dipole moment is defined to point from negative charge to positive charge.

Substituting the definition of the potential into Gauss' Law and assuming a homogeneous polarizibility gives *Poisson's equation*

$$\nabla^2 \Phi(\vec{x}) = -\frac{\rho(\vec{x})}{\epsilon} \tag{5.42}$$

For a point charge located at $\vec{x}_0$, $\rho(\vec{x}) = q \, \delta(|\vec{x} - \vec{x}_0|)$, and the potential is given by equation (5.28). Plugging these into Poisson's equation,

$$\nabla^2 \Phi(\vec{x}) = -\frac{\rho(\vec{x})}{\epsilon}$$

$$\nabla^2 \frac{q}{4\pi\epsilon|\vec{x} - \vec{x}_0|} = -\frac{q}{\epsilon} \, \delta(|\vec{x} - \vec{x}_0|)$$

$$\nabla^2 \frac{1}{|\vec{x} - \vec{x}_0|} = -4\pi \, \delta(|\vec{x} - \vec{x}_0|) \quad . \tag{5.43}$$

This relationship can then be used to show that

$$\Phi(\vec{x}) = \frac{1}{4\pi\epsilon} \int \frac{\rho(\vec{x}')}{|\vec{x} - \vec{x}'|} \, d^3x' \tag{5.44}$$

solves Poisson's equation. $1/|\vec{x} - \vec{x}'|$ is a *Green's function* for this problem, relating the field to an integral over its source. A similar solution can be found for $\vec{E}$ by using

$$\nabla \frac{1}{|\vec{x} - \vec{x}'|} = -\frac{\vec{x} - \vec{x}'}{|\vec{x} - \vec{x}'|^3}$$

$$\Rightarrow \vec{E}(\vec{x}) = -\nabla\Phi(\vec{x})$$

$$= \frac{1}{4\pi\epsilon} \int \rho(\vec{x}') \frac{\vec{x} - \vec{x}'}{|\vec{x} - \vec{x}'|^3} \, d^3x' \quad . \tag{5.45}$$

In free space, Poisson's equation reduces to

$$\nabla^2 \Phi(\vec{x}) = 0 \quad . \tag{5.46}$$

This is called *Laplace's equation*, and governs many other phenomena including the profile of a membrane such as a drumhead stretched around a boundary. One of the properties of its solution is that it can take on an extremum only on the boundary, and so electromagnetic particle traps require time-varying fields. The solution of Laplace's equation requires the specification these boundary conditions, usually given by either the distribution of the potential on the boundary (*Dirichlet* boundary conditions) or its normal derivative (*Neumann* boundary conditions).

Note that we found these equations by integrating Coulomb's Law, which involved canceling the r^2 dependence of a three-dimensional surface area with the r^{-2} dependence of the field. In anything other than three dimensions this would not work. Laplace's equation is routinely solved numerically, and to make it tractable it is frequently solved in two dimensions, but it is very important to remember that solving Laplace's equation in two dimensions corresponds to taking Coulomb's law to have a r^{-1} form because the length of a two-dimensional perimeter is proportional to r. This may effectively be correct if the problem has two-dimensional symmetry so that each point corresponds to a line of charge, but it will be incorrect for an arbitrary two-dimensional slice of a three-dimensional problem.

The *capacitance* between two electrodes relates the charge Q on each of them to the potential difference V between them:

$$C = \frac{Q}{V} \quad . \tag{5.47}$$

The relationship to current is found by differentiating:

$$C\frac{dV}{dt} = \frac{dQ}{dt} = I \quad . \tag{5.48}$$

5.2.2 Magnetostatics

Electric fields are produced by stationary charges; magnetic fields are produced by moving charges. The strength of the *magnetic field* $\vec{H}$ due to an infinitesimal section of a wire carrying a current (Figure 5.4) was found experimentally in 1820 to be governed by the *Biot–Savart Law*

$$d\vec{H} = \frac{I\,d\vec{l} \times \hat{r}}{4\pi r^2}, \tag{5.49}$$

or integrated over space

$$\begin{aligned}
\vec{H} &= \int \frac{I\,d\vec{l} \times \hat{r}}{4\pi r^2} \\
&= \frac{1}{4\pi} \int \vec{J}(\vec{x}') \times \frac{\vec{x} - \vec{x}'}{|\vec{x} - \vec{x}'|^3}\, d^3x' \quad (\text{A}/\text{m}) \quad ,
\end{aligned} \tag{5.50}$$

where $\vec{J}$ is the current density. Using the right hand rule, if your thumb points in the direction of current flow then you fingers will curl in the direction of the field.

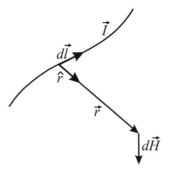

Figure 5.4. The magnetic field due to a differential current element.

The relationship between $\vec{H}$ and $\vec{J}$ can be written more simply by taking the curl,

$$\vec{H}(\vec{x}) = \frac{1}{4\pi} \int \vec{J}(\vec{x}') \times \frac{\vec{x} - \vec{x}'}{|\vec{x} - \vec{x}'|^3} \, d^3 x'$$

$$= \frac{1}{4\pi} \int \vec{J}(\vec{x}') \times \nabla_{\vec{x}} \frac{1}{|\vec{x} - \vec{x}'|} \, d^3 x'$$

$$\nabla_{\vec{x}} \times \vec{H}(\vec{x}) = \frac{1}{4\pi} \int \nabla_{\vec{x}} \times \vec{J}(\vec{x}') \times \nabla_{\vec{x}} \frac{1}{|\vec{x} - \vec{x}'|} \, d^3 x'$$

$$= \frac{1}{4\pi} \int \vec{J}(\vec{x}') \underbrace{\nabla_{\vec{x}}^2 \frac{1}{|\vec{x} - \vec{x}'|}}_{\delta(\vec{x} - \vec{x}')} - \nabla_{\vec{x}} \frac{1}{|\vec{x} - \vec{x}'|} \underbrace{\nabla_{\vec{x}} \cdot \vec{J}(\vec{x}')}_{0} \, d^3 x'$$

$$\nabla \times \vec{H} = \vec{J} \tag{5.51}$$

(using the BAC–CAB rule $\vec{A} \times (\vec{B} \times \vec{C}) = \vec{B}(\vec{A} \cdot \vec{C}) - \vec{C}(\vec{A} \cdot \vec{B})$, Problem 5.1). We will soon see that a term must be added to this equation if the fields are time-varying.

The force on an infinitesimal current element is given in terms of the *magnetic flux density* $\vec{B}$ by

$$d\vec{F} = I d\vec{l} \times \vec{B} \quad , \tag{5.52}$$

or for a single moving charge

$$\vec{F} = q\vec{v} \times \vec{B} \quad . \tag{5.53}$$

The continuum version is

$$\vec{F} = \int \vec{J} \times \vec{B} \, dV \quad . \tag{5.54}$$

Analogous to the relationship between $\vec{D}$ and $\vec{E}$, in a material $\vec{H}$ and $\vec{B}$ are related by the *permeability* μ:

$$\vec{B} = \mu_0(\vec{H} + \vec{M}) = \mu_0(1 + \chi_{\rm m})\vec{H} = \mu_0\mu_{\rm r}\vec{H} = \mu\vec{H} \quad (\text{T}) \quad . \tag{5.55}$$

$\mu_0 = 4\pi \times 10^{-7}$ H/m is the *permeability of free space*, $\vec{M}$ is the *magnetization*, the *relative permeability* $\mu_{\rm r} = 1$ for a vacuum, and $\chi_{\rm m}$ is the *magnetic susceptibility*. $\vec{H}$ is the effective field that results from source currents, and $\vec{B}$ is the physical field that exerts a force on charges. As with $\vec{D} = \epsilon\vec{E}$, the linear relationship $\vec{B} = \mu\vec{H}$ applies only to weak fields; magnetic recording depends on the nonlinear hysteresis in μ. Chapter 12 will explain the origin of this, as well as the reason why $\chi_m < 0$ for diamagnetic materials but $\chi_m > 0$ for paramagnetic, ferromagnetic, and ferrimagnetic materials. For iron $\chi_m \sim 10^3$; for non-magnetic materials $\chi_m \sim 10^{-5}$ and so $\mu_r \sim 1$ for them. In a high-permeability alloy such as *mumetal* ($Fe_{18}Ni_{75}Cu_5Cr_2$) the relative permeability is $\sim 10^5$.

Taking the curl of $\vec{H}$,

$$
\begin{aligned}
\nabla \times \vec{H} &= \vec{J} \\
&= \nabla \times \frac{1}{\mu_0} \vec{B} - \nabla \times \vec{M} \\
&= J_{\text{free}} - J_{\text{induced}} \quad .
\end{aligned}
\tag{5.56}
$$

As with the electrostatic case, the effective current in the material can be decomposed into free and induced currents. The induced current can be understood in terms of the *magnetic moment* of a current distribution, defined to be

$$
\vec{m} = \frac{1}{2} \int \vec{x} \times \vec{J}(\vec{x}) \, dV \quad .
\tag{5.57}
$$

For example, for a circular current loop for radius r,

$$
\begin{aligned}
|\vec{m}| &= \frac{1}{2} r \, I \, 2\pi r \\
&= I \pi r^2 \\
&= I \cdot \text{area} \quad .
\end{aligned}
\tag{5.58}
$$

The magnetic moment associated with the induced current is equal to

$$
\begin{aligned}
\vec{m}_{\text{induced}} &= \frac{1}{2} \int \vec{x} \times \vec{J}_{\text{induced}}(\vec{x}) \, dV \\
&= \frac{1}{2} \int \vec{x} \times (\nabla \times \vec{M}) \, dV \quad ,
\end{aligned}
\tag{5.59}
$$

or for the ith component

$$
m_{i,\text{induced}} = \frac{1}{2} \int \epsilon_{ijk} x_j \epsilon_{klm} \partial_l M_m \, dV \quad .
\tag{5.60}
$$

Since

$$
\partial_l(x_j M_m) = x_j \partial_l M_m + M_m \underbrace{\partial_l x_j}_{\delta_{jl}} \quad ,
\tag{5.61}
$$

this can be rewritten as

$$
m_{i,\text{induced}} = \frac{1}{2} \int \epsilon_{ijk} \epsilon_{klm} \left[\partial_l(x_j M_m) - M_m \delta_{jl} \right] dV \quad .
\tag{5.62}
$$

The integral over all space of the first term in the brackets will vanish. This is because the integral over all space of the magnetization M is bounded, therefore asymptotically M must fall off faster than $1/x$. Hence $\lim_{x \to \infty} x M(x) = 0$, so the integral of the derivative of this quantity is just its value at $\pm\infty$, which is 0. That leaves the remaining term

$$
\begin{aligned}
m_{i,\text{induced}} &= -\frac{1}{2} \int \epsilon_{ijk}\epsilon_{klm} M_m \delta_{jl} \, dV \\
&= -\frac{1}{2} \int \left(\delta_{il}\delta_{jm} - \delta_{im}\delta_{jl} \right) M_m \delta_{jl} \, dV \\
&= -\frac{1}{2} \int \delta_{il}\delta_{jm}\delta_{jl} M_m - \delta_{im}\delta_{jl}\delta_{jl} M_m \, dV \\
&= -\frac{1}{2} \int \delta_{il}\delta_{ml} M_m - \delta_{im} 3 M_m \, dV \\
&= -\frac{1}{2} \int M_i - 3 M_i \, dV \\
&= \int M_i \, dV \quad ,
\end{aligned}
\tag{5.63}
$$

or for all components

$$
\vec{m}_{\text{induced}} = \int \vec{M} \, dV \quad .
\tag{5.64}
$$

The magnetic dipole moment is equal to the integral of the magnetization, therefore the magnetization is the local density of the dipole moment.

As far as we know there is no such thing as a magnetic "charge", called a magnetic monopole, so $\nabla \cdot \vec{B} = 0$. This means that $\vec{B}$ can be written as the curl of a vector field $\vec{A}$ (equation 5.14):

$$
\vec{B} = \nabla \times \vec{A} \quad ,
\tag{5.65}
$$

called the *vector potential*. $\vec{A}$ is related to source currents by

$$
\vec{A}(\vec{x}) = \frac{\mu}{4\pi} \int \frac{\vec{J}(\vec{x}')}{|\vec{x} - \vec{x}'|} \, d^3 x' \quad ,
\tag{5.66}
$$

verified by taking the curl to obtain equation (5.50). This relation holds for static current distributions; Section 7.1 will extend the scalar and vector potentials to time-dependent sources.

In quantum mechanics the vector potential takes on a deep significance beyond this formal definition. The *Aharonov–Bohm effect* considers a particle moving outside an infinite solenoid; the magnetic field vanishes there but the vector potential does not, and this leads to observable quantum interference effects [Sakurai, 1967]. This effect demonstrates the physical reality of the vector potential.

5.2.3 Multipoles

The theory of *multipoles* provides a systematic way to approximate the fields produced by more complex charge and current distributions. One way to understand it is by expanding the inverse distance

$$
\begin{aligned}
\frac{1}{|\vec{r} - \vec{x}|} &= \frac{1}{|\vec{r}|} + \frac{\vec{r} \cdot \vec{x}}{|\vec{r}|^3} + \cdots \\
&= \frac{1}{r} + \frac{\vec{r} \cdot \vec{x}}{r^3} + \cdots \quad ,
\end{aligned}
\tag{5.67}
$$

where $\vec{r}$ is the distance from the source to where the field is being evaluated, $\vec{x}$ is the location within the source relative to its origin, and $\vec{x}$ is assumed to be much smaller than $\vec{r}$. Substituting this series into the potential rewrites it as

$$
\begin{aligned}
\Phi(\vec{r}) &= \frac{1}{4\pi\epsilon} \int \frac{\rho(\vec{x})}{|\vec{r} - \vec{x}|} \, dV \\
&= \frac{1}{4\pi\epsilon} \left(\frac{1}{r} \int \rho(\vec{x}) \, dV + \frac{\vec{r}}{r^3} \cdot \int \rho(\vec{x})\vec{x} \, dV + \cdots \right) \\
&\equiv \frac{1}{4\pi\epsilon} \left(\frac{q}{r} + \frac{\vec{p} \cdot \vec{r}}{r^3} + \cdots \right) \quad .
\end{aligned}
\tag{5.68}
$$

q is the *monopole* term, $\vec{p}$ is the *dipole moment*, and the next term would be the *quadrupole moment*. The corresponding electric field can be found by taking the gradient

$$
\begin{aligned}
\vec{E} &= -\nabla\Phi \\
&= \frac{q\hat{r}}{4\pi\epsilon r^2} + \frac{3\hat{r}(\vec{p} \cdot \hat{r}) - \vec{p}}{4\pi\epsilon r^3} + \cdots \\
&= \frac{q}{4\pi\epsilon r^2}\hat{r} + \frac{2p\cos\theta}{4\pi\epsilon r^3}\hat{r} + \frac{p\sin\theta}{4\pi\epsilon r^3}\hat{\theta} + \cdots \quad ,
\end{aligned}
\tag{5.69}
$$

and the energy associated with the charge distribution is

$$
\begin{aligned}
U &= \int \rho(\vec{x})\Phi(\vec{x}) \, dV \\
&= \int \rho(\vec{x}) \left[\Phi(0) + \vec{x} \cdot \nabla\Phi(\vec{x})|_{\vec{x}=0} + \cdots \right] dV \\
&= q\Phi(0) - \vec{p} \cdot \vec{E}(0) + \cdots \quad .
\end{aligned}
\tag{5.70}
$$

The same expansion can be used with the vector potential,

$$
\begin{aligned}
A_i(\vec{r}) &= \frac{\mu}{4\pi} \int \frac{J_i(\vec{x})}{|\vec{r} - \vec{x}|} \, dV \\
&= \frac{\mu}{4\pi} \left(\frac{1}{r} \int J_i(\vec{x}) \, dV + \frac{\vec{r}}{r^3} \cdot \int J_i(\vec{x})\vec{x} \, dV + \cdots \right) \quad .
\end{aligned}
\tag{5.71}
$$

Because of the vectorial character, finding this field is a bit more tricky. To start, notice that for arbitrary functions f and g, integration by parts shows that

$$
\begin{aligned}
\int f\vec{J} \cdot \nabla g \, dV &= 0 - \int g\nabla \cdot (f\vec{J}) \, dV \\
&= -\int g\vec{J} \cdot \nabla f \, dV - \int fg\nabla \cdot \vec{J} \, dV \quad ,
\end{aligned}
\tag{5.72}
$$

where the integrals are over all space, and the only assumption that's been made is that J vanishes at infinity. Rearranging terms,

$$
\int \left(f\vec{J} \cdot \nabla g + g\vec{J} \cdot \nabla f + fg\nabla \cdot \vec{J} \right) dV = 0 \quad .
\tag{5.73}
$$

If we plug in $\nabla \cdot \vec{J} = 0$, and take $f = 1, g = x_i$, then

$$
\int J_i \, dV = 0
\tag{5.74}
$$

and so the monopole vector potential term vanishes (there are no free magnetic charges). Now taking $f = x_i$ and $g = x_j$,

$$\int \left(x_i J_j + x_j J_i \right) dV = 0 \tag{5.75}$$

or

$$\frac{1}{2} \int \left(x_i J_j - x_j J_i \right) dV = \int x_i J_j \, dV \quad . \tag{5.76}$$

This can be substituted into the dipole term to relate it to the vector potential:

$$A_i(\vec{r}) = \frac{\mu}{4\pi} \frac{\vec{r}}{r^3} \cdot \int \vec{x} J_i(\vec{x}) \, dV$$

$$= \frac{\mu}{4\pi} \frac{1}{r^3} \int r_j x_j J_i \, dV$$

$$= -\frac{1}{2} \frac{\mu}{4\pi} \frac{1}{r^3} \int (r_j x_i J_j - r_j x_j J_i) \, dV$$

$$\vec{A}(\vec{r}) = -\frac{1}{2} \frac{\mu}{4\pi} \frac{\vec{r}}{r^3} \times \int \vec{x} \times \vec{J}(\vec{x}) \, dV$$

$$= \frac{\mu}{4\pi} \frac{\vec{m} \times \vec{r}}{r^3} \quad . \tag{5.77}$$

The magnetic field is found from the curl

$$\vec{B} = \nabla \times \vec{A}$$

$$= \frac{\mu}{4\pi} \nabla \times \frac{\vec{m} \times \vec{r}}{r^3}$$

$$B_i = \frac{\mu}{4\pi} \epsilon_{ijk} \partial_j \frac{1}{r^3} \epsilon_{klm} m_l r_m$$

$$= \frac{\mu}{4\pi} \epsilon_{ijk} \epsilon_{klm} m_l \partial_j \frac{r_m}{r^3}$$

$$= \frac{\mu}{4\pi} \left(\delta_{il} \delta_{jm} - \delta_{im} \delta_{jl} \right) m_l \left(\frac{\delta_{jm}}{r^3} - \frac{3 r_j r_m}{r^5} \right)$$

$$= \frac{\mu}{4\pi} \left(\frac{3 m_i}{r^3} - \frac{m_i}{r^3} - \frac{3 m_i}{r^3} + \frac{3 r_i m_j r_j}{r^5} \right)$$

$$\vec{B} = \frac{\mu}{4\pi} \frac{3 \hat{r}(\vec{m} \cdot \hat{r}) - \vec{m}}{r^3} \quad . \tag{5.78}$$

This is exactly the same as the electrostatic dipole field (equation 5.69).

The force on a magnetic dipole can be derived by applying the substitution used to find the vector potential:

$$\vec{F} = \int \vec{J} \times \vec{B} \, dV$$

$$F_i = \int (\vec{J} \times \vec{B})_i \, dV$$

$$= \epsilon_{ijk} \int J_j B_k \, dV$$

$$= \epsilon_{ijk} \int J_j [B_k(0) + \vec{x} \cdot \nabla B_k(\vec{x})|_{\vec{x}=0}] \, dV$$

$$= 0 + \epsilon_{ijk} \int J_j \vec{x} \cdot \nabla B_k \, dV$$

$$= \epsilon_{ijk} \nabla B_k \cdot \int J_j \vec{x} \, dV$$

$$= \epsilon_{ijk} (\vec{m} \times \nabla B_k)_j$$

$$= \epsilon_{ijk} \epsilon_{jlm} m_l \partial_m B_k$$

$$= (\delta_{im}\delta_{kl} - \delta_{il}\delta_{km}) m_l \partial_m B_k$$

$$= m_k \partial_i B_k - m_i \partial_k B_k$$

$$\vec{F} = \nabla(\vec{m} \cdot \vec{B}) - \vec{m} \underbrace{(\nabla \cdot \vec{B})}_{0} \quad . \tag{5.79}$$

Since a conservative force is the gradient of the potential energy, the energy of a magnetic dipole in a field is

$$U = -\vec{m} \cdot \vec{B}$$

$$= -mB\cos\theta \quad , \tag{5.80}$$

where θ is the angle between the dipole and the local field. There is an angular dependence to this that will seek to align the dipole with the field,

$$\frac{\partial U}{\partial \theta} = mB\sin\theta \quad , \tag{5.81}$$

i.e., there will be a torque about the axis perpendicular to them of

$$\vec{\tau} = \vec{m} \times \vec{B} \quad . \tag{5.82}$$

Note that all of these calculations have assumed that the distance to the point where the field is being evaluated is large compared to the special extent of this source. If the fields are needed closer to the source it's necessary to either use the full distribution or carry the multipole approximation out to a high order.

5.3 DYNAMICS

5.3.1 Maxwell's Equations

We're now ready for Maxwell's contribution. The statics equations tell us that the divergence of $\vec{B}$ and the curl of $\vec{E}$ vanish, and relate the divergence of $\vec{D}$ and the curl of $\vec{H}$ to their sources. Faraday had found that a varying magnetic field induces a current in a wire, and Ampère that a current produces a magnetic field; Maxwell realized that for these equations to be consistent (Problem 5.2) there must also be a time derivative of $\vec{D}$:

$$\nabla \cdot \vec{D} = \rho(\vec{x})$$

$$\nabla \cdot \vec{B} = 0$$

$$\nabla \times \vec{E} = -\frac{\partial \vec{B}}{\partial t}$$

$$\nabla \times \vec{H} = \vec{J}(\vec{x}) + \frac{\partial \vec{D}}{\partial t} \quad . \tag{5.83}$$

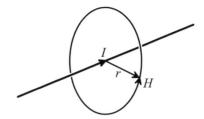

Figure 5.5. The magnetic field around a current-carrying wire.

These are now called *Maxwell's equations*; they show that electric and magnetic fields are connected through a more general theory of electromagnetic phenomena.

Material properties appear in Maxwell's equations through $\vec{D} = \epsilon\vec{E}$ and $\vec{B} = \mu\vec{H}$. In addition, the current $\vec{J}$ is related to the electric field by $\vec{J} = \sigma\vec{E}$. The coefficient σ is the *conductivity*, equal to the inverse of the *resistivity* ρ. In real materials ϵ, μ, and σ can become tensors that depend on direction, and can be complex quantities because of loss mechanisms.

We've already seen the first of Maxwell's equations; integrating it over a volume gives the integral form of Gauss' Law

$$\int_S \vec{D} \cdot d\vec{A} = \int_V \rho \, dV = Q \quad . \tag{5.84}$$

The surface integral of the normal component of $\vec{D}$ is equal to the charge Q enclosed. The second equation lacks a source term and the third one lacks a current term because there are no (known) magnetic monopoles. Integrating the last equation over a surface gives *Stokes' Law*

$$\oint \vec{H} \cdot d\vec{l} = \int_S \left(\vec{J} + \frac{\partial \vec{D}}{\partial t} \right) \cdot d\vec{A} \quad . \tag{5.85}$$

The line integral of the magnetic field around a path is equal to the current crossing an arbitrary surface bounded by the path. The first term on the right hand side of equation (5.85) is the conventional current, and the second term $\partial\vec{D}/\partial t$ is called the *displacement current*. This acts just like a real current, but instead of charge moving it is associated with an electric field changing. The current that flows into and out of a capacitor appears to travel through the space between the capacitor plates; Problem 5.2 shows that this is accounted for by the displacement current. Anyone who has been shocked by a charged capacitor can attest to the reality of this current.

It can be possible to use Stokes' Law to find magnetic fields without direct integration. For example, if a wire is carrying a current I, by symmetry according to the Biot–Savart Law the magnetic field must be directed circumferentially around the wire (Figure 5.5). This means that the line integral is just the field strength times the circumference, and the surface integral of the current density is equal to the current flowing through the wire, therefore

$$2\pi r H = I \Rightarrow H = \frac{I}{2\pi r} \quad . \tag{5.86}$$

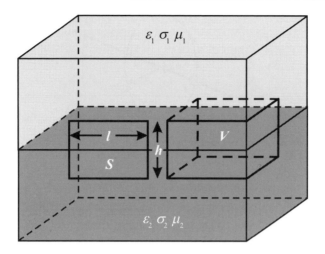

Figure 5.6. Loop and volume used for evaluating boundary conditions.

5.3.2 Boundary Conditions

The integral forms of Maxwell's equations can be used to find the conditions that the fields must satisfy at interfaces between materials, as specified by the dielectric constant ϵ, the conductivity σ, and the permeability μ. Start by integrating Gauss' Law over the volume V in Figure 5.6:

$$\int \nabla \cdot \vec{D} \, dV = \int \vec{D} \cdot d\vec{A} = \int \rho \, dV \quad . \tag{5.87}$$

In the limit that the height of the box $h \rightarrow 0$, the only contributions to the surface integral will come from the top and bottom. If the box if infinitesimal then the fields can be taken to be constant, and so the surface integral is just the normal component of the field times the area:

$$(\vec{D}_1 - \vec{D}_2) \cdot \hat{n} A = \int_V \rho \, dV \quad , \tag{5.88}$$

where $\hat{n}$ is a unit vector normal to the interface. There is a sign change between the integrals over the top and the bottom because the surface normal changes directions. If there is charge at the interface with an areal density ρ_s, then

$$\int_V \rho \, d\vec{V} = \rho_s A \quad . \tag{5.89}$$

Therefore,

$$(\vec{D}_1 - \vec{D}_2) \cdot \hat{n} = \rho_s \quad . \tag{5.90}$$

The change in the normal component of $\vec{D}$ across the interface is equal to the charge density at the interface. An applied field will create such surface charge to match the boundary conditions.

Next, integrate the curl of $\vec{E}$ over the surface S in Figure 5.6:

$$\int \nabla \times \vec{E} \cdot d\vec{A} = \oint \vec{E} \cdot d\vec{l} = -\int_S \frac{\partial \vec{B}}{\partial t} \cdot d\vec{A} \quad . \tag{5.91}$$

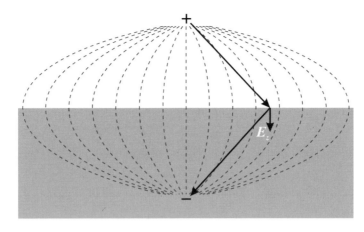

Figure 5.7. Image charge solution for the field of a charge above a ground plane.

As before, if the height h of the loop goes to zero then the only contribution to the line integral comes from the top and bottom, and the integral over the surface on the right hand side vanishes. Therefore

$$(\vec{E}_1 - \vec{E}_2) \times \hat{n} \, l = 0 \tag{5.92}$$

or

$$(\vec{E}_1 - \vec{E}_2) \times \hat{n} = 0 \quad . \tag{5.93}$$

The tangential component of $\vec{E}$ is continuous across the interface.

Since $\vec{J} = \sigma \vec{E}$, and in a perfect conductor $\sigma = \infty$, there can be no $\vec{E}$ field in an ideal conductor otherwise there would be an infinite current. According to equation (5.93), this means that there can be no tangential component on either side of the interface. Equation (5.90) does permit a normal component outside the interface, but it must be screened by a surface charge $\rho_s = D_1$. We see that the electric field must be perpendicular to the surface of a perfect conductor, and that a surface charge is induced to screen the interior from the field.

Integrating the divergence of $\vec{B}$ over the volume as we did for the divergence of $\vec{D}$ shows that

$$(\vec{B}_1 - \vec{B}_2) \cdot \hat{n} = 0 \quad . \tag{5.94}$$

The normal component of the $\vec{B}$ field is continuous across the interface. Similarly, integrating the curl of $\vec{H}$ over the surface instead of the curl of $\vec{E}$ gives

$$(\vec{H}_1 - \vec{H}_2) \times \hat{n} = \vec{J}_s \quad , \tag{5.95}$$

where $\vec{J}_s$ is the density of current at the surface. The tangential component of $\vec{H}$ changes by the surface current density across the interface.

The solution to Laplace's equation is unique. This means that *any* solution that satisfies the boundary conditions, no matter how it is found, is *the* solution. This observation leads to the useful *method of images*. Consider a point charge about an infinite ground plane, shown in Figure 5.7. A fictitious *image charge* is shown equidistant below the plane. By symmetry, the electric field lines are perdendicular to the surface. This is exactly the

Table 5.2. *MKS and CGS governing equations.*

MKS	CGS
$\vec{D} = \epsilon_0 \vec{E} + \vec{P}$	$\vec{D} = \vec{E} + 4\pi\vec{P}$
$\vec{H} = \dfrac{1}{\mu_0}\vec{B} - \vec{M}$	$\vec{H} = \vec{B} - 4\pi\vec{M}$
$\nabla \cdot \vec{D} = \rho$	$\nabla \cdot \vec{D} = 4\pi\rho$
$\nabla \cdot \vec{B} = 0$	$\nabla \cdot \vec{B} = 0$
$\nabla \times \vec{E} = -\dfrac{\partial \vec{B}}{\partial t}$	$\nabla \times \vec{E} = -\dfrac{1}{c}\dfrac{\partial \vec{B}}{\partial t}$
$\nabla \times \vec{H} = \vec{J} + \dfrac{\partial \vec{D}}{\partial t}$	$\nabla \times \vec{H} = \dfrac{4\pi}{c}\vec{J} + \dfrac{1}{c}\dfrac{\partial \vec{D}}{\partial t}$
$\vec{F} = q\vec{E} + q\vec{v} \times \vec{B}$	$\vec{F} = q\vec{E} + q\dfrac{\vec{v}}{c} \times \vec{B}$

Table 5.3. *Conversion between MKS and CGS units. All prefactors of 3 are actually 2.99792..., from the speed of light.*

Quantity	Symbol	MKS	CGS
charge	q	1 coulomb	3×10^9 statcoulombs
current	I	1 ampere	3×10^9 statamps
potential	V	1 volt	$1/300$ statvolt
polarization	P	1 coulomb/m^2	3×10^5 dipole moment/cm^3
electric field	E	1 volt/m	$\frac{1}{3} \times 10^{-4}$ statvolt/cm
displacement	D	1 coulomb/m^2	$4\pi \times 3\times10^5$ statvolt/cm
resistance	R	1 ohm	$\frac{1}{3^2}\times10^{-11}$ s/cm
capacitance	C	1 farad	$3^2\times10^{11}$ cm
magnetic flux	φ	1 weber	10^8 gauss·cm^2 (maxwells)
magnetic induction	B	1 tesla	10^4 gauss
magnetic field	H	1 ampere/m	$4\pi\times10^{-3}$ oersted
magnetization	M	1 ampere/m	10^{-3} magnetic moment/cm^3
inductance	L	1 henry	$\frac{1}{3^2}\times10^{-11}$ stathenry

boundary condition that must be satisfied for a perfect conductor, hence to find the field above the plane the continuous surface charge distribution can be replaced by the single image charge. There are some geometries like this for which it's possible to use an image charge to easily solve for the fields; in more complex geometries image charges are still useful as an expansion technique based on iterating the induced image charge produced by the source image charge.

5.3.3 Electromagnetic Units

There are two common systems of electromagnetic units: MKS that we've used here, and *Gaussian* or CGS. MKS uses familiar macroscopic quantities (volts, amps, ohms) and hence is most suitable for macroscopic phenomena and is commonly used in engineering;

CGS is better matched to microscopic phenomena and is commonly used in physics. Table 5.2 gives the governing equations in these two systems, and in Table 5.3 the (non-obvious) conversion factors are summarized. In addition to these conventional definitions there are many other possible systems offering endless opportunities to go astray, ranging from minor variations of these to the theorist's favorite system in which all fundamental constants are set equal to 1, with units being put back in at the end of a calculation based on dimensional grounds. [Jackson, 1999] has a thorough discussion of the logic of, and the relationships among, the systems.

5.4 RADIATION AND ENERGY

5.4.1 Waves

Perhaps the single most remarkable feature of Maxwell's equations is that they contain a wave solution. To see this, start with the equations for free space

$$\nabla \cdot \vec{D} = 0$$

$$\nabla \cdot \vec{B} = 0$$

$$\nabla \times \vec{E} = -\mu_0 \frac{\partial \vec{H}}{\partial t}$$

$$\nabla \times \vec{H} = \epsilon_0 \frac{\partial \vec{E}}{\partial t} \quad . \tag{5.96}$$

Take the curl of the last two equations:

$$\nabla \times \nabla \times \vec{E} = -\mu_0 \frac{\partial}{\partial t} \nabla \times \vec{H}$$

$$\nabla \times \nabla \times \vec{H} = \epsilon_0 \frac{\partial}{\partial t} \nabla \times \vec{E} \quad . \tag{5.97}$$

These can be simplified with the identity (Problem 5.1)

$$\nabla \times \nabla \times \vec{E} = \nabla(\nabla \cdot \vec{E}) - \nabla^2 \vec{E} \tag{5.98}$$

to give

$$-\nabla^2 \vec{E} = -\mu_0 \frac{\partial}{\partial t} \nabla \times \vec{H}$$

$$-\nabla^2 \vec{H} = \epsilon_0 \frac{\partial}{\partial t} \nabla \times \vec{E} \quad , \tag{5.99}$$

noting that the divergences will be 0 for $\vec{E}$ and $\vec{B}$. Substituting in the curls from equation (5.96),

$$\nabla^2 \vec{E} = \mu_0 \epsilon_0 \frac{\partial^2 \vec{E}}{\partial t^2}$$

$$\nabla^2 \vec{H} = \mu_0 \epsilon_0 \frac{\partial^2 \vec{H}}{\partial t^2} \quad . \tag{5.100}$$

These are wave equations, solved by a plane wave for the electric field

$$\vec{E}(\vec{x}, t) = \vec{E}_0 e^{i(\vec{k} \cdot \vec{x} - \omega t)} \quad , \tag{5.101}$$

with the wave vector $\vec{k}$ pointing in the direction of propagation with a magnitude $|k| = 2\pi/\lambda$, and the velocity $c = \omega/|k| = (\mu_0\epsilon_0)^{-1/2}$. Let there be light.

If we take $\vec{E} = \vec{E}_0 e^{i(\vec{k}\cdot\vec{x}-\omega t)}$ and $\vec{H} = \vec{H}_0 e^{i(\vec{k}\cdot\vec{x}-\omega t)}$,

$$\nabla \times \vec{E} = -\mu_0\frac{\partial\vec{H}}{\partial t}$$

$$i\vec{k} \times \vec{E} = i\omega\mu_0\vec{H}$$

$$\frac{k}{\omega\mu_0}\hat{k} \times \vec{E} = \vec{H}$$

$$\sqrt{\frac{\epsilon_0}{\mu_0}}\hat{k} \times \vec{E} = \vec{H} \quad . \tag{5.102}$$

Similarly,

$$-\sqrt{\frac{\mu_0}{\epsilon_0}}\hat{k} \times \vec{H} = \vec{E} \quad . \tag{5.103}$$

The electric and magnetic fields of a plane electromagnetic wave are perpendicular to each other and to the direction of travel $\hat{k}$. Their ratio

$$\frac{|\vec{E}|}{|\vec{H}|} = \sqrt{\frac{\mu_0}{\epsilon_0}} \approx 377\ \Omega \tag{5.104}$$

has the units of resistance and defines the *impedance of free space*. It will return in Chapter 7 in the effective impedance of antennas.

Since it's reasonable to assume that a wave travels in a medium, the recognition of the wave solution to Maxwell's equations led to a search for the "ether" that supports the wave. The failure of the Michelson–Morely experiment in 1887 to detect the motion of the Earth through the ether helped plant the seeds for the discovery of quantum mechanics, and special relativity. The resolution of the paradox is that electromagnetic waves are carried by photons, which are particles that travel in free space but which also act like waves. One way to understand electromagnetic propagation is to remember that information cannot travel faster than the speed of light. If a charge is moved instantaneously, there is a "kink" in its electric field that travels out at the speed of light: that is an electromagnetic wave packet. Moving the charge periodically creates a wave.

5.4.2 Electromagnetic Energy

If electromagnetic waves can propagate, and if an electromagnetic field can accelerate a charge that is initially at rest, then it must be possible to store and transmit energy in the fields. In this section we will calculate that energy.

A charge in an electric field feels a force $\vec{F} = q\vec{E}$. If the charge moves a distance $d\vec{x}$, work $dW = q\vec{E} \cdot d\vec{x}$ is done against this force. If the charge is moving at a velocity $\vec{v}$ then the rate at which work is being done, or power is being consumed, is $W = q\vec{E} \cdot \vec{v}$. If there is a continuous current density $\vec{J}$, the total rate of work is this quantity integrated over space

$$W = \int_V \vec{E} \cdot \vec{J}\ dV \quad . \tag{5.105}$$

There is no work done by a magnetic field alone on a charge, because the magnetic force is perpendicular to the velocity:

$$\vec{F} \cdot \vec{v} = (q\vec{v} \times \vec{B}) \cdot \vec{v} = 0 \quad . \tag{5.106}$$

Since

$$\nabla \times \vec{H} = \vec{J} + \frac{\partial \vec{D}}{\partial t} \quad \Rightarrow \quad \vec{J} = \nabla \times \vec{H} - \frac{\partial \vec{D}}{\partial t} \quad , \tag{5.107}$$

equation (5.105) can be rewritten as

$$W = \int_V \left[\vec{E} \cdot (\nabla \times \vec{H}) - \vec{E} \cdot \frac{\partial \vec{D}}{\partial t} \right] dV \quad . \tag{5.108}$$

This in turn can be rewritten by using the vector identity

$$\nabla \cdot (\vec{E} \times \vec{H}) = \vec{H} \cdot (\nabla \times \vec{E}) - \vec{E} \cdot (\nabla \times \vec{H}) \tag{5.109}$$

as

$$W = \int_V \left[\vec{H} \cdot (\nabla \times \vec{E}) - \nabla \cdot (\vec{E} \times \vec{H}) - \vec{E} \cdot \frac{\partial \vec{D}}{\partial t} \right] dV \quad . \tag{5.110}$$

Plugging in $\nabla \times \vec{E} = -\partial \vec{B}/\partial t$,

$$W = -\int_V \left[\nabla \cdot (\vec{E} \times \vec{H}) + \vec{E} \cdot \frac{\partial \vec{D}}{\partial t} + \vec{H} \cdot \frac{\partial \vec{B}}{\partial t} \right] dV \quad . \tag{5.111}$$

Since $\vec{D} = \epsilon \vec{E}$,

$$\frac{\partial}{\partial t}(\vec{E} \cdot \vec{D}) = \frac{\partial \vec{E}}{\partial t} \cdot \vec{D} + \vec{E} \cdot \frac{\partial \vec{D}}{\partial t} \tag{5.112}$$

$$= 2\vec{E} \cdot \frac{\partial \vec{D}}{\partial t}$$

and similarly

$$\frac{\partial}{\partial t}(\vec{B} \cdot \vec{H}) = 2\vec{H} \cdot \frac{\partial \vec{B}}{\partial t} \quad . \tag{5.113}$$

Therefore, if we define

$$U = \frac{1}{2}(\vec{E} \cdot \vec{D} + \vec{B} \cdot \vec{H}) \quad \left(\frac{J}{m^3} \right) \tag{5.114}$$

then equation (5.111) becomes

$$W = -\int_V \left[\nabla \cdot (\vec{E} \times \vec{H}) + \frac{\partial U}{\partial t} \right] dV \tag{5.115}$$

with

$$\frac{\partial U}{\partial t} = \vec{E} \cdot \frac{\partial \vec{D}}{\partial t} + \vec{H} \cdot \frac{\partial \vec{B}}{\partial t} \quad . \tag{5.116}$$

Further defining

$$\vec{P} = \vec{E} \times \vec{H} \quad \left(\frac{J}{m^2 \cdot s} \right) \quad , \tag{5.117}$$

the first term can be turned into a surface integral:

$$W = - \int_S \vec{P} \cdot d\vec{A} - \int_V \frac{\partial U}{\partial t} \, dV \quad . \tag{5.118}$$

This has a very natural interpretation. The first term represents an energy flux transported across the boundary of the integration volume by the field, and the second term represents the change in the energy stored in the volume by the field. $\vec{P}$ is called the *Poynting vector*, and U is the energy density. Note that the Poynting vector $\vec{P}$ has nothing to do with the polarization vector $\vec{P}$, they just use the same symbol by convention. Integrating P over an area gives the energy being carried by an electromagnetic wave; integrating U over a volume gives the energy stored in a static field.

Since the energy stored in an electric or magnetic field is equal to the volume integral of the square of the field strength, field lines behave like "*furry rubber bands*" in finding the lowest-energy configuration. They want to be as short as possible to minimize the volume of the integral, and they want to be as far apart from each other as possible to minimize the field density and hence the quadratic energy density.

5.5 SELECTED REFERENCES

[Jackson, 1999] Jackson, John David. (1999). *Classical Electrodynamics*. 3rd edn. New York: Wiley.

> The definitive electrodynamics reference.

[Heald & Marion, 1995] Heald, Mark A., & Marion, Jerry B. (1995). *Classical Electromagnetic Radiation*. 3rd edn. Fort Worth: Saunders.

> Less depth than Jackson, but a more accessible introduction to electrodynamics.

5.6 PROBLEMS

(5.1) Prove the BAC–CAB rule

$$\vec{A} \times (\vec{B} \times \vec{C}) = \vec{B}(\vec{A} \cdot \vec{C}) - \vec{C}(\vec{A} \cdot \vec{B}) \tag{5.119}$$

by writing it out in the summations convention, and use it to show that

$$\nabla \times (\nabla \times \vec{E}) = \nabla(\nabla \cdot \vec{E}) - \nabla^2 \vec{E} \quad . \tag{5.120}$$

(5.2) (*a*) Use Gauss' Law to find the capacitance between two parallel plates of area A at a potential difference V and with a spacing d. Neglect the fringing fields by assuming that this is a section of an infinite capacitor.

(*b*) Show that when a current flows through the capacitor, the integral over the internal displacement current is equal to the external electrical current.

(*c*) Integrate the energy density to find the stored energy at a fixed potential. The answer should be expressed in terms of the capacitance.

(*d*) Batteries are rated by amp-hours, the current they can supply at the design voltage for an hour. Consider a 10 V laptop battery that provides $10 \text{ A} \cdot \text{h}$. Assuming a plate spacing of 10^{-6} m $\equiv 1$ μm and a vacuum dielectric, what area would a capacitor need to be able to store this amount of energy? If such plates were 10 cm on a side and stacked vertically, how tall would the stack have to be to provide this total area?

(5.3) (a) Use Stokes' Law to find the magnetic field of an infinite solenoid carrying a current I with n turns/meter.

(b) Integrate the energy density to find the energy stored in a solenoid of radius r and length l, once again neglecting fringing fields.

(c) Consider a 10 T MRI magnet (Section 9.4) with a bore diameter of 1 m and a length of 2 m. What is the outward force on the magnet? Remember – force is the gradient of potential for a conservative force.

(5.4) Calculate the force per meter between two parallel wires one meter apart, each carrying a current of one ampere (this is the geometry used to define the ampere).

(5.5) (a) Assume that sunlight has a power energy density of 1 kW/m^2 (this is a peak number; the typical average value in the continental USA is $\sim$ 200 W/m^2). Estimate the electric field strength associated with this radiation.

(b) If 1 W of power is focused in a laser beam to a square millimeter, what is the field strength? What about if it is focused to the diffraction limit of $\sim$ 1 μm^2?

6 Circuits, Transmission Lines, and Waveguides

Electric and magnetic fields contain energy, which can propagate. These are the ingredients needed for communications; in this chapter we will look at how electromagnetic energy can be guided. We will start with low-frequency circuits, then progress through transmission lines to high-frequency waveguides.

6.1 CIRCUITS

The elements of an electrical circuit must satisfy Maxwell's equations. In the low-frequency limit this provides a fundamental explanation for the familiar circuit equations. These simple relationships will hold as long as the frequencies are low enough for the size of the circuit to be much smaller than the electromagnetic wavelength. Above this there is a tricky regime in which the entire circuit acts like a distributed antenna, and then when the wavelength becomes small compared to the size of the circuit things become simpler again (this is the subject of Chapter 8 on optics).

6.1.1 Current and Voltage

The *voltage* or *potential* difference between two parts of a circuit is defined by the line integral of the electric field

$$V = - \int \vec{E} \cdot d\vec{l} \quad .$$

(6.1)

As long as $d\vec{B}/dt = 0$ then $\nabla \times \vec{E} = 0$, which implies that the electric field is the gradient of a potential and the value of its line integral is independent of the path; it can go through wires or free space as needed and will always give the same answer. Conversely, if there are time-varying magnetic fields then the potential difference does depend on path and can no longer be defined as a function of position alone.

The electric field is defined to point from positive to negative charge so that the potential increases along a path from negative to positive charge. A charge q such as an electron in a wire feels a force $\vec{F} = q\vec{E}$, and so according to these definitions electrons flow from low to high potentials (Figure 6.1). The *current* $\vec{I}$, in amperes, at a point in a wire is equal to the number of coulombs of charge passing that point per second. It is defined to be in the same direction as the electric field and hence opposite to the direction in which electrons travel. The current density $\vec{J}$ is equal to the current divided by its cross-sectional area $\vec{J} = \vec{I}/A$.

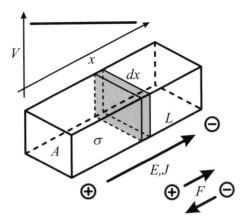

Figure 6.1. A resistive element, showing the relations among the voltage, electric field, current, and charge motion.

6.1.2 Kirchhoff's Laws

There are two *Kirchhoff Laws* that can be used to analyze the current flow in a circuit:

- *The sum of currents into and out of a circuit node must be zero.*
  If multiple wires meet at a point, the sum of all their currents must be equal to zero. This is just a statement of the conservation of charge.
- *The sum of voltages around a circuit must vanish.*
  This follows because the line integral of the electric field around a closed path

$$V = - \oint \vec{E} \cdot d\vec{l} = - \int \nabla \times \vec{E} \cdot d\vec{A} = \frac{\partial}{\partial t} \int \vec{B} \cdot d\vec{A} \qquad (6.2)$$

will vanish if there is no time-varying magnetic flux linking the circuit.

6.1.3 Resistance

In an isotropic conductor the current and electric field are related by

$$\vec{J} = \sigma \vec{E} \ , \qquad (6.3)$$

where σ is the material's *conductivity*. For very large fields there may be nonlinear deviations from this linear relationship, and in a complex material the conductivity may be a tensor that depends on direction. The voltage drop across the resistor in Figure 6.1 with length L, cross-sectional area A, conductivity σ, and carrying a current I is therefore

$$V = - \int_{-}^{+} \vec{E} \cdot d\vec{x} = - \int_{-}^{+} \frac{\vec{J}}{\sigma} \cdot d\vec{x} = \int_{-}^{+} \frac{I}{\sigma A} \, dx = \frac{IL}{\sigma A} \equiv IR \quad . \qquad (6.4)$$

Remember that the integral goes from low to high potential, but that current flows from high to low potentials, so $-\vec{J} \cdot d\vec{x} = J \, dx = I \, dx/A$. This is just *Ohm's Law*, and it defines the *resistance*

$$R = \frac{L}{\sigma A} = \frac{\rho L}{A} \qquad (6.5)$$

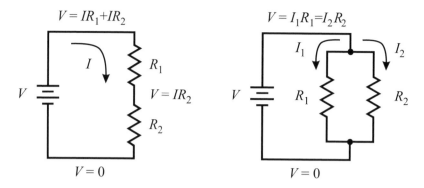

Figure 6.2. Series and parallel circuits.

in terms of the conductivity σ, which has units of siemens per meter (S/m), or the *resistivity* ρ, which has units of ohm-meters ($\Omega\cdot$m). For a two-dimensional film of thickness T, the resistance of a region of length L and width W is

$$
\begin{aligned}
R &= \rho\frac{L}{A} \\
&= \rho\frac{L}{TW} \\
&= \frac{\rho}{T}\frac{L}{W} \\
&\equiv R_\square\frac{L}{W} \quad .
\end{aligned}
$$
(6.6)

This defines the *sheet resistivity* $R_\square$ ("R square"). Since L/w is dimensionless, $R_\square$ has units of resistance without any other length.

Kirchhoff's Laws can be used to simplify the circuits in Figure 6.2. For the series circuit on the left,

$$V = IR_1 + IR_2$$
(6.7)

or

$$I = \frac{V}{R_1 + R_2} \quad ,$$
(6.8)

therefore the total resistance is

$$R_{\text{total}} = R_1 + R_2 \quad .$$
(6.9)

Series resistances simply add. For the parallel circuit on the right, the voltage drop across both legs must be equal since potential is independent of path,

$$V = I_1 R_1 = I_2 R_2 \quad ,$$
(6.10)

and the current in both legs must add up to the total current

$$I_1 + I_2 = \frac{V}{R_{\text{total}}} \quad .$$
(6.11)

Therefore

$$\frac{V}{R_1} + \frac{V}{R_2} = \frac{V}{R_{\text{total}}} \tag{6.12}$$

or

$$\frac{1}{R_{\text{total}}} = \frac{1}{R_1} + \frac{1}{R_2}$$

$$R_{\text{total}} = \frac{R_1 R_2}{R_1 + R_2} \qquad . \tag{6.13}$$

Parallel resistances add inversely. More complex networks of resistances can always be simplified to a single effective resistance by repeated application of these rules.

6.1.4 Power

Now let's now consider a slab of charge of cross-sectional area A and thickness dx moving through the resistor in Figure 6.1. If the charge density is ρ_q, the total charge in this slab is $Q = \rho_q dx A$ and it feels a net force $\vec{F} = Q\vec{E}$. Because a current is flowing, charge is moving relative to this force and so work is being done. The work associated with the slab moving from one end of the resistor to the other is equal to the integral of the force times the displacement:

$$dW = \int_-^+ \vec{F} \cdot d\vec{x} = -Q \int_-^+ \vec{E} \cdot d\vec{x} = -QV = -\rho_q dx A V \tag{6.14}$$

for a negative charge. This decrease in energy is dissipated in the resistor; the *power* is equal to the rate at which work is being done

$$P = -\frac{dW}{dt} = \rho_q \frac{dx}{dt} AV = JAV = IV = I^2 R \qquad . \tag{6.15}$$

The power dissipated in a resistor is equal to the current flowing through it times the voltage drop across it, which by Ohm's Law is also equal to the square of the current times the resistance. This appears as heat in the resistor.

6.1.5 Capacitance

There will be an electric field between an electrode that has a charge of $+Q$ on it and one that has a charge of $-Q$, and hence a potential difference between the electrodes. *Capacitance* is defined to be the ratio of the charge to the potential difference:

$$C = \frac{Q}{V} \qquad . \tag{6.16}$$

The MKS unit is the *farad*, F. Capacitances range from picofarads in circuit components up to many farads in *supercapacitors* based on electrochemical effects [Conway, 1991].

The current across a capacitor is given by

$$C\frac{dV}{dt} = \frac{dQ}{dt} = I \qquad . \tag{6.17}$$

A capacitor is a device that stores energy in an electric field by storing charge on its plates; in Problem 5.2 we saw that this stored energy is equal to $CV^2/2$. The current flowing

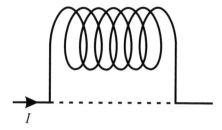

Figure 6.3. A solenoid; the dotted line closes the integration path.

across a capacitor is a *displacement current*: from the point of view of the overall circuit it is a real current, but it arises from the time-varying electric field associated with the capacitor plates storing or releasing charge rather than from real charge passing through the capacitor.

If the applied voltage is $V = e^{i\omega t}$, then the current is

$$I = C\frac{dV}{dt} = i\omega C e^{i\omega t} \quad . \tag{6.18}$$

The *impedance* (complex resistance) is defined to be the ratio of the voltage and current at a fixed frequency,

$$Z = \frac{V}{I} = \frac{e^{i\omega t}}{i\omega C e^{i\omega t}} = \frac{1}{i\omega C} \quad . \tag{6.19}$$

The current leads the voltage by a phase shift of $i = 90°$. When $\omega = 0$ the impedance is infinite (no current flows at DC), and when $\omega = \infty$ the impedance is 0 (the capacitor acts like a wire).

6.1.6   Inductance

An *inductor* stores energy in a magnetic field arising from current flowing through a coil. The inductance is defined to be the ratio of the magnetic flux

$$\Phi = \int \vec{B} \cdot d\vec{A} \tag{6.20}$$

linking a circuit to the current that produces it:

$$L = \frac{\Phi}{I} \quad . \tag{6.21}$$

The MKS unit of inductance is the *henry*, H.

In Figure 6.3 the electric field vanishes along the dotted line for an ideal solenoid, therefore the line intergral of the electric field along the dotted line and around the solenoid is equal to the voltage drop across the solenoid. And the magnetic field vanishes outside the solenoid if it is approximated to be a section of an infinite solenoud, therefore the integral of the magnetic field across the surface bounded by the path is equal to the flux linking the solenoid times the number of turns of the coil. This lets us relate the flux to the potential. If this ideal solenoid is taken to have just a single turn then

$$-V = \oint \vec{E} \cdot d\vec{l} = -\frac{\partial}{\partial t}\int_S \vec{B} \cdot d\vec{A} = -\frac{\partial \Phi}{\partial t} = -\frac{\partial}{\partial t}(LI) = -L\frac{dI}{dt} \tag{6.22}$$

and so

$$V = L\frac{dI}{dt} \quad . \tag{6.23}$$

Extra turns add in series: if an inductor has N turns, then the inductance is N times that due to one turn, assuming that the flux linking all the turns is the same. Since the field of a solenoid of radius r and length l with n turns/meter is $H = nI$, the inductance is

$$L = \frac{\Phi}{I} = \frac{N}{I} \int \vec{B} \cdot d\vec{A} = \frac{nl}{I} \mu n I \pi r^2 = \mu n^2 l \pi r^2 \quad . \tag{6.24}$$

Problem 5.3 showed that the energy stored in a solenoid is $LI^2/2$.

If the current flowing through an inductor is $I = e^{i\omega t}$ then the voltage drop across it is $V = Li\omega e^{i\omega t}$, and so the impedance is

$$Z = \frac{Li\omega e^{i\omega t}}{e^{i\omega t}} = i\omega L \quad . \tag{6.25}$$

The current lags the voltage by $90°$ $(-i)$.

6.2 WIRES AND TRANSMISSION LINES

We have been considering conduction in the low-frequency limit; in this section we will use Maxwell's equations to look at how AC fields penetrate conductors and are guided by them at higher frequencies.

6.2.1 Skin Depth

Assume that the conductor is described by $\vec{J} = \sigma\vec{E}, \vec{D} = \epsilon\vec{E}, \vec{B} = \mu\vec{H}$. If the electric field is periodic as $\vec{E}(\vec{x},t) = \vec{E}(\vec{x})e^{i\omega t}$ then the curl of the magnetic field is

$$\nabla \times \vec{H} = \vec{J} + \frac{\partial \vec{D}}{\partial t}$$

$$\nabla \times \vec{B} = \mu\sigma\vec{E} + \mu\epsilon\frac{\partial \vec{E}}{\partial t}$$

$$\nabla \times \vec{B}(\vec{x}) = (\mu\sigma + i\omega\mu\epsilon)\vec{E}(\vec{x}) \quad . \tag{6.26}$$

Since the divergence of a curl vanishes,

$$\nabla \cdot \nabla \times \vec{B} = 0 = (\mu\sigma + i\omega\mu\epsilon)\nabla \cdot \vec{E}$$

$$\Rightarrow \nabla \cdot \vec{E} = \frac{\rho}{\epsilon} = 0 \quad . \tag{6.27}$$

The linear response coefficients require that there be no free charge.

Now taking the curl of the curl of $\vec{E}$,

$$\nabla \times \vec{E} = -\frac{\partial \vec{B}}{\partial t}$$

$$\nabla \times \nabla \times \vec{E} = -\frac{\partial}{\partial t}\nabla \times \vec{B}$$

$$\underbrace{\nabla(\nabla \cdot \vec{E})}_{0} - \nabla^2\vec{E} = -\mu\sigma\frac{\partial \vec{E}}{\partial t} - \mu\epsilon\frac{\partial^2 \vec{E}}{\partial t^2} \quad . \tag{6.28}$$

The first term on the right hand side is due to real conduction, and the second term is due to the displacement current. Since σ is very large in a good conductor, up to very high (optical) frequencies the displacement current term can be dropped:

$$\nabla^2 \vec{E} = \mu\sigma \frac{\partial \vec{E}}{\partial t} \quad . \tag{6.29}$$

This is now a diffusion equation instead of a wave equation. For a periodic electric field, the spatial part satisfies

$$\nabla^2 \vec{E}(\vec{x}) = i\omega\mu\sigma\vec{E}(\vec{x}) \equiv k^2 \vec{E}(\vec{x}) \quad . \tag{6.30}$$

Since

$$\sqrt{i} = \frac{1+i}{\sqrt{2}} \tag{6.31}$$

(try squaring it),

$$k = \sqrt{i\omega\mu\sigma}$$
$$= (1+i)\sqrt{\frac{\omega\mu\sigma}{2}}$$
$$\equiv \frac{1+i}{\delta} \quad . \tag{6.32}$$

This defines the *skin depth*

$$\delta = \sqrt{\frac{2}{\omega\mu\sigma}} = \frac{1}{\sqrt{\pi\nu\mu\sigma}} \tag{6.33}$$

in terms of the frequency ν, the permeability μ, and the conductivity σ.

Consider the solution to equation (6.30) for a plane wave incident on the surface of a conductor, so that by symmetry we need consider only the distance z into the conductor

$$\frac{d^2 E}{dz^2} = k^2 E \quad . \tag{6.34}$$

E is the magnitude of the electric field, which for a plane wave is transverse to the direction of z. This is solved by

$$E(x) = E_0 e^{-kz} = E_0 e^{-z/\delta} e^{-iz/\delta} \quad , \tag{6.35}$$

where E_0 is the amplitude at the surface, and we've ignored the unphysical possible solution e^{kz}. The total current per unit width that is produced by this field is found by integrating the current density over the depth

$$I = \int_0^\infty J \, dz = \int_0^\infty \sigma E \, dz = \int_0^\infty \sigma E_0 e^{-kz} \, dz = \frac{\sigma E_0}{k} \quad . \tag{6.36}$$

Therefore

$$E_0 = \frac{kI}{\sigma} = \frac{1+i}{\sigma\delta} I = \left(\frac{1}{\sigma\delta} + i\frac{1}{\sigma\delta} \right) I$$
$$\equiv (R_s + \omega L_s)I \quad . \tag{6.37}$$

The total current is proportional to the applied field at the surface; the real part of this defines an effective *surface resistance* R_s, and the imaginary part defines the *surface inductance* L_s. Associated with this current there is dissipation; in a small volume of cross-sectional area A and length along the surface L the dissipation per volume is

$$\frac{I_{\text{volume}}^2 R}{AL} = \frac{1}{AL} J^2 A^2 \frac{L}{\sigma A} = \frac{J^2}{\sigma} \quad . \tag{6.38}$$

If the current is periodic, taking a time average introduces another factor of $\langle \sin^2 \rangle = 1/2$:

$$\left\langle \frac{I^2 R}{AL} \right\rangle = \frac{|J|^2}{2\sigma} \quad . \tag{6.39}$$

Integrating this from $z = 0$ to ∞ gives the energy dissipated by the field in the material per surface area

$$\int_0^\infty \frac{|J|^2}{2\sigma} \, dz = \int_0^\infty \frac{\sigma^2 E_0^2}{2\sigma} e^{2z/\delta} \, dz = \frac{\sigma E_0^2 \delta}{4} \quad . \tag{6.40}$$

The amplitude of the field and current are falling off exponentially with a length scale equal to the skin depth. For example, pure copper at room temperature has a conductivity of 5.8×10^7 S/m and so $\delta \sim 7$ cm at 1 Hz, 2 mm at 1 kHz, 70 μm at 1 MHz, and 2 μm at 1 GHz. Since the skin depth is so small at even fairly low frequencies, very little thickness is needed to screen a field. This is why it is a good approximation to assume that fields vanish at the surface of a conductor, which we have already found to be the boundary condition for a perfect conductor. The part of the field that does leak into the conductor causes a current to flow, and this current leads to resistive dissipation, therefore in making this approximation we are leaving out the mechanism that damps fields around conductors. This is very important in resonant electromagnetic cavities that are designed to have a high Q (low damping rate).

Because of the skin depth, a bundle of fine wires has a smaller AC resistance than a single fat wire because there is more surface area for the current to penetrate into and the overall resistance will be inversely proportional to the effective cross-secional area of the bundle. This is why wires carrying high frequency signals are stranded rather than solid.

6.2.2 Transmission Lines

While electromagnetic fields cannot penetrate far into good conductors, they can be guided long distances by them. Distributed objects can have energy stored in electric fields through capacitance, and in magnetic fields by inductance; the interplay between these can give rise to an energy flow. As a first example such a *transmission line*, consider the coaxial cable in Figure 6.4). Other important transmission line geometries include parallel wires or strips (Problem 6.4), and a strip above a ground plane (called a *stripline*).

Because a transmission line is operated in a closed circuit there is no net charge transfer between either end. We will assume here that any current I in the inner conductor must be matched by a return current $-I$ in the outer conductor; the next section will study higher-frequency modes for which this is no longer true. There is an electric field between the inner and outer conductors, giving rise to a distributed capacitance between them. Current flowing in the inner conductor also produces a magnetic field around it, and

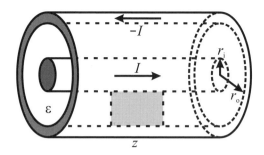

Figure 6.4. A coaxial cable field with a dielectric ϵ.

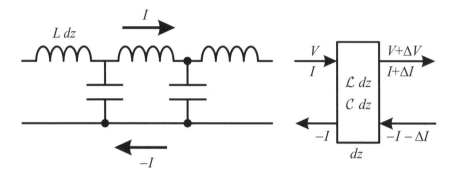

Figure 6.5. Effective circuit model for a transmission line, and a differential element.

hence a distributed inductance along it. As long as the frequency is not so large that the wavelength is comparable to the cross-sectional size, the coaxial cable therefore acts like an extended series inductor and parallel capacitor (Figure 6.5). This circuit model is applicable to arbitrary transmission lines; its solution will reappear in the next section as the fundamental mode for Maxwell's equation in a cylindrical geometry.

From Stokes' Law, the magnetic field between the conductors is

$$H = \frac{I}{2\pi r} \quad , \tag{6.41}$$

and the field vanishes outside of the outer conductor because the net current is then zero. Integrating this field over the surface of length z shown in Figure 6.4 to find the flux between the conductors,

$$\Phi = \int \vec{B} \cdot d\vec{A} = z \int_{r_i}^{r_o} \mu_0 \frac{I}{2\pi r} \, dr = z \frac{\mu_0 I}{2\pi} \ln \frac{r_o}{r_i} \quad . \tag{6.42}$$

Since the dielectric is non-magnetic we can take $\mu_r \sim 1$. Therefore, the inductance per length is

$$\mathcal{L} = \frac{\Phi}{zI} = \frac{\mu_0}{2\pi} \ln \frac{r_o}{r_i} \quad \left(\frac{H}{m}\right) \quad . \tag{6.43}$$

Similarly, from Gauss' Law the electric field between the conductors is

$$E = \frac{Q}{2\pi\epsilon r} \quad , \tag{6.44}$$

where Q is the charge per unit length and the field vanishes outside of the outer conductor. Integrating to find the potential difference,

$$V = - \int_{r_i}^{r_o} \vec{E} \cdot d\vec{l} = \frac{Q}{2\pi\epsilon} \ln \frac{r_o}{r_i} \quad , \tag{6.45}$$

which gives the capacitance per unit length

$$\mathcal{C} = \frac{Q}{V} = \frac{2\pi\epsilon}{\ln(r_o/r_i)} \quad \left(\frac{\mathrm{F}}{\mathrm{m}}\right) \quad . \tag{6.46}$$

6.2.3 Wave Solutions

Now consider the little differential length of the transmission line dz shown in Figure 6.5, with parallel capacitance $\mathcal{C}\,dz$ and series inductance $\mathcal{L}\,dz$. If there is an increase in the current flowing across it

$$\Delta I = \frac{\partial I}{\partial z} dz \tag{6.47}$$

there must be a corresponding decrease in the charge stored in the capacitance

$$\Delta I = -\mathcal{C} dz \frac{\partial V}{\partial t} \quad . \tag{6.48}$$

Therefore

$$\frac{\partial I}{\partial z} dz = -\mathcal{C} dz \frac{\partial V}{\partial t} \tag{6.49}$$

or

$$\frac{\partial I}{\partial z} = -\mathcal{C} \frac{\partial V}{\partial t} \quad . \tag{6.50}$$

Similarly, if there is an increase in the potential across the element

$$\Delta V = \frac{\partial V}{\partial z} dz \tag{6.51}$$

there must be a decrease in the current flowing through the inductance

$$\Delta V = -\mathcal{L} dz \frac{\partial I}{\partial t} \quad . \tag{6.52}$$

Current flows from high to low potential, so an increasing potential drop across the inductor has the opposite sign from the decreasing current. Equating these expressions,

$$\frac{\partial V}{\partial z} = -\mathcal{L} \frac{\partial I}{\partial t} \quad . \tag{6.53}$$

Now take a time derivative of equation (6.50)

$$\frac{\partial^2 I}{\partial t \partial z} = -\mathcal{C} \frac{\partial^2 V}{\partial t^2} \tag{6.54}$$

and a z derivative of equation (6.53)

$$\frac{\partial^2 V}{\partial z^2} = -\mathcal{L} \frac{\partial^2 I}{\partial z \partial t} \tag{6.55}$$

and interchange the order of differentiation to equate the mixed terms (which is permitted for well-behaved functions):

$$\frac{\partial^2 V}{\partial z^2} = \mathcal{LC}\frac{\partial^2 V}{\partial t^2} \equiv \frac{1}{v^2}\frac{\partial^2 V}{\partial t^2} \quad , \tag{6.56}$$

where

$$v \equiv \frac{1}{\sqrt{\mathcal{LC}}} \quad . \tag{6.57}$$

This is a wave equation for the voltage in the transmission line. It is solved by an arbitrary distribution traveling with a velocity $\pm v$

$$V(z,t) = f(z - vt) + g(z + vt) \tag{6.58}$$
$$= V_+ + V_- \quad .$$

If we follow a fixed point in the distribution $f(0)$, $z - vt = 0 \Rightarrow z = vt$. The V_+ solution travels to the right, and V_- to the left. For a sinusoidal wave $V = e^{i(kz-\omega t)}$, $k = \omega/v$.
    Taking derivatives in the opposite order gives a similar equation for the current:

$$\frac{\partial^2 I}{\partial z^2} = \frac{1}{v^2}\frac{\partial^2 I}{\partial t^2} \quad . \tag{6.59}$$

To relate the voltage to the current, substitute equation (6.58) into equation (6.50)

$$\frac{\partial I}{\partial z} = -\mathcal{C}[-vf'(z - vt) + vg'(z + vt)] \tag{6.60}$$

and integrate over z

$$I = \mathcal{C}v[f(z - vt) - g(z + vt)]$$
$$\equiv \frac{1}{Z}[f(z - vt) - g(z + vt)]$$
$$= \frac{1}{Z}[V_+ - V_-]$$
$$= I_+ + I_- \quad , \tag{6.61}$$

where

$$Z = \frac{1}{\mathcal{C}v} = \sqrt{\frac{\mathcal{L}}{\mathcal{C}}} \quad (\Omega). \tag{6.62}$$

The current is proportional to the voltage, with the sign difference in the two terms coming from the difference between the solutions traveling in the right and left directions. The constant of proportionality is the *characteristic impedance* of the transmission line Z. The velocity and impedance of a transmission line are simply related to the capacitance and inductance per unit length. In a real cable, different frequencies are damped at different rates, changing the pulse shape as it travels, and if there are nonlinearities then different frequencies can travel at different rates causing *dispersion*: a sharp pulse will spread out. The dispersion sets a limit on how close pulses can be and still remain separated after traveling a long distance.

6.2.4 Reflections and Terminations

Consider a transmission line with a characteristic impedance Z_0 terminated by a load impedance Z_L. The load might be a resistor, or it could be another transmission line. For a resistor the impedance is associated with energy dissipated by ohmic heating, and for a transmission line the impedance is associated with energy that is transported away, but in both cases the voltage drop across the element is equal to the current applied to it times its impedance.

The incoming transmission line can support signals traveling in both directions (equation 6.58), therefore the voltage at the discontinuity is the sum of these:

$$V_L(t) = V_+(t) + V_-(t) \quad . \tag{6.63}$$

Similarly, the current across the load is

$$I_L(t) = I_+(t) + I_-(t) \quad . \tag{6.64}$$

The current across the termination must equal the current in the transmission line immediately before the termination:

$$\frac{V_L}{Z_L} = \frac{V_+}{Z_0} - \frac{V_-}{Z_0} \quad . \tag{6.65}$$

Eliminating variables between this and equation (6.63) gives the ratio of the incoming and reflected voltages, called the *reflection coefficient*

$$R = \frac{V_-}{V_+} = \frac{Z_L - Z_0}{Z_L + Z_0} \quad , \tag{6.66}$$

and the ratio of the incoming and the transmitted signals is equal to the *transmission coefficient*

$$T = \frac{V_L}{V_+} = \frac{2Z_L}{Z_L + Z_0} \quad . \tag{6.67}$$

Because of the load, V_+ and V_- can no longer be arbitrarily chosen but must satisfy the boundary conditions. These reflection and transmission coefficients have a number of interesting properties. If the load impedance is 0 (a short), $R = -1$ and so there is a reflected pulse of the same shape but opposite sign. If the load resistance is infinite (it is open), $R = 1$ and the reflected pulse has the same sign. These reflections are used in a *Time Domain Reflectometer* (*TDR*) to locate cable faults by measuring the time for a return pulse to arrive. Finally, if $Z_L = Z_0$ then $R = 0$: there is no reflection at all! This is why cables carrying high-frequency signals are terminated with resistors that match the cable's characteristic impedance. Such terminations are particularly important to eliminate clutter from reflected pulses in computer networks and buses.

If $V_+(z) = V_0 e^{ikz}$ going into the load,

$$
\begin{aligned}
V(z) &= V_+(z) + V_-(z) \\
&= V_0 \left(e^{ikz} + R e^{-ikz} \right) \quad .
\end{aligned}
\tag{6.68}
$$

As a function of z, the positive- and negative-going waves will periodically add to and

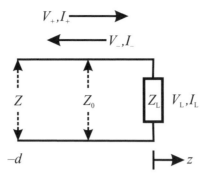

Figure 6.6. A transmission line with impedance Z terminated by a load Z_L.

subtract from each other. Taking the ratio of the maximum to the minimum value for this sum defines the *Voltage Standing-Wave Ratio (VSWR)*

$$\text{VSWR} \equiv \frac{V_{\text{max}}}{V_{\text{min}}}$$

$$= \frac{1 + |R|}{1 - |R|} \tag{6.69}$$

or

$$|R| = \frac{\text{VSWR} - 1}{\text{VSWR} + 1} . \tag{6.70}$$

The VSWR is one of the most important measurements in an RF system, used to ensure that impedances are matched so that all of the power goes in the intended direction.

Now consider the impedance of a transmission line as viewed by a periodic source a distance d from the termination, shown in Figure 6.6:

$$Z(-d) = \frac{V(-d)}{I(-d)}$$

$$= \frac{V_+ e^{-ikd} + V_- e^{ikd}}{Z_0^{-1} \left(V_+ e^{ikd} - V_- e^{-ikd} \right)}$$

$$= \frac{V_+ \left(e^{-ikd} + R e^{ikd} \right)}{V_+ Z_0^{-1} \left(e^{ikd} - R e^{-ikd} \right)}$$

$$= Z_0 \frac{\left(e^{-ikd} + R e^{ikd} \right)}{\left(e^{ikd} - R e^{-ikd} \right)} . \tag{6.71}$$

Normalizing this by the characteristic impedance of the transmission line,

$$\frac{Z(-d)}{Z_0} = \frac{e^{-ikd} + R e^{ikd}}{e^{-ikd} - R e^{ikd}}$$

$$= \frac{1 + R e^{i2kd}}{1 - R e^{i2kd}}$$

$$r + ic \equiv \frac{1 + (x + iy)}{1 - (x + iy)}$$

$$= \frac{1 - (x^2 + y^2)}{(1 - x)^2 + y^2} + i \frac{2y}{(1 - x)^2 + y^2} , \tag{6.72}$$

relates the real and complex parts of the input impedance

$$r + ic = \frac{Z}{Z_0}$$ (6.73)

to those of the round-trip reflection coefficient

$$x + iy = Re^{i2kd} \quad . $$ (6.74)

The real equation can be rewritten suggestively as

$$r = \frac{1 - (x^2 + y^2)}{(1 - x)^2 + y^2}$$

$$\frac{r(1 - x)^2 + x^2}{1 + r} + y^2 = \frac{1}{1 + r}$$

$$x^2 - 2x\frac{r}{1 + r} + \frac{r}{1 + r} + y^2 = \frac{1}{1 + r}$$

$$x^2 - 2x\frac{r}{1 + r} + \left(\frac{r}{1 + r}\right)^2 + y^2 = \frac{1}{1 + r} + \left(\frac{r}{1 + r}\right)^2 - \frac{r}{1 + r}$$

$$\left(x - \frac{r}{1 + r}\right)^2 + y^2 = \frac{1}{(1 + r)^2} \quad . $$ (6.75)

In the complex (x, y) plane, the reflection coefficient lies on a circle of radius $1/(1 + r)$ with a center at $(r/(1 + r), 0)$ set by the real part of the input impedance r. Similary, the complex equation can be rewritten as

$$c = \frac{2y}{(1 - x)^2 + y^2}$$

$$(1 - x)^2 + y^2 = \frac{2y}{c}$$

$$(1 - x)^2 + y^2 - 2y\frac{1}{c} + \frac{1}{c^2} = \frac{1}{c^2}$$

$$(1 - x)^2 + \left(y - \frac{1}{c}\right)^2 = \frac{1}{c^2} \quad . $$ (6.76)

This restricts the reflection coefficient to a circle of radius $1/c$ located at $(1, \pm1/c)$ given by the complex part of the input impedance c. The intersection of these two circles relates the input impedance to the reflection coefficient, conveniently found graphically on a *Smith chart* (Figure 6.7).

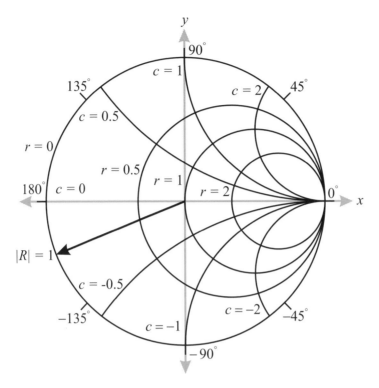

Figure 6.7. The Smith chart.

6.3 WAVEGUIDES

As the wavelength of a signal in a transmission line becomes comparable to the transverse size of the line, more complicated excitations become possible and the circuit model used in the last section no longer applies. A complete solution of Maxwell's equations is then required. Some of these new modes will prove to be desirable, and some will not. *Waveguides*, not surprisingly, guide electromagnetic waves. Depending on the geometry they may or may not be able to transmit a steady current because it is possible to guide waves without a DC return path. Waveguides usually have some symmetry about their long axis; a rectangular pipe is a common type.

6.3.1 Governing Equations

Start with the wave form of Maxwell's equations without any sources:

$$\nabla^2 \vec{E} = \mu\epsilon\frac{\partial^2 \vec{E}}{\partial t^2} \quad \nabla^2 \vec{H} = \mu\epsilon\frac{\partial^2 \vec{H}}{\partial t^2} \quad . \tag{6.77}$$

We are looking for waves that travel along the axis of the waveguide periodically as $e^{i\omega t - \gamma z}$. The real part of γ is the decay rate of the wave and the complex part is the wave vector $2\pi/\lambda$. Cancelling out the time dependence,

$$\nabla^2 \vec{E} = -\omega^2 \mu\epsilon\vec{E} \equiv -k^2 \vec{E} \quad \nabla^2 \vec{H} = -k^2 \vec{H} \quad . \tag{6.78}$$

The Laplacian can be separated into components that are transverse to the waveguide axis and that are axial, taken here to be in the $\vec{z}$ direction:

$$\nabla^2 \vec{E} = \nabla_{\mathrm{T}}^2 \vec{E} + \frac{\partial^2 \vec{E}}{\partial z^2}$$
$$= \nabla_{\mathrm{T}}^2 \vec{E} + \gamma^2 \vec{E} \quad . \tag{6.79}$$

This turns equations (6.78) into *Helmholtz' equations* for the transverse dependence of the field

$$\nabla_{\mathrm{T}}^2 \vec{E} = -(\gamma^2 + k^2)\vec{E} \equiv -k_{\mathrm{c}}^2 \vec{E} \qquad \nabla_{\mathrm{T}}^2 \vec{H} = -k_{\mathrm{c}}^2 \vec{H} \quad , \tag{6.80}$$

defining the characteristic wave vector k_c. Along with this, the curl equation for a periodic signal

$$\nabla \times \vec{E} = -i\omega\mu\vec{H} \tag{6.81}$$

has transverse components

$$\frac{\partial E_z}{\partial y} - \underbrace{\frac{\partial E_y}{\partial z}}_{-\gamma E_y} = -i\omega\mu H_x \qquad \frac{\partial H_z}{\partial y} - \underbrace{\frac{\partial H_y}{\partial z}}_{-\gamma H_y} = i\omega\mu E_z$$

$$\underbrace{\frac{\partial E_x}{\partial z}}_{-\gamma E_x} - \frac{\partial E_z}{\partial x} = -i\omega\mu H_y \qquad \underbrace{\frac{\partial H_x}{\partial z}}_{-\gamma H_x} - \frac{\partial H_z}{\partial x} = i\omega\mu E_y \tag{6.82}$$

which can be rearranged as

$$E_x = -\frac{1}{k_{\mathrm{c}}^2}\left(\gamma\frac{\partial E_z}{\partial x} + i\omega\mu\frac{\partial H_z}{\partial y}\right) \qquad H_x = \frac{1}{k_{\mathrm{c}}^2}\left(i\omega\epsilon\frac{\partial E_z}{\partial x} - \gamma\frac{\partial H_z}{\partial y}\right)$$

$$E_y = \frac{1}{k_{\mathrm{c}}^2}\left(-\gamma\frac{\partial E_z}{\partial x} + i\omega\mu\frac{\partial H_z}{\partial y}\right) \qquad H_y = -\frac{1}{k_{\mathrm{c}}^2}\left(i\omega\epsilon\frac{\partial E_z}{\partial x} + \gamma\frac{\partial H_z}{\partial y}\right) \quad . \tag{6.83}$$

If the axial components E_z, H_z are found from equations (6.80), they completely determine the transverse components through equations (6.83).

This set of equations admits three kinds of solutions: *Transverse Electric (TE)* with $E_z = 0$, *Transverse Magnetic (TM)* with $H_z = 0$, and *Transverse Electromagnetic (TEM)* with $E_z = H_z = 0$. For the TEM case, because the numerator in equations (6.83) vanishes, the only way the transverse components can be non-zero is for the denominator $k_{\mathrm{c}}^2 = \gamma^2 + k^2$ to also vanish. This means that $\gamma = \pm ik = \pm i\omega\sqrt{\mu\epsilon} = \pm i\omega/c$, therefore TEM waves travel at the speed of light in the medium. $k_{\mathrm{c}} = 0$ also reduces Helmholtz' equations to Laplace's equation, giving the static field solutions we used when studying transmission lines. Because in a hollow conductor the boundary is an equipotential, Laplace's equation implies that the field must vanish everywhere in the interior, therefore a TEM wave cannot be supported. Adding another conductor, such as the center lead in a coaxial cable, makes a TEM solution possible.

6.3.2 Rectangular Waveguides

Now consider a rectangular waveguide with width w in the x direction and height h in the y direction. The transverse equation for a TM wave is

$$\nabla_{\rm T}^2 E_z = \frac{\partial^2 E_z}{\partial x^2} + \frac{\partial^2 E_z}{\partial y^2} = -k_{\rm c}^2 E_z \quad . \tag{6.84}$$

Solving this subject to the boundary condition that the field must vanish at the conducting surfaces at $x = 0, w$ and $y = 0, h$ gives

$$E_z = A \sin(k_x x) \sin(k_y y) \quad , \tag{6.85}$$

where

$$\begin{aligned}
k_{\rm c}^2 &= k_x^2 + k_y^2 \\
k_x w &= m\pi \\
k_y h &= n\pi
\end{aligned} \tag{6.86}$$

index the possible modes as a function of integers m and n. If we define a characteristic frequency $\omega_{\rm c}$ associated with each mode by

$$\omega_{\rm c}(m,n) = \frac{k_{\rm c}(m,n)}{\sqrt{\mu\epsilon}} = \frac{1}{\sqrt{\mu\epsilon}} \left[\left(\frac{m\pi}{w}\right)^2 + \left(\frac{n\pi}{h}\right)^2 \right]^{1/2} \quad , \tag{6.87}$$

then we can find the propagation constant

$$\begin{aligned}
\gamma^2 &= k_{\rm c}^2 - k^2 \\
&= k_{\rm c}^2 \left(1 - \frac{k^2}{k_{\rm c}^2} \right) \\
&= k_{\rm c}^2 \left(1 - \frac{\omega^2 \mu\epsilon}{\omega_{\rm c}^2 \mu\epsilon} \right) \\
&= k_{\rm c}^2 \left(1 - \frac{\omega^2}{\omega_{\rm c}^2} \right) \quad .
\end{aligned} \tag{6.88}$$

Therefore

$$\begin{aligned}
\gamma &= k_{\rm c}(m,n) \left[1 - \frac{\omega^2}{\omega_{\rm c}(m,n)^2} \right]^{1/2} \quad &\omega < \omega_{\rm c}(m,n) \\
\gamma &= i k_{\rm c}(m,n) \left[\left(\frac{\omega}{\omega_{\rm c}(m,n)}\right)^2 - 1 \right]^{1/2} \quad &\omega > \omega_{\rm c}(m,n) \quad .
\end{aligned} \tag{6.89}$$

When ω is less than the cutoff frequency $\omega_{\rm c}$ for a mode, or equivalently when the wavelength λ is greater than the cut-off wavelength $\lambda_{\rm c}$, γ is pure real and so the mode decays exponentially. When ω is greater than the cutoff frequency for a mode, γ is pure imaginary and the mode propagates. These modes are labeled TM$_{mn}$. Repeating this analysis for the TE wave by starting with the transverse equation for H_z shows that the TE and TM waves are degenerate with the same cutoff frequencies. At low frequencies nothing propagates; as the frequency is raised more and more modes can be excited, with the distribution of energy among them depending on how the waveguide is driven.

6.3.3 Circular Waveguides

For a waveguide with cylindrical symmetry, the transverse Laplacian for a TM mode is

$$\nabla_{\mathrm{T}}^2 E_z = \frac{1}{r}\frac{\partial}{\partial r}\left(r\frac{\partial E_z}{\partial r}\right) + \frac{1}{r^2}\frac{\partial^2 E_z}{\partial \theta^2} = -k_c^2 E_z \quad, \tag{6.90}$$

which is solved by Bessel functions of the first (J_n) and second (N_n) kind [Gershenfeld, 1999a]:

$$E_z(r,\theta) = [AJ_n(k_c r) + BN_n(k_c r)][C\cos(n\theta) + D\sin(n\theta)] \quad . \tag{6.91}$$

The modes TM_{nl} are indexed by the order of the Bessel function n, and the root l of the Bessel function needed to make the field vanish at the boundaries. Although these frequencies can no longer be solved for analytically, for a coaxial cable a rough approximation for the TM modes is to ask that the wavelength be a multiple of radial spacing

$$\lambda_c \approx \frac{2}{n}(r_o - r_i) \quad n = 1, 2, 3, \dots \quad , \tag{6.92}$$

and for a TE mode that there be an integer number of azimuthal cycles

$$\lambda_c \approx \frac{2\pi}{n}\frac{a+b}{2} \tag{6.93}$$

[Ramo $et\ al.$, 1994]. In the section on transmission lines we studied the fundamental TEM mode. Because these higher-order modes have different velocities, if they are excited they will spread out the signal and hence limit the usefulness of the cable. This is why waveguides are usually designed to be operated with a single mode.

6.3.4 Dielectric Waveguides and Fiber Optics

Fortunately for telecommunications, waves can be guided by dielectric rather than conducting waveguides. The surface resistance that we saw in Section 6.2.1 represents a significant drag on a wave traveling in a waveguide, limiting the distance over which it is useful. Also, the requirement that the transverse dimension of a guide be comparable to the wavelength that is carried is easily met at microwave frequencies from $\sim$1 to 100 GHz ($\sim$10 cm to 1 mm), but it becomes impractical at higher frequencies to work with macroscopic objects with microscopic dimensions. Both of these problems can be addressed by carrying light in a glass fiber instead of RF in a metal box.

To see how a wave can be guided by dielectrics, consider the slab geometry shown in Figure 6.8. We'll look for a mode confined in the y direction with a periodic z dependence of $e^{-\gamma z} \equiv e^{-i\beta z}$. Starting with the TE mode, the transverse equation for H_z becomes

$$\frac{d^2 H}{dy^2} = -(\gamma^2 + k^2)H_z = (\beta^2 - k^2)H_z \tag{6.94}$$

because there is no variation in the x direction. Depending on the relative magnitudes of β and k this can have oscillatory or exponential solutions. For the solution to be confined, and reflect the symmetry of the structure, we require the wave to be exponentially damped

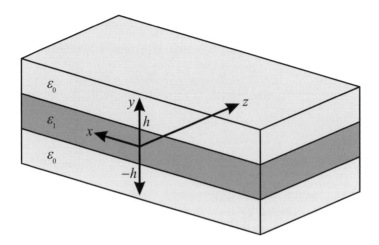

Figure 6.8. A dielectric slab waveguide.

outside of the slab, and periodic across it:

$$\frac{d^2 H_z}{dy^2} = \begin{cases} \left(\beta^2 - k_0^2\right) H_z & (|y| > h) \\ -\left(k_1^2 - \beta^2\right) H_z & (|y| < h) \end{cases} \quad . \tag{6.95}$$

The symmetric solution to this is

$$H_z = \begin{cases} Ae^{-\left(\beta^2 - k_0^2\right)^{1/2}(|y| - h)} \equiv Ae^{-a(|y| - h)} & (|y| > h) \\ B\cos\left((k_1^2 - \beta^2)^{1/2} y\right) \equiv B\cos(by) & (|y| < h) \end{cases} \quad . \tag{6.96}$$

Now the boundary conditions require continuity of the field at the interfaces, hence

$$A = B\cos(bh) \quad . \tag{6.97}$$

The transverse components are found from equations (6.83), which for E_x is

$$\begin{aligned} E_x &= -i\frac{\omega\mu}{k_c^2}\frac{\partial H_z}{\partial y} \\ &= -i\frac{\omega\mu}{k^2 - \beta^2}\frac{\partial H_z}{\partial y} \\ &= \begin{cases} -i\frac{\omega\mu}{a} Ae^{-a(y-h)} & (y > h) \\ i\frac{\omega\mu}{b} B\sin(by) & (|y| < h) \\ i\frac{\omega\mu}{a} Ae^{-a(-y-h)} & (y < -h) \end{cases} \end{aligned} \quad . \tag{6.98}$$

Equating these again at the boundaries,

$$\frac{A}{a} = \frac{B}{b}\sin(bh) \quad . \tag{6.99}$$

Now divide equation (6.99) by (6.97) to find

$$\frac{1}{a} = \frac{1}{b}\tan(bh) \quad . \tag{6.100}$$

This is a transcendental equation relating a and b, with multiple branches because of the periodicity of $\tan(bh)$. A second relationship comes from the definitions of a and b

$$a^2 = \beta^2 - k_0^2$$
$$b^2 = k_1^2 - \beta^2$$
$$\Rightarrow a^2 + b^2 = k_1^2 - k_0^2 \quad . \tag{6.101}$$

a and b are restricted to a circle, with a radius given by the difference of the squares of $k^2 = \omega^2 \mu \epsilon$ in the media. For a and b to be real, the central slab must have the higher dielectric constant. The intersections of these circles with the branches of equation (6.100), found graphically or numerically, give the modes of the waveguide.

The analysis is similar for rectangular slabs that confine modes in both directions and for circular dielectric waveguides, although the imposition of these boundary conditions becomes a more difficult calculation [Yariv, 1991]. The result for the circular geometry is that there are two modes with axial H and E components, one called the HE with H dominant, and an EH mode with E dominant.

Dielectric waveguides for confining light are produced by depositing core doping material on the inside of a cladding glass tube and then drawing it down to a thin *optical fiber*. The first ones were *multi-mode* fibers that had core diameters many times the optical wavelength, resulting in very dispersive communications. In the next chapter we'll see that this can be understood as many different path lengths reflecting at the core–cladding interface. Because they're easier to make and connect to, these are still are used for short links and for many *optical sensors* that measure light coupling into or out of a fiber to determine local material properties [Merzbacher *et al.*, 1996], but long-haul communications uses *single-mode* fibers. The minimum absorption in optical glasses occurs at infrared wavelengths; by using very pure materials this has been reduced below 0.2 dB/km at 1.55 μm [Miya *et al.*, 1979; Takahashi, 1993]. This corresponds to a loss of 10^{-3} over 150 km, making long links possible without active repeaters.

So far we've been considering *step-index* fibers that have a constant dielectric constant in the core. By varying the core doping as a function of thickness it's possible to make *graded-index* fibers that use the radial profile to shape the modes. And an asymmetrical blank when drawn down produces a *polarization-preserving* fiber that retains the polarization of the light [Galtarossa *et al.*, 1994]. We've also assumed that the medium is linear, but the intense fields in the small fiber cores can excite nonlinear effects. We'll see more of this in the Chapter 8, but one of the most important applications is to the creation of *solitons* [Zabusky, 1981]. These are pulses that balance the material's intrinsic frequency-dependent dispersion with a nonlinear response that narrows the pulse, resulting in a stable shape that can propagate for long distances without changing. These can be sent across ocean-scale distances at Gbits/second without errors [Nakazawa *et al.*, 1993; Mollenauer *et al.*, 1996]. Using all of these tricks, fiber links have been demonstrated at speeds above 1 Tbit/second, approaching the limit of 1 bit/second per hertz of optical bandwidth [Ono & Yano, 1998; Cowper, 1998].

6.4 SELECTED REFERENCES

[Ramo *et al.*, 1994] Ramo, Simon, Whinnery, John R., & Duzer, Theodore Van. (1994). *Fields and Waves in Communication Electronics*. 3rd edn. New York: Wiley.

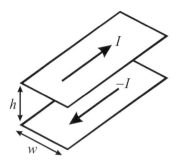

Figure 6.9. Transmission line for Problem 6.4.

A nice introduction to applied electromagnetics.

[Hagen, 1996] Hagen, Jon B. (1996). *Radio-Frequency Electronics: Circuits and Applications*. New York: Cambridge University Press.

[Danzer, 1999] Danzer, Paul (ed). (1999). *The ARRL Handbook for Radio Amateurs*. 76th edn. Newington, CT: American Radio Relay League.

Practical details for all aspects of RF design.

6.5 PROBLEMS

(6.1) Cables designed to carry a low-frequency signal with minimum pickup of interference often consist of a twisted pair of conductors surrounded by a grounded shield. Why the twist? Why the shield?

(6.2) Salt water has a conductivity ~ 4 S/m. What is the skin depth at 10^4 Hz?

(6.3) Integrate Poynting's vector $\vec{P} = \vec{E} \times \vec{H}$ to find the power flowing across a cross-sectional slice of a coaxial cable, and relate the answer to the current and voltage in the cable.

(6.4) Find the characteristic impedance and signal velocity for a transmission line consisting of two parallel strips with a width w and a separation h (Figure 6.4). You can ignore fringing fields by assuming that they are sections of conductors infinitely wide.

(6.5) The most common coaxial cable, RG58/U, has a dielectric with a relative permittivity of 2.26, an inner radius of 0.406 mm, and an outer radius of 1.48 mm.

  (*a*) What is the characteristic impedance?
  (*b*) What is the transmission velocity?
  (*c*) If a computer has a clock speed of 1 ns, how long can a length of RG58/U be and still deliver a pulse within one clock cycle?
  (*d*) It is often desirable to use thinner coaxial cable to minimize size or weight but still match the impedance of RG58/U (to minimize reflections). If such a cable has an outer diameter of 30 mils (a mil is a thousandth of an inch), what is the inner diameter?
  (*e*) For RG58/U, at what frequency does the wavelength become comparable to the diameter?

(6.6) Consider a 10 Mbit/s ethernet signal traveling in a RG58/U cable.

    (*a*) What is the physical length of a bit?

    (*b*) Now consider what would happen if a "T" connector was used to connect one ethernet coaxial cable to two other ones. Estimate the reflection coefficient for a signal arriving at the T.

7 Antennas

In Chapter 5 we found that electromagnetic waves can propagate in free space, and in Chapter 6 we saw that they can be guided in circuits. Now we need to make the important connection between these descriptions: how do electrons in a circuit excite waves in free space, and *vice versa*? This is the essential job of an *antenna*.

There are almost as many antenna designs as there are antenna designers. In theory, we've covered everything that's needed to understand them. In practice, as the size of a circuit grows to become comparable to the wavelenth of the signal that it carries, the explanation of its function can appear to grow from trivial to impenetrable because of the many tricks and approximations that are needed in this difficult regime. As we progress, it will be helpful to keep track of the purpose of an antenna: to match the impedance of a circuit with that of free space, for radiation with the desired frequency, orientation, and polarization. If this is done properly the antenna will efficiently couple the signal across rooms or solar systems; if it is not, then the antenna will serve as a reflector to heat up the signal generator.

7.1 TIME-DEPENDENT POTENTIALS

The electromagnetic potentials will be convenient for studying antennas, but we'll need to understand their behavior for fields that vary in time. Plugging the definition of the vector potential $\vec{B} = \nabla \times \vec{A}$ into $\nabla \times \vec{E} = -\partial \vec{B}/\partial t$ gives

$$\nabla \times \left(\vec{E} + \frac{\partial \vec{A}}{\partial t} \right) = 0 \quad . \tag{7.1}$$

Equation (5.13) showed that this will hold if the argument is the gradient of a function

$$\vec{E} + \frac{\partial \vec{A}}{\partial t} = -\nabla \Phi \quad , \tag{7.2}$$

or

$$\vec{E} = -\nabla \Phi - \frac{\partial \vec{A}}{\partial t} \quad . \tag{7.3}$$

For a time-dependent field, we have to add the derivative of the vector potential to the gradient of the scalar potential to find $\vec{E}$.

Since $\vec{B}$ is the curl of $\vec{A}$, we can add the gradient of another potential to $\vec{A}$ without changing the magnetic field:

$$\vec{A} \rightarrow \vec{A} + \nabla \psi \quad . \tag{7.4}$$

This will change $\vec{E}$, however, but that can be canceled if Φ is replaced with

$$\Phi \rightarrow \Phi - \frac{\partial \psi}{\partial t} \quad . \tag{7.5}$$

This is called a *gauge transformation* and can be used to simplify the vector and scalar potentials.

In terms of the potentials, Poisson's equation $\nabla \cdot \vec{D} = \rho$ becomes in free space

$$\nabla \cdot \epsilon_0 \left(-\nabla \Phi - \frac{\partial \vec{A}}{\partial t} \right) = \rho$$

$$\nabla^2 \Phi + \frac{\partial}{\partial t} \nabla \cdot \vec{A} = -\frac{1}{\epsilon_0} \rho \quad . \tag{7.6}$$

A second relationship between $\vec{A}$ and Φ comes from the Maxwell equation for the curl of $\vec{H}$, which in free space is

$$\frac{1}{\mu_0} \nabla \times \vec{B} = \vec{J} + \epsilon_0 \frac{\partial E}{\partial t}$$

$$\underbrace{\nabla \times (\nabla \times \vec{A})}_{\nabla(\nabla \cdot \vec{A}) - \nabla^2 \vec{A}} = \mu_0 \vec{J} + \underbrace{\mu_0 \epsilon_0}_{c^{-2}} \frac{\partial}{\partial t} \left(-\nabla \Phi - \frac{\partial \vec{A}}{\partial t} \right)$$

$$\nabla^2 \vec{A} - \frac{1}{c^2} \frac{\partial^2 \vec{A}}{\partial t^2} - \nabla \left(\nabla \cdot \vec{A} + \frac{1}{c^2} \frac{\partial \Phi}{\partial t} \right) = -\mu_0 \vec{J} \quad . \tag{7.7}$$

If we transform to the *Lorentz gauge* in which the potentials are related by

$$\nabla \cdot \vec{A} + \frac{1}{c^2} \frac{\partial \Phi}{\partial t} = 0 \quad , \tag{7.8}$$

equations (7.6) and (7.7) take on the satisfying form of wave equations excited by the charge and current respectively,

$$\nabla^2 \Phi - \frac{1}{c^2} \frac{\partial^2 \Phi}{\partial t^2} = -\frac{1}{\epsilon_0} \rho$$

$$\nabla^2 \vec{A} - \frac{1}{c^2} \frac{\partial^2 \vec{A}}{\partial t^2} = -\mu_0 \vec{J} \quad . \tag{7.9}$$

These are linear partial differential equations. If we take a periodic time dependence $\Phi(\vec{x}, t) = \Phi(\vec{x}) e^{i\omega t}$, with likewise for $\vec{A}$, ρ, and $\vec{J}$, the time dependence cancels out:

$$\nabla^2 \Phi(\vec{x}) + k^2 \Phi(\vec{x}) = -\frac{1}{\epsilon_0} \rho(\vec{x})$$

$$\nabla^2 \vec{A}(\vec{x}) + k^2 \vec{A}(\vec{x}) = -\mu_0 \vec{J}(\vec{x}) \quad , \tag{7.10}$$

where $k^2 = \omega^2 / c^2$. The solution for arbitrary time dependence can be found by Fourier superposition.

In Chapter 5 we solved Laplace's equation with a Green's function; we can use the same technique to find a Green's function $G(\vec{x}, \vec{x}')$ that solves *Helmholtz' equation*

$$\nabla^2 G(\vec{x}, \vec{x}') + k^2 G(\vec{x}, \vec{x}') = -4\pi\delta(|\vec{x} - \vec{x}'|) \quad . \tag{7.11}$$

Define $r = |\vec{x} - \vec{x}'|$. Since the delta function depends only on r, by symmetry G will not depend on the angle between $\vec{x}$ and $\vec{x}'$, leaving only the radial term in the expansion of the Laplacian in spherical coordinates (equation 5.20)

$$\frac{1}{r}\frac{\partial^2}{\partial r^2}(rG) + k^2 G = -4\pi\delta(r) \quad . \tag{7.12}$$

For $r \neq 0$ this reduces to the homogeneous equation

$$\frac{\partial^2}{\partial r^2}(rG) + k^2 rG = 0 \quad , \tag{7.13}$$

which can immediately be solved to find

$$rG = \alpha e^{ikr} + \beta e^{-ikr} \tag{7.14}$$

or

$$G = \alpha\frac{e^{ikr}}{r} + \beta\frac{e^{-ikr}}{r} \tag{7.15}$$

for arbitrary constants α and β. In the limit $r \to 0$ the Green's function reduces to

$$G = \frac{\alpha + \beta}{r} \quad . \tag{7.16}$$

Since we know that

$$\nabla^2 \frac{1}{r} = -4\pi\delta(r) \tag{7.17}$$

from equation (5.43), this means that the coefficients must satisfy $\alpha + \beta = 1$. The potentials are then found by integrating the Green's function over the source distributions

$$\Phi(\vec{r}) = \frac{1}{4\pi\epsilon_0}\int \frac{\rho(\vec{x})\left(\alpha e^{ik|\vec{r}-\vec{x}|} + \beta e^{-ik|\vec{r}-\vec{x}|}\right)}{|\vec{r} - \vec{x}|}\,d\vec{x}$$

$$\vec{A}(\vec{r}) = \frac{\mu_0}{4\pi}\int \frac{\vec{J}(\vec{x})\left(\alpha e^{ik|\vec{r}-\vec{x}|} + \beta e^{-ik|\vec{r}-\vec{x}|}\right)}{|\vec{r} - \vec{x}|}\,d\vec{x} \quad . \tag{7.18}$$

The exponentials in the numerators represent a phase shift in the propagation of a spherical wave from the sources. Equating these with the time delay $e^{\pm ik\Delta x} = e^{i\omega\Delta t}$ shows that $\Delta t = \pm k\Delta x/\omega = \pm\Delta x/c$. There's a problem here: the positive solution corresponds to a wave that travels backwards rather than forwards in time. Since as far as we know that's not possible, we'll drop the *advanced potential* solution and stick with the causal *retarded potential* solution with $\beta = 1$ [Anderson, 1992]:

$$\Phi(\vec{r}) = \frac{1}{4\pi\epsilon_0}\int \frac{\rho(\vec{x})\,e^{-ik|\vec{r}-\vec{x}|}}{|\vec{r} - \vec{x}|}\,d\vec{x}$$

$$\vec{A}(\vec{r}) = \frac{\mu_0}{4\pi}\int \frac{\vec{J}(\vec{x})\,e^{-ik|\vec{r}-\vec{x}|}}{|\vec{r} - \vec{x}|}\,d\vec{x} \quad . \tag{7.19}$$

In homogeneous media these are modified by including the relative permittivity and permeability.

Substituting the periodic time dependence of $\vec{A}$ and Φ into the Lorentz gauge relates them by

$$\nabla \cdot \vec{A} = -\mu_0 \epsilon_0 \frac{\partial \Phi}{\partial t}$$
$$= -i\omega \mu_0 \epsilon_0 \Phi \quad . \tag{7.20}$$

This in turn means that a periodic electric field can be written in terms of the vector potential alone,

$$\vec{E} = -\nabla \Phi - \frac{\partial \vec{A}}{\partial t}$$
$$= -\nabla \Phi - i\omega \vec{A}$$
$$= \frac{1}{i\omega \mu_0 \epsilon_0} \nabla (\nabla \cdot \vec{A}) - i\omega \vec{A} \quad . \tag{7.21}$$

Therefore if we can solve equation (7.19) for $\vec{A}$ then everything else can be found from it.

There's one final subtlety in using time-dependent electromagnetic potentials. We've written them with impunity as complex quantities $A = A_0 e^{i\theta_A} e^{i\omega t}$, with the phase angle keeping track of the *sin* and *cos* components independently. This trick works for linear transformations like addition and integration, but it fails for nonlinear operations like multiplication which will mix the real and imaginary parts. Consider the time-average of the real parts of two complex quantities A and B, which can be found from

$$\langle \text{Re}[A] \text{Re}[B] \rangle = \left\langle \frac{(A + A^*)}{2} \frac{(B + B^*)}{2} \right\rangle$$
$$= \frac{1}{4} \langle AB + A^*B + AB^* + A^*B^* \rangle$$
$$= \frac{1}{4} A_0 B_0 \left\langle e^{i(\theta_A + \theta_B + 2\omega t)} + e^{-i(\theta_A + \theta_B + 2\omega t)} + e^{i(\theta_A - \theta_B)} + e^{-i(\theta_A - \theta_B)} \right\rangle$$
$$= \frac{1}{2} A_0 B_0 [\cos(\theta_A - \theta_B) + \underbrace{\langle \cos(\theta_A + \theta_B + 2\omega t) \rangle}_{0}]$$
$$= \frac{1}{2} A_0 B_0 \cos(\theta_A - \theta_B) \quad . \tag{7.22}$$

This does not equal $\text{Re}\langle AB \rangle$, but notice that it is the same as

$$\frac{1}{2} \text{Re}[A^*B] = \frac{1}{4} (A^*B + AB^*)$$
$$= \frac{1}{4} A_0 B_0 \left(e^{i(-\theta_A + \theta_B)} + e^{i(\theta_A - \theta_B)} \right)$$
$$= \frac{1}{2} A_0 B_0 \cos(\theta_A - \theta_B)$$
$$= \langle \text{Re}[A] \text{Re}[B] \rangle$$
$$= \frac{1}{2} A_0 B_0 \cos(\theta_B - \theta_A)$$
$$= \frac{1}{2} \text{Re}[AB^*] \quad . \tag{7.23}$$

This means that we can continue our practice of finding observable values by taking the real part at the end of a calculation involving complex quantities if the time average of a product is replaced with half of one factor times the complex conjugate of the other. The most important place where this will be needed is in evaluating the Poynting vector

$$\langle \vec{P} \rangle = \langle \vec{E} \times \vec{H} \rangle$$
$$= \frac{1}{2} \text{Re} \left[\vec{E} \times \vec{H}^* \right] \tag{7.24}$$

for periodic fields.

7.2 DIPOLE RADIATION

7.2.1 Infinitesimal Length

We're now equipped to find the fields radiated by oscillating charges and currents. The simplest case is a thin wire carrying a periodic current with a constant amplitude $\vec{J} = I_0 \delta(x, y) \hat{z}$. If we further assume that the wire has an infinitesimal length d in the $\hat{z}$ direction, the vector potential can be read off from equation (7.19) to be

$$\vec{A}(r) = \mu_0 \frac{I_0 d e^{-ikr}}{4\pi r} \hat{z} \quad , \tag{7.25}$$

with $r = |\vec{x} - \vec{x}'|$ the distance from the source. This is called a *Hertz dipole* because the current must be associated with periodically-varying point charges at the ends of the wire. Since in spherical coordinates (r, θ, φ) the unit normal $\hat{z} = \cos\theta \, \hat{r} - \sin\theta \, \hat{\theta}$, the vector potential around the wire is

$$A_r = \mu_0 \frac{I_0 d e^{-ikr}}{4\pi r} \cos\theta \quad A_\theta = -\mu_0 \frac{I_0 d e^{-ikr}}{4\pi r} \sin\theta \quad . \tag{7.26}$$

The only non-zero term in equation (5.21) is

$$\vec{B} = \nabla \times \vec{A}$$
$$= \frac{1}{r} \left[\frac{\partial}{\partial r}(r A_\theta) - \frac{\partial}{\partial \theta} A_r \right] \hat{\varphi} \quad , \tag{7.27}$$

giving

$$B_\varphi = \frac{\mu_0 I_0 d}{4\pi} e^{-ikr} \left(\frac{ik}{r} + \frac{1}{r^2} \right) \sin\theta \quad . \tag{7.28}$$

Equation (7.21) can be used to find the corresponding electric field (Problem 7.1),

$$E_\theta = \frac{I_0 d}{4\pi} e^{-ikr} \left(\frac{i\omega\mu_0}{r} + \frac{1}{r^2}\sqrt{\frac{\mu_0}{\epsilon_0}} + \frac{1}{i\omega\epsilon_0 r^3} \right) \sin\theta$$
$$E_r = \frac{I_0 d}{4\pi} e^{-ikr} \left(\frac{2}{r^2}\sqrt{\frac{\mu_0}{\epsilon_0}} + \frac{2}{i\omega\epsilon_0 r^3} \right) \cos\theta \quad . \tag{7.29}$$

There are three exponents for the radial dependence. The r^{-3} term will dominate as $r \to 0$, reducing to the electric field of a static dipole with $I_0 d = p$ (equation 5.69). This is the *near-field* or *static* zone. The r^{-1} term will dominate as $r \to \infty$, the *far-field* or *radiation* zone. And the r^{-2} is significant in the intermediate *induction* zone. Taking only the r^{-1} terms significant in the far field, the Poynting vector is

$$\langle \vec{P} \rangle = \frac{1}{2} \text{Re}[\vec{E} \times \vec{H}^*]$$

$$= \hat{r} \frac{1}{2} \text{Re}[E_\theta H_\varphi^*]$$

$$= \hat{r} \frac{I_0^2 k^2 d^2}{32\pi^2 r^2} \sqrt{\frac{\mu_0}{\epsilon_0}} \sin^2 \theta \quad . \tag{7.30}$$

This can be integrated over a sphere to find the total energy radiated,

$$W = \int \vec{P} \cdot d\vec{A}$$

$$= \int_0^{2\pi} \int_0^{\pi} P_r r^2 \sin \theta \, d\theta \, d\varphi$$

$$= \int_0^{\pi} 2\pi P_r r^2 \sin \theta \, d\theta$$

$$= \int_0^{\pi} 2\pi \frac{I_0^2 k^2 d^2}{32\pi^2 r^2} \sqrt{\frac{\mu_0}{\epsilon_0}} \sin^2 \theta r^2 \sin \theta \, d\theta$$

$$= \frac{I_0^2 k^2 d^2}{16\pi} \sqrt{\frac{\mu_0}{\epsilon_0}} \int_0^{\pi} \sin^3 \theta \, d\theta$$

$$= \frac{I_0^2 k^2 d^2}{12\pi} \sqrt{\frac{\mu_0}{\epsilon_0}}$$

$$= \frac{I_0^2 \pi}{3} \sqrt{\frac{\mu_0}{\epsilon_0}} \left(\frac{d}{\lambda}\right)^2 \quad . \tag{7.31}$$

The r^{-1} dependence of the leading terms in $\vec{E}$ and $\vec{H}$ give an r^{-2} decay that is canceled by the r^2 surface area, leaving the total radiated power over all directions independent of the distance.

In a periodically driven resistor, the power consumption is

$$W = \langle I^2 R \rangle$$

$$= I_0^2 \underbrace{\langle \sin^2(\omega t) \rangle}_{1/2} R$$

$$= \frac{I_0^2 R}{2} \quad . \tag{7.32}$$

If we turn this around to define a resistance, the power radiated from the dipole is related to the magnitude of the current that excites it by

$$R_{\text{rad}} = \frac{2W}{I_0^2}$$

$$= \frac{2\pi}{3} \sqrt{\frac{\mu_0}{\epsilon_0}} \left(\frac{d}{\lambda}\right)^2 \quad . \tag{7.33}$$

This is called the *radiation resistance*. If it is small, the real impedances in the circuit

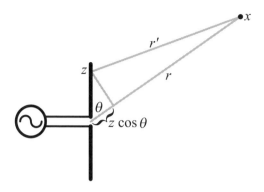

Figure 7.1. Center-fed dipole antenna.

feeding the antenna will dominate, leaving most of the energy dissipated there rather than radiated as desired. That will certainly be true here because d is infinitesimal.

7.2.2 Finite Length

A more effective antenna drives a resonant mode of a longer dipole so that the oscillatory energy can stay in the antenna. Following the resonant modes we found in the last chapter, let's now take

$$\vec{J}(\vec{x}) = I_0 \, \delta(x, y) \sin \left[k \left(\frac{d}{2} - |z| \right) \right] \hat{z} \qquad (7.34)$$

for the current distribution in a center-fed dipole of length d, having a maximum at the center and vanishing at the ends. This neglects the small corrections to the current distribution due to radiation damping and the finite transverse thickness of the wire.

The geometry is shown in Figure 7.1. We'll be interested in the far-field radiation pattern, for which $r \gg z$, therefore the path difference as a function of position along the antenna can be approximated by $r' \approx r - z \cos \theta$. This difference will be sigificant in the phase of the spherical wave $e^{ikr'}$ over the antenna, but not in the decay of the amplitude r'^{-1} which will be a small change in the large radius. With these approximations equation (7.19) can easily be integrated to find

$$\vec{A} = \frac{\mu_0}{4\pi} \int_{-d/2}^{d/2} I_0 \sin \left[k \left(\frac{d}{2} - |z| \right) \right] \frac{e^{ik(r - z \cos \theta)}}{r} \, dz$$

$$= \frac{\mu_0}{2\pi} I_0 \frac{e^{ikr}}{kr} \frac{\cos \left(\frac{kd}{2} \cos \theta \right) - \cos \left(\frac{kd}{2} \right)}{\sin^2 \theta} \hat{z} \quad . \qquad (7.35)$$

Referring back to the curl in spherical coordinates, equation (5.21), all of the terms have an r^{-1} dependence and hence can be ignored except for two:

$$\vec{B} = \nabla \times \vec{A} \rightarrow -\frac{1}{r} \frac{\partial}{\partial r} (r A_\varphi) \, \hat{\theta} + \frac{1}{r} \frac{\partial}{\partial r} (r A_\theta) \, \hat{\varphi} \quad . \qquad (7.36)$$

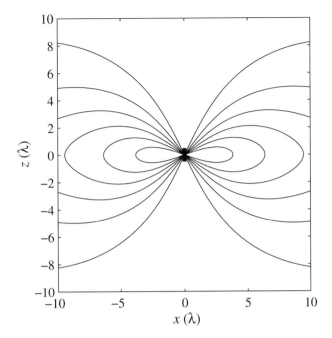

Figure 7.2. Radiation from a half-wave dipole.

Writing $\hat{z} = \cos\theta\,\hat{r} - \sin\theta\,\hat{\theta}$ and taking the curl,

$$\vec{B} = -\frac{\mu_0}{2\pi}I_0\frac{ie^{ikr}}{r}\frac{\cos\left(\frac{kd}{2}\cos\theta\right) - \cos\left(\frac{kd}{2}\right)}{\sin\theta}\,\hat{\varphi}\quad. \tag{7.37}$$

From this the electric field can be found,

$$\begin{aligned}
\vec{E} &= -\sqrt{\frac{\mu_0}{\epsilon_0}}\hat{k}\times\vec{H}\\
&= -\frac{1}{2\pi}\sqrt{\frac{\mu_0}{\epsilon_0}}I_0\frac{ie^{ikr}}{r}\frac{\cos\left(\frac{kd}{2}\cos\theta\right) - \cos\left(\frac{kd}{2}\right)}{\sin\theta}\,\hat{\theta}\quad,
\end{aligned} \tag{7.38}$$

and the Poynting vector

$$\begin{aligned}
\langle\vec{P}\rangle &= \frac{1}{2}\mathrm{Re}[\vec{E}\times\vec{H}^*]\\
&= \frac{1}{8\pi^2}\sqrt{\frac{\mu_0}{\epsilon_0}}\frac{I_0^2}{r^2}\left[\frac{\cos\left(\frac{kd}{2}\cos\theta\right) - \cos\left(\frac{kd}{2}\right)}{\sin\theta}\right]^2\,\hat{r}\quad.
\end{aligned} \tag{7.39}$$

Contours of constant $\vec{P}$ are plotted in cross-sectional slice in Figure 7.2 for $kd/2 = \pi/2$, a *half-wave* dipole antenna. The difference between these curves and the circles that would describe spherical waves represents *directivity*: the antenna sends more energy in a desired direction at the expense of less in others.

The total power emitted is

$$
\begin{aligned}
W &= \int \vec{P} \cdot d\vec{A} \\
&= \int_0^{2\pi} \int_0^\pi P_r r^2 \sin\theta \; d\theta \; d\varphi \\
&= \int_0^\pi 2\pi P_r r^2 \sin\theta \; d\theta \\
&= \frac{I_0^2}{4\pi} \sqrt{\frac{\mu_0}{\epsilon_0}} \int_0^\pi \frac{\cos\left(\frac{kd}{2}\cos\theta\right) - \cos\left(\frac{kd}{2}\right)}{\sin\theta} \; d\theta \quad .
\end{aligned}
\tag{7.40}
$$

The integral can be written in terms of special functions or evaluated numerically; for $kd = \pi$ it is

$$
W \approx \frac{I_0^2}{4\pi} \sqrt{\frac{\mu_0}{\epsilon_0}} \; 1.22 \quad .
\tag{7.41}
$$

This gives a radiation resistance of

$$
\begin{aligned}
R_{\text{rad}} &= \frac{2W}{I_0^2} \\
&= \frac{2.44}{4\pi} \sqrt{\frac{\mu_0}{\epsilon_0}} \\
&\approx 73 \; \Omega \quad ,
\end{aligned}
\tag{7.42}
$$

quite a difference from the infinitesimal case. This value explains the prevalence of 75 Ω transmission lines, although the radiation resistance will vary from this value because of loading from nearby objects (such as the Earth), and have an imaginary component due to the antenna's non–infinitesimal dimensions.

7.3 DUALITY AND RECIPROCITY

Maxwell's equations in free space, equation (5.96), have an interesting symmetry: they are left unchanged if $\vec{E}$ is replaced with $\vec{H}$, $\vec{H}$ replaced with $-\vec{E}$, and μ and ϵ are interchanged. This *duality* relationship means that any free-space solution immediately provides a second dual one with the electric and magnetic fields swapped. Since we've already seen that the static field of an electrical dipole (equation 5.69) has the same form as the static field of a magnetic dipole (equation 5.78), this means that the roles of the electric and magnetic fields can be exchanged in equation (7.29) to find the radiation from a magnetic dipole source.

A deeper symmetry starts with the rather formal observation that any two sets of fields

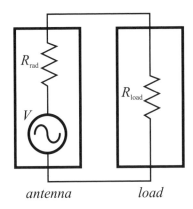

Figure 7.3. Effective circuit for a receiving antenna.

$(\vec{E}_1, \vec{H}_1)$ and $(\vec{E}_2, \vec{H}_2)$ that are periodic with a frequency ω must satisfy

$$\nabla \cdot \left(\vec{E}_1 \times \vec{H}_2 - \vec{E}_2 \times \vec{H}_1 \right)$$
$$= \vec{H}_2 \cdot \left(\nabla \times \vec{E}_1 \right) - \vec{E}_1 \cdot \left(\nabla \times \vec{H}_2 \right) - \vec{H}_1 \cdot \left(\nabla \times \vec{E}_2 \right) + \vec{E}_2 \cdot \left(\nabla \times \vec{H}_1 \right)$$
$$= \vec{H}_2 \cdot \left(-\frac{\partial \vec{B}_1}{\partial t} \right) - \vec{E}_1 \cdot \left(\vec{J}_2 + \frac{\partial \vec{D}_2}{\partial t} \right) - \vec{H}_1 \cdot \left(-\frac{\partial \vec{B}_2}{\partial t} \right) + \vec{E}_2 \cdot \left(\vec{J}_1 + \frac{\partial \vec{D}_1}{\partial t} \right)$$
$$= -i\omega \vec{H}_2 \cdot \vec{B}_1 - \vec{E}_1 \cdot \vec{J}_2 - i\omega \vec{E}_1 \cdot \vec{D}_2 + i\omega \vec{H}_1 \cdot \vec{B}_2 + \vec{E}_2 \cdot \vec{J}_1 + i\omega \vec{E}_2 \cdot \vec{D}_1$$
$$= -i\omega \mu \vec{H}_2 \cdot \vec{H}_1 - \vec{E}_1 \cdot \vec{J}_2 - i\omega \epsilon \vec{E}_1 \cdot \vec{E}_2 + i\omega \mu \vec{H}_1 \cdot \vec{H}_2 + \vec{E}_2 \cdot \vec{J}_1 + i\omega \epsilon \vec{E}_2 \cdot \vec{E}_1$$
$$= \vec{E}_2 \cdot \vec{J}_1 - \vec{E}_1 \cdot \vec{J}_2 \qquad (7.43)$$

for sources $\vec{J}_1$ and $\vec{J}_2$. If both sides are integrated over the volume of a sphere,

$$\int \left(\vec{E}_2 \cdot \vec{J}_1 - \vec{E}_1 \cdot \vec{J}_2 \right) dV = \int \nabla \cdot \left(\vec{E}_1 \times \vec{H}_2 - \vec{E}_2 \times \vec{H}_1 \right) dV \quad , \qquad (7.44)$$

Gauss' Law turns the right hand side into a surface integral,

$$\int \left(\vec{E}_2 \cdot \vec{J}_1 - \vec{E}_1 \cdot \vec{J}_2 \right) dV = \int \left(\vec{E}_1 \times \vec{H}_2 - \vec{E}_2 \times \vec{H}_1 \right) \cdot d\vec{A} \quad , \qquad (7.45)$$

and in the limit that the radius goes to infinity the near-field solutions will disappear, leaving $\vec{E}$ and $\vec{H}$ as the transverse components of an outgoing spherical wave with the local wave vector $\vec{k} = k\hat{r}$ pointing radially so that

$$\int \left(\vec{E}_2 \cdot \vec{J}_1 - \vec{E}_1 \cdot \vec{J}_2 \right) dV = \int \left[\vec{E}_1 \times \left(\sqrt{\frac{\epsilon_0}{\mu_0}} \vec{k} \times \vec{E}_2 \right) - \vec{E}_2 \times \left(\sqrt{\frac{\epsilon_0}{\mu_0}} \vec{k} \times \vec{E}_1 \right) \right] \cdot d\vec{A}$$
$$= 0 \quad . \qquad (7.46)$$

This is one form of the *Lorentz Reciprocity Theorem*, which provides a connection between the transmitting and receiving properties of an antenna.

The effective circuit of an antenna used as a receiver is shown in Figure 7.3. The radiation induces a voltage V across the antenna terminals, which appears to the load as in ideal generator in series with the antenna's radiation resistance. Problem 7.3 will show that a maximum power of $W = |V|^2 / 8R_{\text{load}}$ is delivered to the load if $R_{\text{load}} = R_{\text{rad}}$.

To find how V relates to the source current in the transmitting antenna, equation (7.46) can be applied to a pair of antennas. The integral will vanish except where the currents are non–zero, which can easily be evaluated for an infinitesimal dipole,

$$\int \vec{E}_2 \cdot \vec{J}_1 \, dV = \int \vec{E}_1 \cdot \vec{J}_2 \, dV$$

$$\int \vec{E}_2(\vec{x}_1) \cdot \vec{dl} \, J_1 \, dA = \int \vec{E}_1(\vec{x}_2) \cdot \vec{dl} \, J_2 \, dA$$

$$V_2(\vec{x}_1)I_1 = V_1(\vec{x}_2)I_2$$

$$\frac{I_1}{V_1(\vec{x}_2)} = \frac{I_2}{V_2(\vec{x}_1)} \qquad . \tag{7.47}$$

The ratio of the source current I_1 in antenna 1 to the voltage $V_1(\vec{x}_2)$ it induces in antenna 2 is equal to the ratio with the radiation going in the opposite direction. This result, called *reciprocity*, can be extended to apply to arbitrary pairs of antennas. This is surprising: antennas usually receive far-field plane-wave radiation, but they radiate solutions with much more complex near-field patterns; there's no *a priori* reason to expect any symmetry between these processes.

Reciprocity provides useful connections among many antenna properties. The *gain G* of an antenna is defined to be the maximum value of the Poynting vector, evaluated on the surface of a unit sphere, divided by the total power radiated over the area of that sphere

$$G \equiv \max_{\theta,\varphi} \frac{P(r = 1, \theta, \varphi)}{W/4\pi} \qquad . \tag{7.48}$$

Directional antennas can have gains much larger than 1, the value for an isotropic radiator.

If an antenna transmits a power W_1, the received far-field power at a second antenna W_2 will be

$$W_2 = A_2 \frac{G_1}{4\pi r^2} W_1 \qquad . \tag{7.49}$$

The transmitted power decreases as a spherical wave but is increased by the gain, and A defines the receiving antenna's *area* which is the effective cross-section it presents to capture the incoming power density. The ratio of the transmitted and received power is thus

$$\frac{W_2}{W_1} = \frac{1}{4\pi r^2} A_2 G_1 \qquad . \tag{7.50}$$

If antenna 1 receives and antenna 2 transmits the roles are reversed:

$$\frac{W_1}{W_2} = \frac{1}{4\pi r^2} A_1 G_2 \qquad , \tag{7.51}$$

but because of reciprocity these ratios must be equal:

$$\frac{W_1}{W_2} = \frac{W_2}{W_1}$$

$$A_1 G_2 = A_2 G_1$$

$$\frac{A_1}{G_1} = \frac{A_2}{G_2} \qquad . \tag{7.52}$$

Since these antennas are arbitrary, we conclude that the ratio of an antenna's area to its

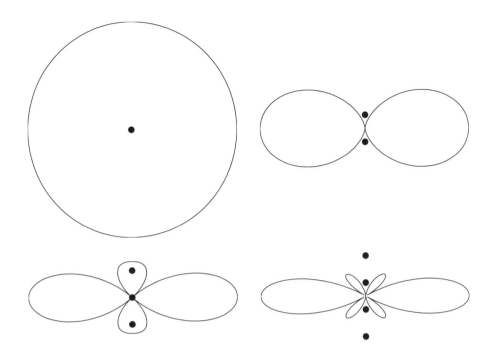

Figure 7.4. Looking down on antennas in a broadside array, showing the location of the antennas and contours of the field strength.

gain is a constant independent of its design. In Problem 7.4 we'll see that this value is $\lambda^2/4\pi$.

These concepts of gain and area are essential to the design of useful RF links: the overall SNR can be improved by increasing the transmitter power, the transmitting antenna's gain, the receiving antenna's area, or the receiving amplifier's sensitivity. These present very different costs in the system's size/weight/dissipation/complexity/expense that must be allocated to where they are best borne.

7.4 ANTENNA TYPES

Antenna gain serves to direct energy where it is wanted, saving power and reducing interference. But how is it increased? One approach is shown in Figure 7.4, which plots the in-plane far-field radiation pattern for arrays of dipoles, viewed from above. A single dipole radiates a spherical wave, with field components that depend on distance as e^{ikr}/r. If there is more than one dipole, and the spacing between them is small compared to the distance r, then as in Figure 7.1 the dependence of the amplitude on the relative distances can be ignored and the phase dependence approximated by summing over their locations y_n as $\sum_n e^{ik(r-y_n\cos\theta)}/r$. Normal to the axis of the array the dipoles add in phase, but if they are placed a half-wavelength apart they will interfere destructively in the transverse direction, resulting in increasingly directional radiation patterns as antennas

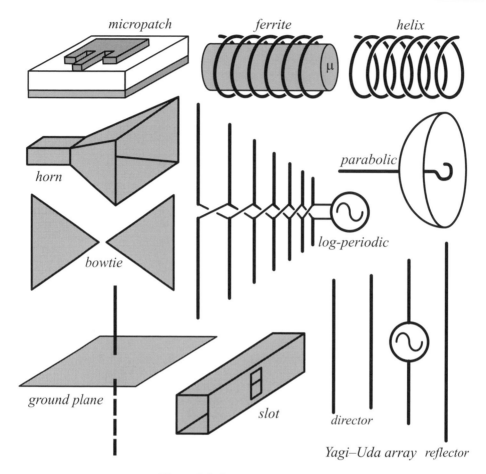

Figure 7.5. Some antenna types.

are added. This is called a *broadside* array, because the beam comes from the broad side. Alternatively, if the relative phases of the antennas match the phase shift over their separation, then they will add constructively along their axis and destructively in the normal direction, giving an *end-fire* array. AM radio stations usually use one of these configurations, so the antenna orientation points to the market that they are aimed at.

The *Yagi–Uda* array shown in Figure 7.5 cleverly obtains directionality with just one driven antenna, by parasitically exciting the rest of the array. Depending on the spacing, these secondary antennas can be *directors* that add constructively in a desired direction, or *reflectors* that add destructively in an undesired one.

Beyond dipoles, the enormous range of antenna designs reflects their many applications. A *micropatch* antenna uses a stripline to drive a resonant mode of a conductor over a ground plane. Because they have a planar structure, these are used where antennas must be conformal to a surface. The slots in the micropatch in Figure 7.5 provide impedance matching between the patch and the driving line. A *horn* matches the impedance between a waveguide and free space by increasing the area so gradually that there is no appreciable scattering. In the limit that the wavelength is small compared to the size of the antenna, the waves can be directed geometrically, as is done in the *parabolic* antennas that provide the large areas needed for weak satellite signals. Another way to increase the effective size of an antenna that's useful at low frequencies is to pull in more of the magnetic flux with a *ferrite* that provides permeability without conductivity (Chapter 12). These are found in AM receivers. At low frequencies the coil serves just to collect magnetic flux, but at much higher frequencies a traveling wave will follow it around. This is used in a *helical* antenna to launch and detect circularly-polarized radiation, eliminating the signal dependence on the relative orientation between the transmitter and the receiver, and providing two independent polarization channels that can be used in a single band.

Antennas are frequently called upon to operate over a range of frequencies. To be independent of wavelength, their shape must be invariant if it is rescaled. One kind of broadband antenna is the *bow-tie*, which unlike a dipole has a continuous spectrum of modes that extend further out as the frequency is decreased. A variant is the *log-peridioc* antenna, familiar on roofs as TV aerials, which is a finite segment from an infinite series of dipoles with a geometrical scaling of their size and spacing, connected in an alternating sequence so that the gain is directed towards the tapered end.

Antennas can also be defined by the imposition of boundary conditions. If a monopole antenna is placed above a ground plane, the induced image charge will match the radiation from a dipole. This can be used to synthesize extra antenna elements, and to shield an antenna from interfering materials (like the Earth). And if a hole is cut into a waveguide a wave will leak out, which can be understood as the radiation from the currents matching the boundary conditions. These are used in distributed phased *slot* antennas.

Perhaps the most interesting antennas of all are the ones that make no (apparent) sense. These give a nonlinear search algorithm free reign to optimize a structure to meet desired specifications, using numerical simulations and experimental tests as error metrics to guide the search. The resulting shapes can have no recognizable logic, but nevertheless beat the performance of conventional antenna designs [Johnson & Rahmat-Samii, 1997].

7.5 SELECTED REFERENCES

[Ramo *et al.*, 1994] Ramo, Simon, Whinnery, John R., & Duzer, Theodore Van. (1994). *Fields and Waves in Communication Electronics*. 3rd edn. New York: Wiley.

[Balanis, 1997] Balanis, Constantine. (1997). *Antenna Theory: Analysis and Design*. 2nd edn. New York: Wiley.

> Everything you ever wanted to know about antennas, and more.

7.6 PROBLEMS

(7.1) Find the electric field for an infinitesimal dipole radiator.

(7.2) What is the magnitude of the Poynting vector at a distance of 1 km from an antenna radiating 1 kW of power, assuming that it is an isotropic radiator with a wavelength much less than 1 km? What is the peak electric field strength at that distance?

(7.3) For what value of R_{load} is the maximum power delivered to the load in Figure 7.3?

(7.4) For an infinitesimal dipole antenna, what are the gain and the area, and what is their ratio?

8 Optics

This chapter will complete our tour of Maxwell's equations by looking at electromagnetic waves with frequencies so high that their wavelength becomes small compared to the structures that manipulate them. The familiar equations of geometrical optics will naturally emerge from matching boundary conditions in this regime, with applications from microscopy to optical information processing. The chapter will close by considering some of the obstacles and opportunities associated with relaxing the assumptions of short wavelengths in linear media through the study of Gaussian and nonlinear optics. Although the focus will be on light (pun not intended), these same ideas are used at lower frequencies with *quasi-optical* RF components, at higher frequencies with reflection *X-ray optics*, and with magnetic lenses in *electron optics*.

8.1 REFLECTION AND REFRACTION

In Chapter 5 we saw that a plane TEM electromagnetic wave in a homogeneous medium can propagate with the following properties:

- $\vec{E} = \vec{E}_0 e^{i(\vec{k}\cdot\vec{r}-\omega t)}$, where the direction of the electric field $\hat{E}_0$ is perpendicular to the direction of travel $\hat{k}$.
- The magnetic field is proportional and transverse to the electric field, $\vec{H} = \sqrt{\epsilon/\mu}\,\hat{k}\times\vec{E}$.
- The velocity $v = 1/\sqrt{\mu\epsilon}$, and the wave number is $k = \omega/v = \omega\sqrt{\mu\epsilon} = 2\pi/\lambda$.

Materials are conveniently described by n, the *index of refraction*, which is the ratio of the speed of light in vacuum to its speed in the material:

$$n = \frac{c}{v} = \frac{\sqrt{\mu_0\mu_r\epsilon_0\epsilon_r}}{\sqrt{\mu_0\epsilon_0}} = \sqrt{\mu_r\epsilon_r} \quad . \tag{8.1}$$

A typical value for glass is $n = 1.5$. Notice that the ratio

$$\frac{k}{n} = \frac{2\pi}{\lambda n} = \frac{\omega}{v}\frac{v}{c} = \frac{\omega}{c} \tag{8.2}$$

is independent of the material and is determined solely by the frequency. This means that as light moves through materials with different indices of refraction the frequency won't change because it must still oscillate at the same rate, but the wavelength will.

In Section 5.3.2 we saw that at an interface between two media a and b the boundary conditions in the absence of free surface charges and currents are

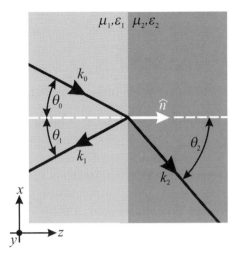

Figure 8.1. Reflection and refraction at an interface.

- Normal component of $\vec{D}$ and $\vec{B}$ continuous

$$(\vec{D}_a - \vec{D}_b) \cdot \hat{n} = 0 \quad (\vec{B}_a - \vec{B}_b) \cdot \hat{n} = 0 \tag{8.3}$$

- Tangential component of $\vec{E}$ and $\vec{H}$ continuous

$$(\vec{E}_a - \vec{E}_b) \times \hat{n} = 0 \quad (\vec{H}_a - \vec{H}_b) \times \hat{n} = 0 \quad, \tag{8.4}$$

where $\hat{n}$ is normal to the interface.

Now consider a plane wave incident on an interface between two insulating dielectric materials, and let's allow for reflected and transmitted waves that might have different angles and wave vectors as shown in Figure 8.1.

The continuity of the tangential component of the electric field requires

$$[\vec{E}_0 e^{i(\vec{k}_0 \cdot \vec{r} - \omega_0 t)} + \vec{E}_1 e^{i(\vec{k}_1 \cdot \vec{r} - \omega_1 t)} - \vec{E}_2 e^{i(\vec{k}_2 \cdot \vec{r} - \omega_2 t)}] \times \hat{n} = 0 \quad . \tag{8.5}$$

This must hold for all times, therefore

$$\omega_0 = \omega_1 = \omega_2 \quad, \tag{8.6}$$

and it must hold everywhere on the boundary, which implies that all three waves must have the same periodicity variation along the boundary

$$\vec{k}_0 \cdot \hat{x} = \vec{k}_1 \cdot \hat{x} = \vec{k}_2 \cdot \hat{x} \tag{8.7}$$

$$k_0 \sin \theta_0 = k_1 \sin \theta_1 = k_2 \sin \theta_2 \quad .$$

Since the magnitudes of the wave vectors k_0 and k_1 are equal because their wavelength is the same, we find that the angle of incidence equals the angle of reflection:

$$k_0 \sin \theta_0 = k_0 \sin \theta_1 \tag{8.8}$$

$$\theta_0 = \theta_1 \quad .$$

A second condition comes from $k_1/n_1 = k_2/n_2$ (equation 8.2):

$$k_1 \sin\theta_1 = k_2 \sin\theta_2 \qquad (8.9)$$

$$\frac{k_1}{k_2} = \frac{\sin\theta_2}{\sin\theta_1}$$

$$\frac{n_1}{n_2} = \frac{\sin\theta_2}{\sin\theta_1} \quad .$$

This is *Snell's Law*, discovered experimentally by Willebrord Snell around 1621. Light rays bend when they cross the interface between media with different indices of refraction. The index of refraction can depend on wavelength and so different colors are bent in different directions; this gives rise to *chromatic dispersion* in a prism, which is usually good, and to *chromatic aberration* in lenses, which is usually bad.

We've found the directions of the rays; now we will find the amplitudes from the continuity equations

$$(\vec{E}_0 + \vec{E}_1) \times \hat{n} = \vec{E}_2 \times \hat{n} \qquad (8.10)$$

and

$$(\vec{H}_0 + \vec{H}_1) \times \hat{n} = \vec{H}_2 \times \hat{n} \quad . \qquad (8.11)$$

The $\vec{H}$ and $\vec{E}$ components of the wave are related by

$$\vec{H} = \sqrt{\frac{\epsilon}{\mu}}\, \hat{k} \times \vec{E}$$

$$= \sqrt{\frac{\epsilon}{\mu}}\, \frac{\vec{k}}{|\vec{k}|} \times \vec{E}$$

$$= \sqrt{\frac{\epsilon}{\mu}}\, \frac{1}{\omega\sqrt{\mu\epsilon}} \vec{k} \times \vec{E}$$

$$= \frac{1}{\omega\mu}\, \vec{k} \times \vec{E}$$

$$\approx \frac{1}{\omega\mu_0}\, \vec{k} \times \vec{E} \qquad (8.12)$$

(since in most dielectric materials $\mu_r \approx 1$). Substituting into equation (8.11),

$$(\vec{k}_0 \times \vec{E}_0 + \vec{k}_1 \times \vec{E}_1) \times \hat{n} = (\vec{k}_2 \times \vec{E}_2) \times \hat{n} \quad . \qquad (8.13)$$

An arbitrary incoming wave can be separated into two components that will be analyzed separately: the component with the electric field perpendicular to the plane of incidence ($\vec{E}$ points in the $\hat{y}$ direction in Figure 8.1) and the component with the electric field in the plane of incidence ($E_y = 0$).

- $\vec{E}$ perpendicular to the plane of incidence

  Since all the electric field vectors point in the $\hat{y}$ direction, the continuity equation

$$(\vec{E}_0 + \vec{E}_1 - \vec{E}_2) \times \hat{n} = 0 \qquad (8.14)$$

  becomes the scalar equation

$$E_0 + E_1 = E_2 \quad . \qquad (8.15)$$

Using the BAC–CAB rule $\vec{A} \times (\vec{B} \times \vec{C}) = \vec{B}(\vec{A} \cdot \vec{C}) - \vec{C}(\vec{A} \cdot \vec{B})$, equation (8.13) expands to

$$[\vec{E}_0(\vec{k}_0 \cdot \hat{n}) - \hat{n}(\vec{k}_0 \cdot \vec{E}_0)] + [\vec{E}_1(\vec{k}_1 \cdot \hat{n}) - \hat{n}(\vec{k}_1 \cdot \vec{E}_1)]$$

$$= [\vec{E}_2(\vec{k}_2 \cdot \hat{n}) - \hat{n}(\vec{k}_2 \cdot \vec{E}_2)] \quad . \tag{8.16}$$

Because this is a TEM wave the $\vec{k} \cdot \vec{E}$ terms vanish, and writing out the dot products gives

$$E_0(\vec{k}_0 \cdot \hat{n}) + E_1(\vec{k}_1 \cdot \hat{n}) = E_2(\vec{k}_2 \cdot \hat{n})$$

$$E_0 k_0 \cos\theta_0 - E_1 k_1 \cos\theta_1 = E_2 k_2 \cos\theta_2$$

$$E_0 \cos\theta_0 - E_1 \cos\theta_1 = \frac{k_2}{k_1} E_2 \cos\theta_2 \quad (k_0 = k_1)$$

$$= \frac{n_2}{n_1} E_2 \cos\theta_2 \quad . \tag{8.17}$$

For the incoming wave we know E_0 and θ_0, and therefore θ_1. From Snell's Law we know θ_2, leaving two unknowns (E_1 and E_2) and two equations (8.1 and 8.17). These can be solved to find the desired relationship between the incoming and reflected amplitudes:

$$E_0 \cos\theta_0 - E_1 \cos\theta_1 = \frac{n_2}{n_1} E_2 \cos\theta_2$$

$$= \frac{\sin\theta_1}{\sin\theta_2} E_2 \cos\theta_2$$

$$= \frac{\sin\theta_1}{\sin\theta_2} (E_0 + E_1) \cos\theta_2 \quad \text{(equation 8.15)}$$

$$E_0 \cos\theta_0 \sin\theta_2 - E_1 \cos\theta_1 \sin\theta_2 = (E_0 + E_1) \cos\theta_2 \sin\theta_1$$

$$E_1 = \frac{\cos\theta_0 \sin\theta_2 - \cos\theta_2 \sin\theta_1}{\cos\theta_2 \sin\theta_1 + \cos\theta_1 \sin\theta_2} E_0$$

$$= \frac{\cos\theta_0 \sin\theta_2 - \cos\theta_2 \sin\theta_0}{\cos\theta_2 \sin\theta_0 + \cos\theta_0 \sin\theta_2} E_0$$

$$= \frac{\sin(\theta_2 - \theta_0)}{\sin(\theta_2 + \theta_0)} E_0 \quad . \tag{8.18}$$

From this we find the transmitted amplitude

$$E_2 = E_0 + E_1$$

$$= \left[1 + \frac{\sin(\theta_2 - \theta_0)}{\sin(\theta_2 + \theta_0)} \right] E_0$$

$$= \frac{\sin(\theta_2 + \theta_0) + \sin(\theta_2 - \theta_2)}{\sin(\theta_2 + \theta_0)} E_0$$

$$= \frac{2 \sin\theta_2 \cos\theta_0}{\sin(\theta_2 + \theta_0)} E_0 \quad . \tag{8.19}$$

Summarizing the results,

$$E_1 = \frac{\sin(\theta_2 - \theta_0)}{\sin(\theta_2 + \theta_0)} E_0 \quad ,$$

$$E_2 = \frac{2 \sin\theta_2 \cos\theta_0}{\sin(\theta_2 + \theta_0)} E_0 \quad . \tag{8.20}$$

- $\vec{E}$ in the plane of incidence

  Continuity of the tangential component of $\vec{E}$ gives

  $$(\vec{E}_0 + \vec{E}_1 - \vec{E}_2) \times \hat{n} = 0 \qquad (8.21)$$

  or

  $$E_0 \cos\theta_0 - E_1 \cos\theta_1 = E_2 \cos\theta_2 \qquad (8.22)$$

since the cross products all point in the $\hat{y}$ direction. Similary, equation (8.13) becomes a scalar equation since $\vec{k} \times \vec{E}$ points in the $\hat{y}$ direction, and then $\hat{y} \times \hat{n}$ points in the $\hat{x}$ direction:

$$k_0 E_0 + k_1 E_1 = k_2 E_2$$
$$E_0 + E_1 = \frac{k_2}{k_1} E_2 \quad (k_0 = k_1)$$
$$= \frac{n_2}{n_1} E_2 \quad . \qquad (8.23)$$

Once again we have two equations for our two unknowns, which can be solved with a bit more algebra to show that

$$E_1 = \frac{\tan(\theta_0 - \theta_2)}{\tan(\theta_0 + \theta_2)} E_0 \quad ,$$
$$E_2 = \frac{2\cos\theta_0 \sin\theta_2}{\sin(\theta_0 + \theta_2)\cos(\theta_0 - \theta_2)} E_0 \quad . \qquad (8.24)$$

Equations (8.20) and (8.24) are the *Fresnel equations*. Notice that E_1 can vanish in equation (8.24) if the numerator vanishes, which will happen when $\theta_0 = \theta_2$. This is trivial: it says that there is no reflection if the materials are the same. E_1 can also vanish when the denominator diverges, which will happen if $\theta_0 + \theta_2 = \pi/2$, which means that the transmitted and reflected beams are perpendicular. This angle is called *Brewster's angle* θ_B, and may be found from Snell's Law to be

$$\frac{n_2}{n_1} = \frac{\sin\theta_B}{\sin[(\pi/2) - \theta_B]} = \tan\theta_B \quad . \qquad (8.25)$$

At this angle incoming radiation with the field pointing in an arbitrary direction will be reflected with no component of the field in the plane of incidence. The reflected radiation will be *linearly polarized* with the field pointing solely parallel to the plane of the interface between the materials. This is how polarizing sunglasses work: since reflected light close to Brewster's angle is nearly linearly polarized, glasses that contain vertically oriented polarizers will block most of the reflected glare [Land, 1951]. We will cover polarization in more detail in Chapter 11.

A second important angle for reflections is the *critical angle* θ_c for which $\theta_2 = \pi/2$:

$$\frac{n_2}{n_1} = \frac{\sin\theta_c}{\sin\theta_2} = \frac{\sin\theta_c}{1} \quad \Rightarrow \quad \theta_c = \sin^{-1}\left(\frac{n_2}{n_1}\right) \quad . \qquad (8.26)$$

Since $\sin\theta_2$ can be no larger than 1, if the light arrives at an angle closer to the surface than θ_c then the boundary conditions cannot be satisfied if there is a transmitted wave and so the wave will be completely reflected. This is called *total internal reflection*, and is used to confine light in display panels, light pipes, and multi-mode optical fibers. The

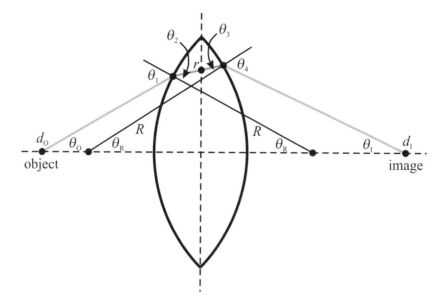

Figure 8.2. A spherical lens.

light will make multiple bounces inside the dielectric material, but as long as the angle is kept below the critical angle it will not leak out.

8.2 GEOMETRICAL OPTICS

Geometrical optics considers the propagation of light when the wavelength is small compared to the relevant length scales in a problem, so that it can be approximated by the reflection and refraction of plane waves from locally straight interfaces. Consider a ray passing through a spherical lens with a radius of curvature R and index of refraction n, shown in cross-section in Figure 8.2.

We will consider only light rays that lie near the axis, called *paraxial* rays, therefore angles from horizontal will be approximated with $\sin\theta \approx \theta$. We will also assume that this is a *thin lens*, so that when a ray passes through the lens its radius r from the axis of the lens is approximately constant. To find out how the lens modifies a ray we must analyze the trigonometry of Figure 8.2 with these approximations. Since the sum of the angles in a triangle must equal $180°$ we can relate the incoming angle, measured from the surface normal, to the radius of curvature of the lens and the distance from the lens by

$$\theta_O + \theta_R + (180° - \theta_1) = 180°$$

$$\theta_1 = \theta_R + \theta_O$$

$$\approx \tan^{-1}\left(\frac{r}{R}\right) + \tan^{-1}\left(\frac{r}{d_O}\right) \quad . \tag{8.27}$$

In the paraxial limit $\tan\theta = \sin\theta/\cos\theta \approx \theta/1 \approx \theta$, and so

$$\theta_1 \approx \frac{r}{R} + \frac{r}{d_O} \quad . \tag{8.28}$$

Similarly, the outgoing angle measured from the surface normal is

$$\theta_4 = \theta_R + \theta_1 \approx \frac{r}{R} + \frac{r}{d_I} \quad . \tag{8.29}$$

Adding equations (8.28) and (8.29),

$$\theta_1 + \theta_4 \approx r \left(\frac{2}{R} + \frac{1}{d_1} + \frac{1}{d_2} \right) \quad . \tag{8.30}$$

These angles are related to the angles inside the lens by Snell's Law

$$\frac{\sin\theta_1}{\sin\theta_2} \approx \frac{\theta_1}{\theta_2} = \frac{n_2}{n_1} = n \qquad \frac{\theta_4}{\theta_3} = n \quad , \tag{8.31}$$

taking $n = 1$ outside the lens. Combining these,

$$\theta_1 + \theta_4 = n(\theta_2 + \theta_3) \quad . \tag{8.32}$$

Finally, the internal angles are related by the included angles:

$$\theta_2 - \theta_R = \theta_R - \theta_3 \tag{8.33}$$

and so

$$\theta_2 + \theta_3 = 2\theta_R \approx 2\frac{r}{R} \quad . \tag{8.34}$$

Substituting (8.32) and (8.34) into (8.30) gives the result

$$2n\frac{r}{R} = r \left(\frac{2}{R} + \frac{1}{d_O} + \frac{1}{d_I} \right)$$

$$\underbrace{(n-1) \left(\frac{2}{R} \right)}_{\dfrac{1}{f}} = \frac{1}{d_O} + \frac{1}{d_I} \quad , \tag{8.35}$$

where f is the *focal length* of the lens. This is the *lens equation*, giving the relationship between where a ray starts on the axis on one side of the lens and where it crosses the axis on the other side. Notice that the angles have dropped out of this equation: all rays starting at the same distance from the lens on one side in the *object plane* are rejoined in a plane on the other side in the *image plane*. Problem 8.4 looks at the magnification associated with this.

If a ray starts in the *focal plane* $d_O = f$, then the lens equation requires that $d_I = \infty$. This means that the outgoing rays are parallel (*collimated*), meeting only at infinity. Lenses can also be described in terms of the *F number*, which is the ratio of the focal length to the diameter. If the focal length is 10 cm, and the diameter of the lens is 5 cm, then the F number is 10/5=2 and is written $f/2$. Another way to characterize a lens is by the *numerical aperture* (*NA*), the sine of the half-divergence angle times the index of refraction of the space the light is traveling in.

8.2.1 Ray Matrices

The calculations leading up to the lens equation were not too difficult, but they will rapidly become awkward in a system with multiple optical elements. This task can be simplified by introducing *ray matrices* that define how an arbitrary optical element transforms a light ray.

At a point along the axis of an axisymmetric optical system, a geometrical optics ray is characterized by the radius from the axis r and the slope $r' = dr/dz$. The action of a linear optical element can be specified in terms of a matrix operating on this state vector:

$$\begin{bmatrix} r_{\text{out}} \\ r'_{\text{out}} \end{bmatrix} = \begin{bmatrix} A & B \\ C & D \end{bmatrix} \begin{bmatrix} r_{\text{in}} \\ r'_{\text{in}} \end{bmatrix} \quad . \tag{8.36}$$

The advantage of this approach is that the action of multiple elements can be found by multiplying their ray matrices. As a simple example, a ray that passes through a homogeneous medium of width w will emerge with the same angle but a new radius of $r_{\text{out}} = r_{\text{in}} + r'w$, and so the corresponding ray matrix is

$$\begin{bmatrix} 1 & w \\ 0 & 1 \end{bmatrix} \quad . \tag{8.37}$$

To find the ray matrix for a lens, first notice that equation (8.35) can be written as

$$\begin{aligned} \frac{1}{f} &= \frac{1}{d_{\text{O}}} + \frac{1}{d_{\text{I}}} \\ &\approx \frac{1}{r}(\theta_{\text{I}} - \theta_{\text{O}}) \\ &\approx \frac{1}{r}(r'_{\text{in}} - r'_{\text{out}}) \quad , \end{aligned} \tag{8.38}$$

with a negative sign because d_{I} is on the opposite side of the lens from d_{O}. This can be rearranged as

$$r'_{\text{out}} = r'_{\text{in}} - \frac{r_{\text{in}}}{f} \quad ,$$

therefore the ray matrix for a lens is

$$\begin{bmatrix} 1 & 0 \\ -1/f & 1 \end{bmatrix} \quad . \tag{8.39}$$

Snell's Law in ray matrices is simply

$$\begin{bmatrix} 1 & 0 \\ 0 & n_1/n_2 \end{bmatrix} \quad . \tag{8.40}$$

As a final example, the ray matrix for a rod of length d with an index of refraction that depends quadratically with radius as $n = n_0[1 - \alpha r^2]$ is [Yariv, 1991]

$$\begin{bmatrix} \cos(\kappa d) & \frac{1}{\kappa}\sin(\kappa d) \\ -\kappa\sin(\kappa d) & \cos(\kappa d) \end{bmatrix} \quad , \tag{8.41}$$

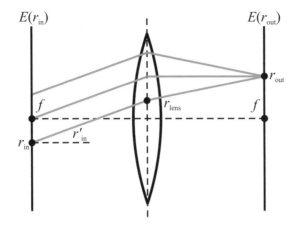

Figure 8.3. A thin lens performing a Fourier transform.

where $\kappa = \sqrt{2\alpha}$. This is called a *GRIN* (GRaded INdex of refraction) lens; by selecting different lengths it can perform a range of useful functions, such as collimating light into or out of an optical fiber.

8.2.2 Optical Transforms

The rays that we have been studying also have a phase that depends on the path lengths and indices of refraction, which leads to interference effects when multiple *coherent* rays are combined that maintain their relative phases. An unexpected consequence of this is shown in Figure 8.3. There is a input field distribution $E(x)$ in the focal plane at a distance f from a thin lens, and we are interested in the resulting distribution in the focal plane on the other side of the lens. The ray matrix for this combined system consists of propagation through free space of width f, the lens, and then more space of width f:

$$\begin{bmatrix} 1 & f \\ 0 & 1 \end{bmatrix} \begin{bmatrix} 1 & 0 \\ -1/f & 1 \end{bmatrix} \begin{bmatrix} 1 & f \\ 0 & 1 \end{bmatrix} = \begin{bmatrix} 0 & f \\ -1/f & 0 \end{bmatrix} \quad . \tag{8.42}$$

This means that the position of a ray in the output focal plane is $r_{\text{out}} = f r'_{\text{in}}$, depending solely on the angle of the ray at the input focal plane and not on its position. Therefore, the field at r_{out} is the sum over the input of all rays with an angle $r'_{\text{in}} = r_{\text{out}}/f$. Because these rays travel over different paths, there will be interference between them that we now calculate.

The position of a ray at the lens for this geometry is

$$r_{\text{lens}} = r_{\text{in}} + r'_{\text{in}} f$$
$$= r_{\text{in}} + \frac{r_{\text{out}}}{f} f$$
$$= r_{\text{in}} + r_{\text{out}} \quad . \tag{8.43}$$

In terms of this, the distance from the input to the lens is

$$
\sqrt{f^2 + (r_{\text{lens}} - r_{\text{in}})^2} = \sqrt{f^2 + r_{\text{out}}^2}
$$

$$
= f\sqrt{1 + r_{\text{out}}^2/f^2}
$$

$$
\approx f\left(1 + \frac{r_{\text{out}}^2}{2f^2}\right)
$$

$$
= f + \frac{r_{\text{out}}^2}{2f} \quad , \tag{8.44}
$$

making the usual paraxial approximation. Similarly, the distance from the lens to the output is

$$
\sqrt{f^2 + (r_{\text{lens}} - r_{\text{in}})^2} \approx f + \frac{r_{\text{in}}^2}{2f} \quad . \tag{8.45}
$$

In addition to the phase shift associated with traveling over these distances there will also be a phase shift in the lens. If the lens has a thickness $2d_0$ on the optical axis and a radius of curvature R, in the paraxial approximation the thickness of the lens at a distance r from the axis is

$$
[(R - d_0) + d]^2 + r^2 = R^2
$$

$$
R - d_0 + d = \sqrt{R^2 - r^2}
$$

$$
\approx R - \frac{r^2}{2R}
$$

$$
d = d_0 - \frac{r^2}{2R}
$$

$$
2d = 2d_0 - \frac{r^2}{R} \quad . \tag{8.46}
$$

Assuming that the index of refraction outside the lens is $n = 1$, the wave vector in the lens is found from

$$
\frac{k_{\text{lens}}}{k_{\text{air}}} = \frac{n}{1} \quad \Rightarrow \quad k_{\text{lens}} = n k_{\text{air}} \quad . \tag{8.47}
$$

The extra phase associated with traveling through the lens, compared to the phase if the lens was not there, is

$$
kn2d - k2d = k(n - 1)2d \tag{8.48}
$$

$$
= k(n - 1)\left(2d_0 - \frac{r^2}{R}\right) \quad .
$$

Since by definition $R = 2f(n - 1)$,

$$
k(n - 1)2d = k(n - 1)2d_0 - \frac{kr^2}{2f} \quad . \tag{8.49}
$$

We're almost done. The field at the output is found by adding up the phase shifts, integrated over the input:

$$E(r_{out}) = \int_{-\infty}^{\infty} e^{ik(f+r_{out}^2/2f)} e^{ik[(n-1)2d_0 - r_{lens}^2/2f]} e^{ik(f+r_{in}^2/2f)} \, E(r_{in}) \, dr_{in}$$

$$= \int_{-\infty}^{\infty} e^{ik(f+r_{out}^2/2f)} e^{ik[(n-1)2d_0 - (r_{in}+r_{out})^2/2f]} e^{ik(f+r_{in}^2/2f)} \, E(r_{in}) \, dr_{in}$$

$$= e^{i2k[f+(n-1)d_0]} \int_{-\infty}^{\infty} e^{ik[r_{out}^2 - (r_{in}+r_{out})^2 + r_{in}^2]/2f} \, E(r_{in}) \, dr_{in}$$

$$= e^{i2k[f+(n-1)d_0]} \int_{-\infty}^{\infty} e^{-ikr_{in}r_{out}/f} \, E(r_{in}) \, dr_{in} \quad . \tag{8.50}$$

This should look familiar: it is the Fourier transform of the input distribution! Given coherent illumination at its input focal plane, a thin lens produces the Fourier transform at its output focal plane multiplied by an extra phase factor. Because this is computed, literally, at the speed of light, and works just as fast with two-dimensional inputs, optical transforms are appealing for high-speed signal and image processing. Many algorithms that can be expressed in terms of Fourier transforms such as convolution and filtering have been implemented in such optical computers (although "computer" is really a misnomer, because there is no nonlinear interaction among different paths).

8.3 BEYOND GEOMETRICAL OPTICS

According to geometrical optics, a plane wave can be focused down to a spot of infinitesimal size and hence infinite energy density. But this is of course impossible, and does not happen because the plane-wave approximation is no longer justified when dimensions become comparable to the wavelength. It's then necessary to return the wave equation.

Consider a point source emitting a spherical wave e^{ikr}/r. If the wave is run backwards so that it travels towards the source, radiation arrives from all directions. Because of the uniqueness of solutions to partial differential equations, if the radiation is limited to a divergence angle fixed by the finite diameter of a lens then it can't match these boundary conditions for a point source and so there must be some width to the focus.

To relate this size to the divergence angle, instead of a point source now assume that spherical waves are emitted uniformly over an aperture of width w. For convenience, this will be analyzed in 2D, corresponding to a distribution of line- rather than point-sources; the conclusion will be unchanged in the full three-dimensional *scalar diffraction theory* [Heald & Marion, 1995]. This is similar to the geometry of Figure 7.1, but by optics convention θ will be measured relative to the surface normal, and x will be the position along the aperture from the z axis. We'll be concerned with the field distribution far from the source, once again making the approximation that the distance from a source at $(z = 0, x)$ to a field location at (r, θ) relative to the origin is $r - x \cos(90 - \theta) = r - x \sin \theta$, and will include the x dependence in the relative phases but not the amplitude over the

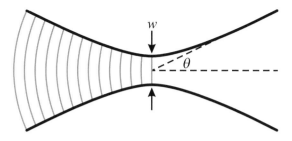

Figure 8.4. Wave fronts around a Gaussian focus.

aperture. Then integrating e^{ikr} over the source gives

$$\int_{-w/2}^{w/2} e^{ikr(x)}\,dx = \int_{-w/2}^{w/2} e^{ik(r - x\sin\theta)}\,dx$$

$$= e^{ikr} \int_{-w/2}^{w/2} e^{-ikx\sin\theta}\,dx$$

$$= 2e^{ikr} \frac{\sin\left(k\frac{w}{2}\sin\theta\right)}{k\sin\theta}$$

$$\approx e^{ikr} w \cos\left(k\frac{w}{2}\sin\theta\right) \quad . \tag{8.51}$$

This last line follows by taking the derivative of the numerator and denominator for the paraxial limit $\sin\theta \to 0$. If we ask for the boundary where the argument of the cosine equals 1,

$$1 = k\frac{w}{2}\sin\theta \approx k\frac{w}{2}\theta = \frac{2\pi n}{\lambda}\frac{w}{2}\theta$$

$$\theta = \frac{\lambda}{\pi n w} \quad . \tag{8.52}$$

The divergence angle is proportional to the wavelength, and inversely proportional to the source aperture width and the index of refraction. Problem 8.5 shows that this presents a serious constraint on the resolution of optical devices, and is why microscopes use oil immersion lenses, telescopes are big, and blue lasers are desired for optical storage.

To find the field distribution everywhere, it's necessary to solve Maxwell's equations in cylindrical coordinates for a diverging wave. This is a surprisingly non-trivial calculation [Yariv, 1991], but the result drawn in Figure 8.4 for the fundamental radial and azimuthal mode TEM_{00} is close to what we just found. At the focus, the phase fronts are parallel and the transverse amplitude has a Gaussian dependence with a standard deviation of w, the *beam waist*. That is why this limit is called *Gaussian optics*. Then, after a transition region, the beam diverges with spherical phase fronts in a cone given by equation (8.52). Light with this distribution is said to be *diffraction limited*. It has the narrowest beam waist possible; an imperfect lens will produce a broader distribution. While diffraction-limited optics were once restricted to specialized scientific instruments, the inexpensive plastic lens in a CD player now attains this limit.

Gaussian optics tames the infinite energy density associated with the infinitesimal focus of geometrical optics, but as Problem 5.5 showed the field strength in a tightly-focused optical beam can still be very large. It can in fact exceed the strength of the intra-atomic fields, and can therefore be used to drive atoms into novel states [Weinacht *et al.*, 1999]. While reaching this limit does require powerful lasers, well before then the assumption of a linear material response breaks down. The lowest-order correction to the polarization is

$$\vec{D} = \epsilon_0 \vec{E} + \vec{P} = \epsilon_0 \epsilon_r \vec{E} + \epsilon_0 \vec{E}^T \cdot \mathbf{d}_2 \cdot \vec{E} \quad . \tag{8.53}$$

The nonlinear optical coefficient **d** will usually depend on the orientation relative to the axes of the medium. For the simplest scalar case,

$$D = \epsilon_0 \epsilon_r E + \epsilon_0 d_2 E^2 \quad . \tag{8.54}$$

Because the optical coefficient is small, on the order of 10^{-12} m/V in typical nonlinear optical materials, it usually can be considered to be a small perturbation on the polarization. Repeating the derivation of the wave equation (5.100) with this nonlinear polarization adds a term

$$\nabla^2 E - \mu_0 \epsilon_0 \epsilon_r \frac{\partial^2 E}{\partial t^2} = \mu_0 \epsilon_0 d_2 \frac{\partial^2 E^2}{\partial t^2} \quad . \tag{8.55}$$

If the electric field has a time dependence $Ee^{i\omega t}$, the right hand side squares this to become $-\omega^2 \mu_0 \epsilon_0 d_2 e^{2i\omega t}$. This acts as a forcing function, generating a wave at twice the excitation frequency. Microscopically, it corresponds to exciting transitions with two photons rather than one. This is called *second harmonic generation*, and is used to make short-wavelength light from more powerful longer-wavelength sources. A common combination is to use KDP (KH_2PO_4) to double the 1064 nm fundamental of a $Nd:YAG$ ($Nd^{3+}:Y_3Al_5O_{12}$) laser up to 532 nm. Weaker, higher harmonics can be generated with still shorter wavelengths. And if two laser beams are incident on a nonlinear crystal, the quadratic nonlinearity will generate sum- and difference-frequencies. These can be used for *parametric up-* and *down-conversion* to shift frequencies, and for gain in an *Optical Parametric Amplifier* (*OPA*).

One more restrictive assumption that we've been making is to assume that our optical elements are axisymmetric and passive. Microelectronic fabrication techniques can be used to create lenses with arbitrary shapes, called *binary optics* [Stern, 1996]. For example, a lens can be made with multiple focal lengths by interleaving the profiles for the individual lenses. *Holographic Optical Elements* (*HOEs*) are planar structures that offer the same flexibility for coherent light, using the diffractive effects to be covered in the next chapter.

The performance of a perfect optical system will still be degraded by fluctuations in the ambient environment, which sets a severe limit on ground-based telescopes and long-range optical links. But by analyzing the degradation of a point source it is possible to reconstruct a model of the atmospheric perturbations and then significantly decrease their influence by correcting for them by continuously deforming the shape of *active optical* elements [Bortoletto *et al.*, 1999].

Even better, consider what happens in an inhomogeneous medium to an arbitrary paraxial wave, which can be written as a modulated plane wave $E(\vec{x}) = f_+(\vec{x})e^{i(kz-\omega t)}$. Substituting this into the wave equation and separating out the transverse and axial parts

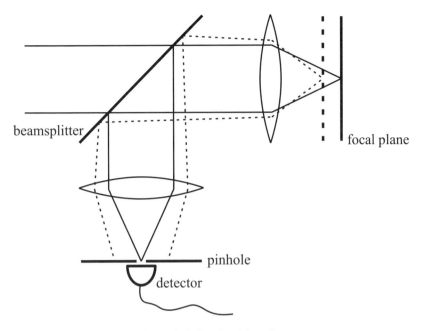

Figure 8.5. Confocal imaging.

of the Laplacian,

$$0 = \nabla^2 E + \omega^2 \mu\epsilon(\vec{x})E$$

$$= \nabla_T^2 E + \frac{\partial^2 E}{\partial z^2} + \omega^2 \mu\epsilon(\vec{x})E$$

$$= \nabla_T^2 f_+ + \frac{\partial^2 f_+}{\partial z^2} + 2ikf_+ - k^2 f_+ + \omega^2 \mu\epsilon(\vec{x})f_+ \quad . \tag{8.56}$$

This has a complex conjugate

$$0 = \nabla_T^2 f_+^* + \frac{\partial^2 f_+^*}{\partial z^2} - 2ikf_+^* - k^2 f_+^* + \omega^2 \mu\epsilon(\vec{x})f_+^* \tag{8.57}$$

(assuming that the material response is real, without gain or loss). If, instead, we started with a solution traveling in the opposite direction given by $E(\vec{x}) = f_-(\vec{x})e^{i(-kz-\omega t)}$, then the wave satisfies

$$0 = \nabla_T^2 f_- + \frac{\partial^2 f_-}{\partial z^2} - 2ikf_- - k^2 f_- + \omega^2 \mu\epsilon(\vec{x})f_- \quad . \tag{8.58}$$

Comparing equations (8.57) and (8.58), f_- and f_+^* satisfy the same governing equations. This means that if it's possible to propagate a beam through the medium, then sending its complex conjugate back through the medium will undo the distortion [Yariv & Pepper, 1977; Yariv, 1987]. This is called *phase conjugate optics* and can be realized through nonlinear mixing.

Finally, we've seen that lenses can be used to form two-dimensional images, but the world is three-dimensional. Because light that enters a lens from sources away from its focal plane will blur the image, it's not possible to use a lens to see into a three-dimensional object even if light can propagate through it. But consider what happens if

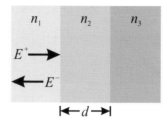

Figure 8.6. Reflection from dielectric interfaces.

the illumination is focused onto a detector through a pinhole in the *confocal* geometry shown in Figure 8.5 [Minsky, 1957; Lichtman, 1994]. Now most of the light that scatters away from the focal plane is deflected away from the pinhole, so that by scanning the sample it is possible to reconstruct a three-dimensional image. The spatial resolution and chemical sensitivity can be further enhanced by using a two-photon process for illumination [Denk *et al.*, 1990]. Such confocal microscopy has become a workhorse for biomedical imaging.

8.4 SELECTED REFERENCES

[Born & Wolf, 1999] Born, Max, & Wolf, Emil. (1999). *Principles of Optics: Electromagnetic Theory of Propagation, Interference and Diffraction of Light*. 7th edn. New York: Cambridge University Press.

The classic optics reference.

[Yariv, 1991] Yariv, Amnon. (1991). *Optical Electronics*. 4th edn. Philadelphia: Saunders College Pub.

A good source for modern passive and active optical systems.

8.5 PROBLEMS

(8.1) Optics (as well as most of physics) can be derived from a global law as well as a local one, in this case *Fermat's Principle:* a light ray chooses the path between two points that minimizes the time to travel between them. Apply this to two points on either side of a dielectric interface to derive Snell's Law.

(8.2) (*a*) Use Fresnel's equations and the Poynting vectors to find the *reflectivity* and *transmissivity* of a dielectric interface, defined by the ratios of incoming and outgoing energy.

(*b*) For a glass–air interface ($n = 1.5$) what is the reflectivity at normal incidence?

(*c*) What is the Brewster angle?

(*d*) What is the critical angle?

(8.3) Consider a wave at normal incidence to a dielectric layer with index n_2 between layers with indices n_1 and n_3 (Figure 8.6).

(*a*) What is the reflectivity? Think about matching the boundary conditions, or about the multiple reflections.

(*b*) Can you find values for n_2 and d such that the reflection vanishes?

(8.4) Consider a ray starting with a height r_0 and some slope, a distance d_1 away from a thin lens with focal length f. Use ray matrices to find the image plane where all rays starting at this point rejoin, and discuss the magnification of the height r_0.

(8.5) Common CD players use an AlGaAs laser with a 790 nm wavelength.

(a) The pits that are read on a CD have a diameter of roughly 1 μm and the optics are diffraction-limited; what is the beam divergence angle?

(b) Assuming the same geometry, what wavelength laser would be needed to read 0.1 μm pits?

(c) How large must a telescope mirror be if it is to be able to read a car's license plate in visible light ($\lambda \sim 600$ nm) from a *Low Earth Orbit (LEO)* of 200 km?

9 Lensless Imaging and Inverse Problems

The previous chapter completed our tour through Maxwell's equations, showing how simple dielectric interfaces can serve as lenses to form images. While the power and simplicity of such a lens is appealing, it is also limiting. A confocal microscope scans the position of its sample, something that is not feasible if an image is being made of, say, a planet. Optics are available to image just a fraction of the range of energies and mechanisms that are useful for probing an object, and even then there is information in the interactions that can reveal information far beyond what is seen in a conventional image.

To address these problems we will now introduce intelligence into the apparatus. This will start by abstracting the operation performed by a lens into something that can be implemented by measuring and processing other kinds of signals, such as microwaves or sound. Then we will look beyond two-dimensional images to explore a number of tomographic techniques to recover three-dimensional structure, closing with an introduction to the general problem of inverting a set of measurements to deduce the signal sources.

9.1 MATCHED FILTERS AND SYNTHETIC LENSES

The first step is synthesizing lenses is to introduce the notion of a *matched filter*, which we will see is the optimal linear detector for a known signal. Matched filters will provide a way to generalize the response of a lens to other domains.

Consider a signal $x(t)$ with Fourier transform $X(\omega)$ passed through a linear filter that has an impulse response $f(t)$ [Gershenfeld, 1999a], or equivalently, a frequency response $F(\omega)$ given by the Fourier transform of the impulse response. The frequency domain response of the output of the filter $Y(\omega)$ will be the product of the Fourier transforms

$$Y(\omega) = X(\omega)F(\omega) \quad , \tag{9.1}$$

and the time domain response will be the convolution

$$y(T) = x(T) * f(T) = \int_0^T x(T - t) f(t) \, dt \quad , \tag{9.2}$$

where the bounds of the integral are the time during which the signal has been applied to the filter.

The size of the output of the filter can be bounded by the integral version of the *Cauchy–Schwarz inequality*: the magnitude of the integral of a product is less than or

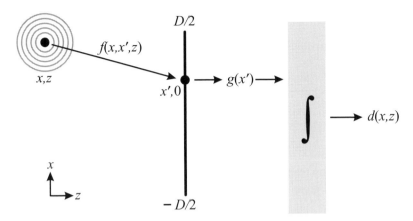

Figure 9.1. Mathematical model of a lens.

equal to the product of the integrals of the magnitude

$$y^2(T) = \left| \int_0^T x(T-t)f(t)\, dt \right|^2$$

$$\leq \int_0^T |x(T-t)|^2\, dt \int_0^T |f(t)|^2\, dt \quad . \tag{9.3}$$

By inspection, this bound will be saturated if

$$f(t) = Ax^*(T-t) \tag{9.4}$$

for arbitrary A. The filter will produce the maximum output for a given input signal if the filter's impulse response is equal to the complex conjugate of that signal reversed in time. This is called a matched filter, and is routinely used to detect and time known signals such as radar echoes.

What do matched filters have to do with lenses? Consider the abstract view of a lens shown in Figure 9.1. The amplitude and phase of a periodic wave that starts at the point (x, z) and arrives at a point on the surface of the lens $(x', 0)$ will be changed by the wave's propagation. This change in amplitude and phase can be described by multiplying the wave by a complex factor $f(x, x', z)$ that depends on the distances. The signal seen by a detector on the far side of the lens will be equal to the integral over the surface of the lens of f multiplied by another term $g(x')$ that contains the amplitude and phase shifts introduced by passing through the lens and then traveling to the detector; let's call the overall factor giving the change in detected amplitude and phase $d(x, z)$.

f will be the product of a phase shift due to the wave traveling the distance from (x, z) to $(x', 0)$ and an amplitude change due to the wave spreading out. Since the amplitude change is a higher-order correction than the phase shift, in this calculation we will include just the phase shift, which is

$$f(x, x', z) = e^{ikr}$$

$$= e^{ik\sqrt{(x-x')^2+z^2}}$$

$$\approx e^{ik[z+(x-x')^2/2z]}$$

$$= e^{ikz}e^{ik(x-x')^2/2z} \quad . \tag{9.5}$$

We have once again made the paraxial approximation $(x - x' \ll z)$ for sources near the axis of the lens. The e^{ikz} term can also be dropped because we will be interested only in the functional form of the transverse dependence of f on x.

The response of the detector is found by integrating over the lens the propagation term f and the lens response function g:

$$d(x, z) = \int_{-D/2}^{D/2} g(x')f(x - x', z) \, dx' \quad . \tag{9.6}$$

This is just a convolution over x', therefore if we want the lens to produce the maximum output for a signal at a given location (i.e., to focus the waves from the source onto the detector) then from the last section we've learned that the lens should be designed to have a response function equal to the complex conjugate of the propagation factor, with the direction reversed

$$
\begin{aligned}
g(x') &= f^*(-x') \\
&= e^{-ik(x+x')^2/2z} \quad .
\end{aligned}
\tag{9.7}
$$

For a source on the axis $(x = 0)$ this reduces to

$$g(x') = e^{-ikx'^2/2z} \quad . \tag{9.8}$$

The matched-filter lens has a quadratic phase dependence on the distance from the axis x'.

Now let's look at how such a lens responds to a signal away from the focus. If the lens is matched to a source on-axis at $(0, z)$ but the source is actually at (x, z), the response is

$$
\begin{aligned}
d(x, z) &= \int_{-D/2}^{D/2} g(x')f(x - x', z) \, dx' \\
&= \int_{-D/2}^{D/2} e^{-ikx'^2/2z} e^{ik(x-x')^2/2z} \, dx' \\
&= e^{ikx^2/2z} \int_{-D/2}^{D/2} e^{-ikxx'/z} \, dx' \\
&= -D e^{ikx^2/2z} \frac{\sin(kxD/2z)}{kxD/2z} \\
&= -D e^{ikx^2/2z} \operatorname{sinc} \frac{xD}{z\lambda} \quad ,
\end{aligned}
\tag{9.9}
$$

where $\operatorname{sinc}(x) \equiv \sin(\pi x)/\pi x$. This is called the *point spread function*; it has a central peak and side lobes, and in the limit $D \to \infty$ it becomes a delta function (Figure 9.2). As we saw with Gaussian optics, the resolution of the lens increases as the aperture increases; Problem 9.2 looks at resolution in more detail.

Matched filter lenses can be implemented for a variety of purposes. By doing the delay and summation electronically rather than physically in a piece of glass it is possible to make lenses for a much broader range of types and wavelengths of radiation, and the properties of such lenses can quickly be modified. In the simplest implementation, the

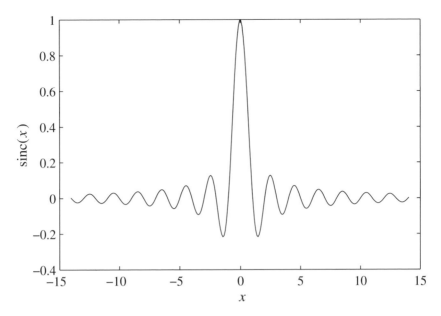

Figure 9.2. $\mathrm{sinc}(x) \equiv \sin(\pi x)/\pi x$.

lens is made up of an array of transmitters and receivers. Doing the calculation that we just did in reverse shows that by introducing delays in the outgoing signals it is possible to produce an electronically steerable beam, and then with the same delays on the return signal it is possible to probe for an object at a particular point. This is how *phased-array radar* works; the number and spacing of the individual transceivers produce *grating lobes* that further blur the overall point spread function that results from the aperture size. Since the resolution of the lens will increase as the size of the aperture D increases it is desirable to make that as large as possible, but physical constraints such as the size of a satellite will limit the maximum aperture. However, a single moving transmitter and receiver can do the work of a much larger array by sending out a signal and collecting the return from many points along a trajectory, such as the satellite's orbit. This is called a *synthetic aperture* [Fitch, 1988], and is used in both satellite radars mapping the Earth and towed sonar transducers mapping the ocean (Problem 9.3).

9.2 COHERENT IMAGING

The preceeding calculation assumed that the radiation was *coherent*, so that interference is possible independent of the distance that a wave travels. We will shortly see the limits of that assumption, but if it is justified then a detector that records the wave's mean square amplitude will throw away the information stored in the phase. That can be recovered by *holography*, which saves the interference rather than just the intensity [Gabor, 1948; Gabor, 1966]

Consider two beams of coherent radiation, such as sound or light, incident on a piece of film or other medium which responds to the intensity in the (x, y) plane (Figure 9.3).

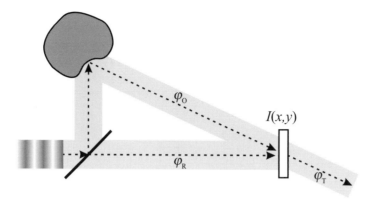

Figure 9.3. Creation of a hologram.

One of the beams is unmodified, and the other one has been scattered from an object. After traveling a distance r, the phase of the reference beam in the plane of the film will be

$$\varphi_R = e^{-i\vec{k}\cdot\vec{r}} = e^{-i(k_x x + k_y y)} \quad , \tag{9.10}$$

and the scattered beam will have its amplitude and phase modified by the object:

$$\varphi_O = A(x,y)e^{-i\theta(x,y)} \quad . \tag{9.11}$$

The recorded intensity on the film is the square magnitude of the sum of these:

$$\begin{aligned}
I(x,y) &= |\varphi_O + \varphi_R|^2 \\
&= (\varphi_O + \varphi_R)(\varphi_O + \varphi_R)^* \\
&= \varphi_O \varphi_R^* + \varphi_O^* \varphi_R + \varphi_O \varphi_O^* + \varphi_R \varphi_R^* \\
&= A(x,y)e^{-i\theta(x,y)}e^{+i(k_x x + k_y y)} + A(x,y)e^{+i\theta(x,y)}e^{-i(k_x x + k_y y)} \\
&\quad + |A|^2 + 1 \quad .
\end{aligned} \tag{9.12}$$

Now, if an identical reference beam is later sent through this film, the transmitted beam will have its intensity modulated by the recorded pattern:

$$\begin{aligned}
\varphi_T &= I(x,y)e^{-i(k_x x + k_y y)} \\
&= A(x,y)e^{-i\theta(x,y)} + \text{(background terms)} \quad .
\end{aligned} \tag{9.13}$$

The first term gives rise to a transmitted beam that is identical to the light arriving from the object, and the other terms give rise to various kinds of background (glare and unfocused virtual images) which can be eliminated by more sophisticated geometries. The transmitted beam that is identical to the arriving light truly is identical. There is no difference between it and what would be seen if the object really was there, up to the limits of the film. The most obvious difference between a hologram and a normal incoherent film image is parallax: the image looks different from different directions.

Instead of creating holograms by recording light bounced off a real object, it is also possible to create them numerically to make synthetic holograms either on film or in real time with a fast acousto–optic modulator (Chapter 11) [Lucente, 1997]. Since holograms can reproduce the light that would be generated by an arbitrary source, it is also possible

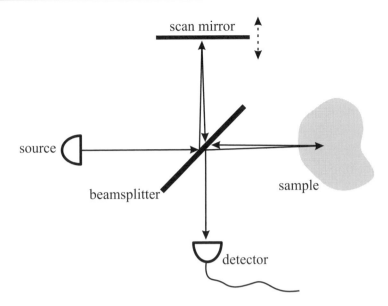

Figure 9.4. Optical Coherence Tomography (the beams have been displaced for clarity).

to design holograms to serve as *Holographic Optical Elements* (*HOEs*), thin sheets that can transform coherent light in ways that are impossible with conventional lenses [Stern, 1996]. HOEs are routinely used in applications such as laser beam scanning, wavelength multiplexing, and heads–up displays [Sweatt, 1979; Rallison, 1984; Herzig & Dandliker, 1987].

Real light sources are of course not perfectly coherent. If we approximate the spectrum as a Gaussian with a width $\Delta\omega$, then the temporal distribution will also have a Gaussian envelope with a width of $\Delta t = 1/\Delta\omega$ because the Fourier transform of a Gaussian is also Gaussian with the variance inverted (equation 3.23). Therefore,

$$\Delta\omega\,\Delta t = 1 \quad , \tag{9.14}$$

or

$$\Delta f\,\Delta t = \frac{1}{2\pi} \quad . \tag{9.15}$$

This is called the *frequency–time uncertainty relation*. Associated with the temporal spread is a distance $c\,\Delta t$, called the *coherence length*. This is the maximum path difference over which interference effects can be seen. The most stable lasers have linewidths below a hertz, giving a coherence length so large that it can be ignored for all but the most sensitive metrology measurements. Incoherent lights sources can have linewidths of tens to hundreds of nanometers, giving coherence lengths of a few microns.

If a hologram is illuminated by broadband light, each wavelength will scatter at a different angle, blurring the image. This can be partially compensated for by using a slit to restrict the beamwidth reflected from the object used to make the hologram, so that when it is read by incoherent light the images associated with different colors will not overlap [Benton, 1969].

Limited coherence becomes a feature rather than a bug when it is used in *Optical Coherence Tomography* (*OCT*) [Huang *et al.*, 1991] . The idea is shown in Figure 9.4. Light from a source with a short coherence length enters a *Michelson interferometer*. One beam reflects off of a mirror, and the other one scatters off of the sample. These are recombined by the beamsplitter and detected by a photodetector. If the two path lengths exactly match the beams will add coherently, giving a strong signal. Because of the short coherence length, if the paths are slightly different then the detector will see an incoherent background. Scanning the mirror then gives a profile of the scattering as a function of depth in the sample; these measurements can be assembled into a three-dimensional image by moving the sample also. OCT is promising for biomedical applications because it can provide internal information with a benign light source.

9.3 COMPUTED TOMOGRAPHY

OCT is limited in the depth that can be probed because photons must be scattered back to the detector, but if they scatter multiple times in the sample the image will degrade. An opposite limit applies to higher-energy radiation such as X-rays: most photons either travel straight through the material or are absorbed. Measurement of this absorption as a function of orientation can reveal the internal structure via *Computed Tomography* (*CT*), also know as *Computerized Axial Tomography* or *Computer-Assisted Tomography* (*CAT*) [Kak & Slaney, 1988].

Let $\rho(x, y)$ be the distribution of the absorption coefficient of the material in a two-dimensional slice (Figure 9.5). Assume that the radiation is generated in a parallel beam inclined at an angle θ to the x axis, and let (s, t) be the coordinates in the direction of, and perpendicular to, the radiation. The total absorption in that direction, $A_\theta(t)$, is found by projecting along the s direction:

$$
\begin{aligned}
A_\theta(t) &= \int_{-\infty}^{\infty} \rho(s, t) \, ds \\
&= \int_{-\infty}^{\infty} \int_{-\infty}^{\infty} \rho(s, t') \, \delta(t' - t) \, ds \, dt' \\
&= \int_{-\infty}^{\infty} \int_{-\infty}^{\infty} \rho(x, y) \, \delta(x \cos\theta + y \sin\theta - t) \, dx \, dy \quad .
\end{aligned}
\tag{9.16}
$$

This is called the *Radon transform* of the distribution [Radon, 1917]; it is a parallel projection onto the t axis.

Now consider the spatial Fourier transform of the absorption density:

$$
R(u, v) = \int_{-\infty}^{\infty} \int_{-\infty}^{\infty} \rho(x, y) \, e^{-i2\pi(ux + vy)} \, dx \, dy \quad .
\tag{9.17}
$$

Expressing the reciprocal space variables u and v in polar coordinates r and θ (but not changing the integral to polar coordinates),

$$
\begin{aligned}
R(r\cos\theta, r\sin\theta) &= \int_{-\infty}^{\infty} \int_{-\infty}^{\infty} \rho(x, y) \, e^{-i2\pi r(x\cos\theta + y\sin\theta)} \, dx \, dy \\
&= \int_{-\infty}^{\infty} \int_{-\infty}^{\infty} \rho(x, y) \times
\end{aligned}
$$

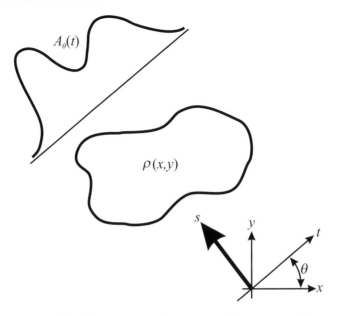

Figure 9.5. The geometry for tomographic reconstruction.

$$\left[\int_{-\infty}^{\infty} e^{-i2\pi rt}\delta(x\cos\theta + y\sin\theta - t)\ dt\right]\ dx\ dy$$

$$= \int_{-\infty}^{\infty}\left[\int_{-\infty}^{\infty}\int_{-\infty}^{\infty} \rho(x,y)\ \delta(x\cos\theta + y\sin\theta - t)\ dx\ dy\right]$$

$$\times\ e^{-i2\pi rt}\ dt$$

$$R(r\cos\theta, r\sin\theta) = \int_{-\infty}^{\infty} A_\theta(t)\ e^{-i2\pi rt}\ dt \quad . \tag{9.18}$$

The Fourier transform of the Radon transform along the projection axis is equal to the Fourier transform of the absorption density distribution expressed in polar coordinates. This is called the *Fourier Slice Theorem*. If the parallel projection is measured for many values of θ, this relationship can then be inverted to find the internal distribution. This is how a CAT scan works. A similar calculation holds for a point source that generates a fan or a cone beam instead of the parallel propagation assumed here.

9.4 MAGNETIC RESONANCE IMAGING

The techniques we've discussed so far give information about where something is, but not what it is. *Magnetic Resonance Imaging* (*MRI*) probes the environment of nuclear spins to provide remarkable chemical as well as spatial information. It is based on *Nuclear Magnetic Resonance* (*NMR*), but was renamed to sound less frightening (even though the energies involved are only a factor of $\sim 10^{13}$ or so less than what is required for a nuclear reaction). To understand NMR we must introduce the basic features of spins; Chapters 10, 12 and 15 will continue this development.

Quantum mechanical magnetic particles and systems possess an intrinsic magnetic moment μ and an angular momentum $\vec{J}$ that are related by the *gyromagnetic ratio* γ:

$$\vec{\mu} = \gamma \vec{J} \quad . \tag{9.19}$$

Such a particle acts just like a spinning top that is also a magnetic dipole, hence this property is called *spin* (even though nothing is actually spinning in the classical sense).

If a magnetic field is applied, it will exert a torque of $\vec{\tau} = \vec{\mu} \times \vec{B}$ on the magnetic moment. This must be equal to the rate of change of the angular momentum,

$$\frac{d\vec{J}}{dt} = \vec{\tau} = \vec{\mu} \times \vec{B}$$

$$\frac{d\vec{\mu}}{dt} = \vec{\mu} \times \gamma \vec{B} \quad . \tag{9.20}$$

Therefore the spin precesses around the field with a frequency γB, called the *Larmor frequency*. Although this is a classical result, it matches the quantum calculation in the limit of non-interacting spins [Slichter, 1992].

The mysterious gyromagnetic ratio relating the magnetic moment and the angular momentum can be estimated semiclassically by considering a particle with a charge q traveling in an orbit of radius r with period T. The angular momentum is

$$J = mvr = m\frac{2\pi r}{T}r \quad , \tag{9.21}$$

the current is

$$I = \frac{q}{T} \quad , \tag{9.22}$$

the magnetic moment is

$$m = IA = \frac{q}{T}\pi r^2 \quad , \tag{9.23}$$

and so the gyromagnetic ratio is

$$m = \gamma J \Rightarrow \gamma = \frac{q}{2m} \quad . \tag{9.24}$$

This is independent of the radius, suggesting that the result might be more generally valid. In fact, this estimate is approximately equal to the correct formulas found from complete atomic and nuclear calculations. The inverse dependence on mass means that there will be many orders-of-magnitude difference in nuclear versus electronic resonance frequencies (Problem 9.4).

In an NMR experiment a strong magnetic field is applied, customarily taken to define the $\hat{z}$ directon. The resolution and sensitivity of the instrument will depend on the strength of this field; the largest spectrometers have static fields on the order of 20 T. In equilibrium the spins in the sample line up around this field with a thermal distribution. The fractional excess of spins aligned with the field, which is $\sim 10^{-5}$ for fields on the order of tesla, produces a magnetization in the sample. Transverse to the static field there is a coil that can apply an RF field to rotate the spins into the (x, y) plane (a *tipping pulse*). After this is turned off the magnetization will then precess around the static field, generating an RF signal back in the coil. This *Free Induction Decay* (*FID*) will continue until the spins thermalize again, on a time scale called T_1 that is typically on the order of seconds.

The signal strength will depend on the number of spins, the gyromagnetic ratio will depend on the spin species, and in a material the local magnetic field will vary due to internal fields. Therefore the RF spectrum provides detailed information about the density and type of spins and their local chemical environments. To find out where they are, a gradient is added to the static field.

Let $f(\omega)$ be the NMR signal as a function of frequency, normalized to 1:

$$\int f(\omega)\, d\omega = 1 \quad , \tag{9.25}$$

and similarly let $\rho(\vec{r})$ be the normalized spatial distribution of the density of the spin being probed

$$\int \rho(\vec{r})\, d^3\vec{r} = 1 \quad . \tag{9.26}$$

If the gradient can be approximated as a constant slope along the z direction,

$$B(z) = B_0 + Gz \quad , \tag{9.27}$$

then the resonance frequency as a function of position is

$$\omega(z) = \gamma B = \gamma B_0 + \gamma Gz \tag{9.28}$$

or

$$d\omega = \gamma G dz \quad . \tag{9.29}$$

Because of this gradient, the spin density integrated over a slice gives the corresponding component of the resonance spectrum

$$f(\omega)\, d\omega = dz \int \int \rho(x, y, z(\omega))\, dx\, dy$$

$$f(\omega)\gamma G = \int \int \rho(x, y, z(\omega))\, dx\, dy \quad . \tag{9.30}$$

Therefore the Fourier transform of the FID provides a map of the spin density as a function of z, and additional gradients permit the full spin distribution to be reconstructed [Lauterbur, 1973; Stehling et al., 1991]. This projection of the spin density in a slice is similar to the Radon transform projection of an absoprtion coefficient along a slice. The spatial resolution depends on the strength of the gradients and the homogeneity of the field; in the best instruments this is on the order of microns.

The spatial and chemical analyses can be performed simultaneously; one of the most dramatic consequences of this has been the development of *functional MRI (fMRI)* [Ogawa et al., 1990; Kwong, 1995]. It is possible to resolve the change in the oxygenation state of hemoglobin to determine the amount of oxygen in the bloodstream, and it was found experimentally that the brain delivers more oxygen to active parts of the brain. This *Blood-Oxygen-Level-Dependent (BOLD)* contrast provides a means to spatially resolve the workings of the brain, and has revealed the location of not only gross motor control but also abstract thoughts as they happen.

9.5 INVERSE PROBLEMS

Each of the techniques we've covered has drawn on insight into a particular domain to use a set of measurements to estimate the sources of the signals. The most general approach to imaging replaces these problem-specific details with a framework for working back from observations to inferences, the study of *inverse problems*.

Let $\vec{y}$ be a vector of measurements (pixel intensities, antenna waveforms, . . .), let $\vec{x}$ be the desired internal state information (mass distribution, airplane location, . . .), and let $\vec{y} = \vec{f}(\vec{x})$ be the *forward model* that relates them (geometrical optics, RF propagation, . . .). It's tempting to simply write $\vec{x} = \vec{f}^{-1}(\vec{y})$, but there are at least two important problems with doing that: $\vec{f}$ may be a terribly complicated function, so that its inverse is not available, and $\vec{x}$ may have more elements than $\vec{y}$. For example, we might want to find a three-dimensional distribution from two-dimensional observations.

If the measurement has noise it's necessary to work with a probabilistic relationship $p(\vec{y}|\vec{x})$. This too must be inverted, because we want to find the most likely value of $\vec{x}$ given a measurement of $\vec{y}$. This can be done via Bayes' Theorem:

$$\max_{\vec{x}} p(\vec{x}|\vec{y}) = \max_{\vec{x}} \frac{p(\vec{y}|\vec{x})\,p(\vec{x})}{p(\vec{y})} \tag{9.31}$$

or

$$\max_{\vec{x}} \log p(\vec{x}|\vec{y}) = \max_{\vec{x}} \log \frac{p(\vec{y}|\vec{x})\,p(\vec{x})}{p(\vec{y})}$$
$$= \max_{\vec{x}}[\log p(\vec{y}|\vec{x}) + \log p(\vec{x}) - \log p(\vec{y})] \quad . \tag{9.32}$$

$p(\vec{x}|\vec{y})$ is called the *posterior*, because it updates our knowledge of the internal state following a measurement of the observable. $p(\vec{y}|\vec{x})$ is the *likelihood*, giving the forward probability for the observations. The $p(\vec{y})$ term, the *evidence*, measures how reliable we think the data are, and has no influence on the maximization over a single data set. The middle term, called a *prior* or *regularizer*, is the important modification. It expresses our beliefs about the internal state in advance of seeing any data, such as whether the distribution should be smooth, or perhaps have sharp features.

Problems that can be written in terms of a Green's function (equation 5.44) are linear in the sources, letting the maximization be solved by matrix inversion (assuming that application of the prior is also linear). This inverse is well conditioned because of the presence of the regularizer. More generally, a nonlinear search is needed to find an acceptable solution; efficient approximate techniques have been developed to do this in high-dimensional parameter spaces [Besag *et al.*, 1995].

The maximization can be done to find the value of $\vec{x}$ at a discrete set of locations, or to find the parameters in a continuous representation of $\vec{x}$. For example, a least-squares likelihood term results from taking the logarithm of a Gaussian error model, and a *maximum entropy* prior finds the distribution that makes the weakest assumption about the data, so if we want to find $(x_1, \ldots, x_N)$ from $(y_1, \ldots, y_M)$ we must solve

$$0 = \frac{\partial}{\partial x_n}\left[-\sum_{m=1}^{M} \frac{[f_m(x_1, \ldots, x_N) - y_m]^2}{\sigma_m^2} - \lambda \sum_{n'=1}^{N} x_{n'} \log x_{n'}\right] \quad . \tag{9.33}$$

This gives a system of equations for the x_n. The measurement error is σ_m, and λ controls the essential tradeoff between trusting the data and trusting our advance beliefs. Note that the Lagrange multiplier could equivalently have been put in front of either term.

While the introduction of a prior might appear to violate the sanctity of the data, most any experimental procedure contains some kind of implicit prior even if it is not written down, and this assumption is what makes it possible to generalize from limited measurements. In theory, the combination of a prior and a search algorithm makes it possible to deduce the most plausible signal sources from any set of measurements. In practice, salvation lies in the details of the implementation, but there are broadly-applicable lessons about search, constraint, and functional approximation that can help guide the development of entirely new approaches to imaging [Gershenfeld, 1999a].

9.6 SELECTED REFERENCES

[Kino, 1987] Kino, Gordon S. (1987). *Acoustic Waves: Devices, Imaging, and Analog Signal Processing*. Englewood Cliffs: Prentice-Hall.

   Advanced acoustics, including synthetic lenses, holography, and microscopy.

[Slichter, 1992] Slichter, Charles P. (1992). *Principles of Magnetic Resonance*. 3rd edn. New York: Springer-Verlag.

   A great introduction to magnetic resonance (and quantum mechanics).

[Shung *et al.*, 1992] Shung, K. Kirk, Smith, Michael B., & Tsui, Benjamin. (1992). *Principles of Medical Imaging*. San Diego: Academic Press.

   CAT, PET, and all that.

[Press *et al.*, 1992] Press, William H., Teukolsky, Saul A., Vetterling, William T., & Flannery, Brian P. (1992). *Numerical Recipes in C: The Art of Scientific Computing*. 2nd edn. New York: Cambridge University Press.

   Recipes has a good practical introduction to inverse problems and transforms.

9.7 PROBLEMS

(9.1) Is a thin spherical lens (like the ones we studied in the last chapter) a matched filter with a response like equation (9.8)?

(9.2) The resolving power of a lens can be defined in terms of the distance between the two points at which the point spread function has decreased from its maximum by 3 dB. Two objects can be resolved if they are separated by this distance. This is also sometimes defined by *Rayleigh's criterion*: the principal maximum of the point spread function from one object is at the first minimum of the point spread function of the other.

   (*a*) For equation (9.9) what is the resolving power in terms of the wavelength, the lens aperture size, and the distance from the lens?

   (*b*) For an ultrasonic signal (100 kHz) in air ($\sim$ 350 m/s), what size aperture is needed to resolve 1 cm at a distance of 1 m?

   (*c*) For a radar satellite (10 GHz) in Low Earth Orbit ($\sim$ 200 km) what size aperture is needed to resolve 1 cm? What is the angle subtended by this aperture relative to the Earth's surface?

(9.3) Work out the delay function $g(x')$ to implement a matched filter in the plane for a paraxial synthetic aperture radar. Assume that the transceiver is moving along a straight line, at each point sending out a spherical wave and accumulating the return signal convolved by $g(x')$, and assume that the transceiver velocity is slow compared to the wave speed.

(9.4) Estimate the typical resonance frequency for a nuclear spin (NMR) and an electronic spin (ESR) in a 1 T field.

(9.5) Consider a point charge about an infinite conducting ground plane.

(*a*) Using the method of images, find the surface charge distribution on the plane.

(*b*) Assume that the plane is divided up into a grid of square electrodes, and analytically integrate the charge density to find the measured charge at each electrode.

(*c*) Numerically evaluate the electrode charge distribution generated by the point charge.

(*d*) Use these measurements to estimate the source charge distribution as a function of height above the center of the surface charge distribution. Take a least-squares likelihood function, and for a regularizer use the sum of squares of the source charges. Plot the charge distribution as the relative weight of the regularizer term is varied, showing the minimum amount of charge compatible with the measurements for a given total error.

10 Semiconductor Materials and Devices

This chapter is the heart of the book. We've learned about how physical phenomena can represent and communicate information, and will learn about how it can be input, stored, and output, but here we turn to the essential electronic devices that transform it. To understand these devices we need to understand how electrons behave in materials. We will start by reviewing the statistical mechanics of quantum systems, and then solve a simplified model of a particle in a periodic potential to introduce the idea of band structure. Based on this quantization of the available electronic states we will study junction and field-effect semiconductor devices, leading up to an introduction to digital logic. The chapter will close by considering some of the fundamental physical limits on making and using these devices.

10.1 QUANTUM STATISTICAL MECHANICS

When statistical mechanics was introduced in Section 3.4 we did not worry about the role of quantum mechanics. Now we must: quantum mechanics is essential in explaining the states available to electrons in a semiconductor. The statistics will need to account for the allowed occupancy, and the variable number of electrons as a function of external fields or internal doping.

Remember that statistical mechanical distributions are found by maximizing their entropy subject to external constraints expressed by Lagrange multipliers. In equation (3.60) we saw that fixing the average energy

$$\sum_i E_i p_i = E \tag{10.1}$$

introduced the temperature $\beta = 1/kT$ and gave the partition function for the canonical distribution

$$\mathcal{Z} = \sum_i e^{-\beta E_i} \quad . \tag{10.2}$$

Varying the number of particles introduces another constraint,

$$\sum_i N_i p_i = N \quad , \tag{10.3}$$

where the sum is over the possible number of particles N_i in each state, and N is the expected total number of particles in the system. Adding this to equation (3.49) and repeating the derivation of the maximum entropy distribution gives the partition function for the *grand canonical distribution*

$$\mathcal{Z} = \sum_i e^{-\beta(E_i - \mu N_i)} \quad , \tag{10.4}$$

where the sum is now over the available states and their possible occupancies and E_i is the total energy of a given configuration. The expected value of any quantity f_i that depends on the state of the system is then found from

$$\langle f \rangle = \frac{1}{\mathcal{Z}} \sum_i f_i \, e^{-\beta(E_i - \mu N_i)} \quad . \tag{10.5}$$

The new parameter that arises from the constraint on the particle number is μ, the *chemical potential*. Once again comparing the microscopic and macroscopic thermodynamic definitions shows that the chemical potential is equal to the rate at which the free energy F grows as particles are added to the sytem [Reif, 1965],

$$\mu = \frac{\partial F}{\partial N} \quad . \tag{10.6}$$

Remember that there are two kinds of quantum systems: *fermions*, which have 1/2-integer spin and which cannot be in the same state because of the *Pauli exclusion principle*, and *bosons*, which have integer spin and can share the same state. Since electrons are fermions, here we will need to solve equation (10.5) subject to the condition that each state can have only one electron. When we consider photons, which are bosons, in Chapter 11 we'll want solutions that allow an unlimited number of particles per state.

The most important statistical quantity will be the expected occupancy of the available quantum states as a function of their energy and the temperature. If there are N_i particles in the ith quantum state, with a single-particle energy E_i, we'll neglect interactions and take the total energy to be $E_i N_i$. Then the expected number of particles in state s is found by summing over all possible configurations,

$$\begin{aligned} \langle N_s \rangle &= \frac{\sum_{N_1} \sum_{N_2} \cdots N_s e^{-\beta(E_1 N_1 + E_2 N_2 + \cdots) + \beta\mu(N_1 + N_2 + \cdots)}}{\sum_{N_1} \sum_{N_2} \cdots e^{-\beta(E_1 N_1 + E_2 N_2 + \cdots) + \beta\mu(N_1 + N_2 + \cdots)}} \\ &= \frac{\sum_{N_s} N_s e^{-\beta E_s N_s + \beta\mu N_s}}{\sum_{N_s} e^{-\beta E_s N_s + \beta\mu N_s}} \quad . \end{aligned} \tag{10.7}$$

The sum over N_s can be pulled out of the other sums, and except for it the numerator and denominator cancel. Equation (10.7) can now be evaluted for the two cases of interest.

- *Fermions*

  Here N_s must be either 0 or 1 because there can't be more than one fermion in a single state:

$$\begin{aligned} \langle N_s \rangle &= \frac{\sum_{N_s=0}^{1} N_s e^{-\beta E_s N_s + \beta\mu N_s}}{\sum_{N_s=0}^{1} e^{-\beta E_s N_s + \beta\mu N_s}} \\ &= \frac{0 + e^{-\beta E_s + \beta\mu}}{1 + e^{-\beta E_s + \beta\mu}} \\ &= \frac{1}{e^{\beta(E_s - \mu)} + 1} \quad . \end{aligned} \tag{10.8}$$

  This is called the *Fermi–Dirac distribution*.

- *Bosons*

For bosons, the sum over N_s runs from 0 to ∞:

$$
\begin{aligned}
\langle N_s \rangle &= \frac{\sum_{N_s=0}^{\infty} N_s e^{-\beta E_s N_s + \beta \mu N_s}}{\sum_{N_s=0}^{\infty} e^{-\beta E_s N_s + \beta \mu N_s}} \\[2mm]
&= \frac{\sum_{N_s=0}^{\infty} N_s C^{N_s}}{\sum_{N_s=0}^{\infty} C^{N_s}} \quad (C \equiv e^{-\beta E_s + \beta \mu}) \\[2mm]
&= \frac{C \frac{d}{dC} \sum_{N_s=0}^{\infty} C^{N_s}}{\sum_{N_s=0}^{\infty} C^{N_s}} \\[2mm]
&= \frac{C \frac{d}{dC} (1 - C)^{-1}}{(1 - C)^{-1}} \\[2mm]
&= \frac{C(1 - C)^{-2}}{(1 - C)^{-1}} \\[2mm]
&= \frac{1}{C^{-1} - 1} \\[2mm]
&= \frac{1}{e^{\beta(E_s - \mu)} - 1} \quad .
\end{aligned}
\tag{10.9}
$$

This is the *Bose–Einstein* distribution. It differs from the Fermi–Dirac distribution only by the minus sign in the denominator, but we will see that this sign difference makes all the difference in the world.

10.2 ELECTRONIC STRUCTURE

In an atom that is part of a crystal, there are *core* electrons that remain tightly bound to the nucleus unless they are knocked out by energetic particles such as photons in *X-ray Photoemission Spectroscopy* (*XPS*) or electrons in *Auger spectroscopy*. There are also less weakly bound *outer-shell* electrons that can move around the crystal. A surprisingly good approximation is to ignore the inner-shell electrons, and treat the outer-shell electrons as a non-interacting ideal gas of fermions traveling in a medium with a spatially periodic potential due to the atoms on the crystal lattice sites [Ashcroft & Mermin, 1976]. From this simple and historically important model we'll find that there are momentum bands of allowed and forbidden electron energies which we will then use to explain the basic features of semiconductors. This is a physicists' approach; similar but complementary insights come from viewing band structure as resulting from the electronic states available in a system with many bonds [Hoffmann, 1988].

Here we'll introduce just enough quantum mechanics to analyze a one-dimensional model of an electron in a periodic potential, saving the general structure of quantum mechanics for Chapter 15. The spatial state of the electron will be described by its *wave function* $\psi(x)$, with the probability of seeing the electron at a point given by $|\psi(x)|^2$. Since the electron must be somewhere the wave function is normalized,

$$
\int_{-\infty}^{\infty} |\psi(x)|^2 \, dx = 1 \quad .
\tag{10.10}
$$

Measurable quantities are given by differential operators that act on the wave function; the most important one being the total energy, called the *Hamiltonian*

$$\mathcal{H}[\psi(x)] = \left[-\frac{\hbar^2}{2m}\frac{d^2}{dx^2} + V(x) \right]\psi(x) \quad , \tag{10.11}$$

where $V(x)$ is the electron's potential energy, and $(-\hbar^2/2m)\,d^2/dx^2$ is the operator corresponding to the kinetic energy. The expected value of the energy associated with the wave function is

$$\langle \psi | \mathcal{H} | \psi \rangle = \int_{-\infty}^{\infty} \psi^*(x)\mathcal{H}[\psi(x)]\,dx \quad . \tag{10.12}$$

The evolution of a wave function is given by the *time-dependent Schrödinger equation*

$$\mathcal{H}[\psi(x)] = i\hbar\frac{\partial\psi}{\partial t} \quad . \tag{10.13}$$

If the Hamiltonian does not depend on time, the allowable energy states are given by wave functions that satisfy the *time-independent Schrödinger equation*

$$\mathcal{H}[\psi_E(x)] = E\psi_E(x) \quad , \tag{10.14}$$

where the $\psi_E(x)$ are the *eigenfunctions* of the Hamiltonian, and the possible values of E are the corresponding *eigenvalues*. If the potential vanishes, this is easily solved to find for free space

$$-\frac{\hbar^2}{2m}\frac{d^2\psi_E(x)}{dx^2} = E\psi_e(x) \quad \Rightarrow \quad \psi(x) = Ae^{ikx} + Be^{-ikx} \quad , \tag{10.15}$$

where A and B are constants that depend on the boundary conditions, and the energy E and wave vector k are related by $E = \hbar^2k^2/2m$.

In Chapter 15 we'll see that if another operator O *commutes* with the Hamiltonian so that

$$\mathcal{H}\{O[\psi(x)]\} = O\{\mathcal{H}[\psi(x)]\} \quad , \tag{10.16}$$

then wave functions can be chosen to simultaneously be eigenfunctions of both operators. In particular, if the potential is periodic so that $V(x + \Delta) = V(\Delta)$, the Hamiltonian will be unchanged by an operator $T_\Delta[\psi(x)] = \psi(x + \Delta)$ that translates the wavefunction by this distance, so T_Δ commutes with H. This means that an energy eigenfunction will also be an eigenfunction of the translation operator, with an eigenvalue λ that can depend on the energy

$$T_\Delta[\psi(x)] = \lambda_\Delta(E)\psi(x) \quad . \tag{10.17}$$

If we compose two translations by two multiples of the period of the potential,

$$T_{\Delta'}\{T_\Delta[\psi(x)]\} = \lambda_{\Delta'}\lambda_\Delta\psi(x) \quad , \tag{10.18}$$

but by definition the translations add so that

$$T_{\Delta'}\{T_\Delta[\psi(x)]\} = \lambda_{\Delta'+\Delta}\psi(x) \quad . \tag{10.19}$$

These equations will be consistent if $\lambda_\Delta(E) = e^{\alpha(E)\Delta}$, where α is a constant that depends on E. A further constraint comes from requiring that after translation the wave function remain normalized

$$\int_{-\infty}^{\infty} |T_\Delta[\psi(x)]|^2\, dx = \int_{-\infty}^{\infty} \left| e^{\alpha\Delta}\psi(x) \right|^2 dx \;\Rightarrow\; |e^{\alpha\Delta}|^2 = 1 \quad. \tag{10.20}$$

This means that the exponent is imaginary, so the translation operator is a multiplication by a phase factor that depends on the energy $T_\Delta = e^{ik(E)\Delta}$, or

$$\psi_E(x+\Delta) = e^{ik(E)\Delta}\psi_E(x) \quad. \tag{10.21}$$

If we write the wave function as a product of two terms $\psi(x) = A_k(x)u_k(x)$, where $u_k(x+\Delta) = u_k(x)$ reflects the periodicity of the potential, then

$$\begin{aligned}
\psi(x+\Delta) &= A_k(x+\Delta)u_k(x+\Delta) \\
&= A_k(x+\Delta)u_k(x) \\
&= T_\Delta[\psi_k(x)] \\
&= e^{ik\Delta}A_k(x)u_k(x) \quad,
\end{aligned} \tag{10.22}$$

which will hold if $A_k(x) = e^{ikx}$, therefore

$$\psi(x) = e^{ikx}u_k(x) \quad. \tag{10.23}$$

This is *Bloch's Theorem*.

The simplest periodic potential ignores the size of the atoms and models them as a sum of delta functions,

$$V(x) = \sum_{n=-\infty}^{\infty} V_0\, \delta(x - n\Delta) \quad, \tag{10.24}$$

called the *Kronig–Penney model*. The atoms will be assumed to be fixed; in a real crystal there are quantized displacement waves called *phonons* [Ashcroft & Mermin, 1976]. In the intervals between these idealized atoms the potential vanishes, therefore the wave function there is just that of a free plane wave

$$\begin{aligned}
\psi(x) &= Ae^{iqx} + Be^{-iqx} \\
\Rightarrow u_k(x) &= Ae^{i(q-k)x} + Be^{-i(q+k)x} \quad,
\end{aligned} \tag{10.25}$$

where $q\hbar = \sqrt{2mE}$, and A and B are unknown constants.

We will now find how q and hence E are related to the k that was introduced by Bloch's Theorem. Requiring that $u_k(0) = u_k(\Delta)$ gives

$$A + B = Ae^{i(q-k)\Delta} + Be^{-i(q+k)\Delta} \quad. \tag{10.26}$$

A second relationship comes from Schrödinger's equation

$$\left[-\frac{\hbar^2}{2m}\frac{d^2}{dx^2} + V \right]\psi = E\psi$$

$$\left[E + \frac{\hbar^2}{2m}\frac{d^2}{dx^2} \right]\psi(x) = \sum_{n=-\infty}^{\infty} V_0\delta(x - n\Delta)\psi(x) \quad. \tag{10.27}$$

The wave function must be continuous across the delta functions, but its slope can change

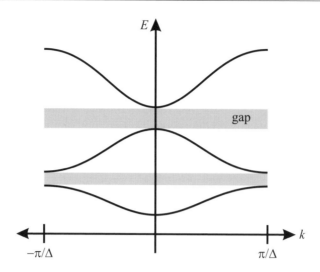

Figure 10.1. Band structure for the Kronig–Penney model.

at them. Integrating from $x = -\epsilon$ to ϵ, taking the limit $\epsilon \to 0$, and using the periodicity of u shows that (Problem 10.1)

$$\frac{\hbar^2}{2m} iq \left(A - B - Ae^{i(q-k)\Delta} + Be^{-i(q+k)\Delta} \right) = V_0(A + B) \quad . \tag{10.28}$$

Now we have two equations in the two unknowns A and B. Eliminating B gives

$$\left[\cos(k\Delta) - \cos(q\Delta) - \frac{mV_0}{q\hbar^2} \sin(q\Delta) \right] A = 0$$

$$\Rightarrow \quad \cos(k\Delta) = \cos(q\Delta) + \frac{mV_0\Delta}{\hbar^2} \frac{\sin(q\Delta)}{q\Delta} \quad . \tag{10.29}$$

In the limit of infinite lattice spacing $\Delta \to \infty$ this reduces to $k = q \Rightarrow E = \hbar^2 k^2/2m$, the free electron case. For non-infinite spacing the relationship is more complicated: since $|\cos(k\Delta)| \le 1$, there will be bands of q values for which there is no k that solves this equation, and hence gaps in the allowable energy $E = \hbar^2 q^2/2m$. The relationship between k and $E(q)$ is plotted in Figure 10.1, with successive bands shifted back to the origin by multiples of $2\pi/\Delta$ (which can be done without changing the value of $\cos(k\Delta)$). Each of the regions shifted back is called a *Brillouin zone*. The symmetries of a real crystal lead to a much more complicated three-dimensional *band structure*, but the basic features are similar. For a free particle the bands are just sections of a parabola, which are bent by the crystal periodicity near the gaps at the zone boundaries. k is called the *crystal momentum*. It indexes the eigenstates, playing a role that is analogous but no longer equal to the real momentum.

Next, assume that the crystal has a finite length of $L = N\Delta$, and to avoid end effects assume periodic boundary conditions $\psi(0) = \psi(L)$. This implies that

$$u_k(0) = e^{ikL} u_k(L) \quad \Rightarrow \quad e^{ikL} = 1 \tag{10.30}$$

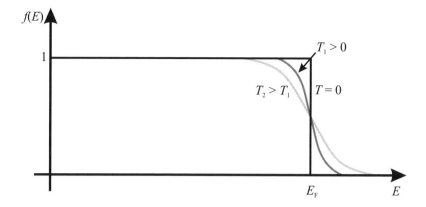

Figure 10.2. The Fermi–Dirac distribution.

because of the periodicity of u_k. This will hold if

$$kL = 2\pi n \quad \Rightarrow \quad k = \frac{2\pi}{L}n = \frac{2\pi}{N\Delta}n \qquad (10.31)$$

for integers n. Since these crystal momentum states are separated by a difference of $2\pi/(N\Delta)$ and each band is $2\pi/\Delta$ wide, there are

$$\frac{2\pi}{\Delta}\frac{N\Delta}{2\pi} = N \qquad (10.32)$$

states per band. Because electrons are spin-1/2 fermions, each of these momentum states can hold two electrons, one "up" and one "down", and the occupation probability as a function of temperature is given by the Fermi–Dirac distribution

$$f(E) = \frac{1}{1 + e^{(E-\mu)/kT}} \qquad . \qquad (10.33)$$

The Fermi–Dirac distribution is shown in Figure 10.2. The chemical potential μ is the change in the free energy when one electron is added. The *Fermi energy* or *Fermi level* E_F is the chemical potential at $T = 0$ K, and if it lies in a band it gives the highest filled state at $T = 0$ K. The Fermi level will be a function of the number of electrons in the crystal and hence the number of states that can be filled. As the temperature is raised, electrons in states below the Fermi energy will be excited above it, which will move the chemical potential relative to the Fermi energy. Because this difference is small at room temperature, we will follow the common (but not quite correct) practice of using them interchangeably.

An applied voltage can move an electron only if there is an electron to be moved, and a state for it to go into. Therefore, for conduction the states far below the Fermi energy don't matter because there are no available nearby states for an electron to move into, and states far above the Fermi energy don't matter because there is little probability of them being occupied. The uppermost filled band is called the *valence band*, and the lowest unfilled band is called the *conduction band*. In an insulator, the valence band is completely full, hence it is not possible for electrons to move unless they are excited over the band gap. The chemical potential for an insulator lies in the middle of the gap because each carrier excited out of the valence band appears in the conduction band. In

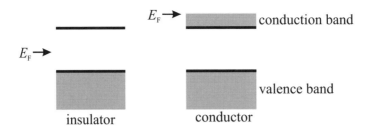

Figure 10.3. Band structure for an insulator and conductor.

a metal the chemical potential lies within the conduction band and so there are plenty of conduction states available (Figure 10.3). The energy difference between a conduction electron and a free electron removed from the metal is called the *work function*.

A *semiconductor* is just an insulator that has an energy gap small enough for there to be an appreciable probability for an electron to be thermally excited across it at room temperature. For example, Ge, Si, and diamond all have full valence bands, but the gap energy of Ge is 0.67 eV, for Si it is 1.11 eV, and for diamond it is 5 eV (kT at room temperature is 0.026 eV). The room temperature resistivity of a good insulator is $\sim 10^{10}$ $\Omega \cdot$ cm, that of a good metal is $\sim 10^{-6}$, and for a semiconductor it is $\sim 10^{6}$.

When a full valence band has a single electron removed it leaves behind one available state. This single state can be viewed as a positively charged particle called a *hole* that moves under an applied field. Actually, all of the electrons in the band are moving in the opposite direction, much like the motion of a bubble trapped in a glass of water. The effective mass m^* of a hole is given by finding the change in its momentum under an applied force, and is equal to $m_p^* = 0.56m_0$ for Si (where m_0 is the free electron mass). Electrons in a crystal also have an effective mass different from that of a free electron because of the curvature of the bands; in Si $m_n^* = 1.1m_0$.

The conduction properties of a material depend sensitively on the location of the Fermi energy relative to the nearest energy gap. Adding doping atoms can add or remove electrons, moving the Fermi energy and changing the character of the material. Such *extrinsic* materials are produced by adding *donor* atoms, such as P or As which have one extra outer electron compared to Si, or *acceptor* atoms such as Al which has one less electron. A donor will raise the Fermi level by giving electrons up to the conduction band, and an acceptor will lower the Fermi level by trapping valence band electrons, thereby producing holes. This ability to selectively move the Fermi level relative to the undoped level in the *intrinsic* material is the key to making semiconductor devices. Materials which are doped so that the dominant conduction is by electrons are called *n-type* (n for negative), and those doped by holes *p-type* (Figure 10.4).

The density of carriers n of the conduction band is found by integrating up from the conduction band edge E_c the product of the density of states $N(E)$, which gives the number of available states per volume in an energy range, times the Fermi distribution $f(E)$, which is their thermodynamic occupancy

$$n = \int_{E_c}^{\infty} f(E)N(E)\,dE \quad . \tag{10.34}$$

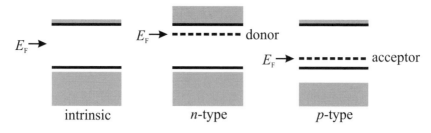

Figure 10.4. Doping of a semiconductor.

Since at room temperature kT is much smaller than gap energies, a good approximation for the Fermi distribution is

$$f(E) = \frac{1}{1 + e^{(E-E_\mathrm{F})/kT}} \simeq e^{-(E-E_\mathrm{F})/kT} \quad . \tag{10.35}$$

The density of states can be approximated with the free-electron distribution. In 3D, equation (10.31) becomes

$$\vec{k} = \frac{2\pi}{L} \left(n_x \hat{x} + n_y \hat{y} + n_z \hat{z} \right) \quad , \tag{10.36}$$

hence

$$\begin{aligned}
E &= \frac{\hbar^2}{2m_n^*} |k|^2 \\
&= \frac{\hbar^2}{2m_n^*} \left(\frac{2\pi}{L} \right)^2 \left(n_x^2 + n_y^2 + n_z^2 \right) \\
&\equiv \frac{\hbar^2}{2m_n^*} \left(\frac{2\pi}{L} \right)^2 r^2 \\
&= \frac{h^2}{2m_n^* V^{2/3}} r^2
\end{aligned} \tag{10.37}$$

or

$$dE = \frac{h^2}{m_n^* V^{2/3}} r \, dr \quad . \tag{10.38}$$

m_n^* is the electron's effective mass, $V = L^3$ is the volume of the crystal, and we're assuming that because there are so many electron states, the sum of the squares of the integers $n_x^2 + n_y^2 + n_z^2$ can be taken to be a continuous variable r^2. The density of states $N(E)$ can equivalently be taken to be a function of this variable $N(r)$. In r-space, states occupy cubes of unit volume, therefore the total number of states in the crystal $V N(r) \, dr$ in an infinitesimal shell around r is given by the area of the shell times its thickness

$$V N(r) \, dr = 2 \cdot 4\pi r^2 \, dr \quad , \tag{10.39}$$

with the factor of 2 coming from spin-up and -down occupancy. Substituting in equations

(10.37) and (10.38),

$$V N(r) \, dr = 8\pi r^2 \, dr$$

$$= 8\pi r \cdot r \, dr$$

$$V N(E) \, dE = 8\pi \frac{V^{1/3}\sqrt{2m_n^* E}}{h} \frac{V^{2/3} m_n^*}{h^2} \, dE$$

$$N(E) \, dE = 8\pi \frac{m_n^{*\,3/2}\sqrt{2E}}{h^3} \, dE$$

$$= \frac{1}{2\pi^2} \left(\frac{2m_n^*}{\hbar^2} \right)^{3/2} \sqrt{E} \, dE \quad . \tag{10.40}$$

Then

$$n = \int_{E_c}^{\infty} f(E) N(E) \, dE$$

$$= \int_{E_c}^{\infty} e^{-(E-E_F)/kT} \frac{1}{2\pi^2} \left(\frac{2m_n^*}{\hbar^2} \right)^{3/2} \sqrt{E} \, dE$$

$$= \frac{1}{2\pi^2} \left(\frac{2m_n^*}{\hbar^2} \right)^{3/2} e^{E_F/kT} \int_{E_c}^{\infty} e^{-E/kT} \sqrt{E} \, dE \quad . \tag{10.41}$$

In doing this integration we're free to choose any energy scale we wish, and so can simplify it by taking $E_c = 0$

$$n = \frac{1}{2\pi^2} \left(\frac{2m_n^*}{\hbar^2} \right)^{3/2} e^{E_F/kT} \int_0^{\infty} e^{-E/kT} \sqrt{E} \, dE$$

$$= \frac{1}{2\pi^2} \left(\frac{2m_n^*}{\hbar^2} \right)^{3/2} e^{E_F/kT} \frac{\sqrt{\pi}}{2} (kT)^{3/2}$$

$$= 2 \left(\frac{m_n^* kT}{2\pi\hbar^2} \right)^{3/2} e^{E_F/kT} \quad . \tag{10.42}$$

Going back to units in which $E_c \neq 0$ requires subtracting the difference off from E_F,

$$n = 2 \left(\frac{m_n^* kT}{2\pi\hbar^2} \right)^{3/2} e^{-(E_c - E_F)/kT}$$

$$\equiv N_n e^{-(E_c - E_F)/kT} \quad . \tag{10.43}$$

Similarly, the hole occupancy in the valence band is given by $1 - f(E)$; integrating this from the valence band edge E_v gives the symmetrical relationship

$$p = 2 \left(\frac{m_p^* kT}{2\pi\hbar^2} \right)^{3/2} e^{-(E_F - E_v)/kT}$$

$$\equiv N_p e^{-(E_F - E_v)/kT} \quad . \tag{10.44}$$

For an intrinsic semiconductor the Fermi level will be in the middle of the gap, and the hole and electron concentrations will be equal $n = p = n_i$. The Fermi energy will move depending on the doping, but the product of the occupancies will be a constant that depends only on the gap energy E_g:

$$np = N_n N_p e^{-(E_c - E_v)/kT} = N_n N_p e^{-E_g/kT} = n_i^2 \quad . \tag{10.45}$$

The carrier densities can be rewritten in terms of the intrinsic density n_i and the intrinsic Fermi energy E_i as

$$n = n_i e^{(E_F - E_i)/kT} \quad p = n_i e^{(E_i - E_F)/kT} \quad . \tag{10.46}$$

If there is no doping then these will be equal.

Now consider what happens to one of these electrons in response to the force of an external electric field

$$F = \frac{dp}{dt} = qE \quad . \tag{10.47}$$

In free space this causes a steady increase in the velocity

$$dp = m \, dv = qE \, dt \quad , \tag{10.48}$$

but in a material the collisions with the lattice and defects will slow it down. *Kinetic theory* [Balian, 1991] makes the rough approximation that after a characteristic time τ a collision occurs which randomizes the electron's velocity, so that the average drift velocity is the expected value of the non-random contribution

$$\langle v \rangle = \frac{q\tau}{m} E \equiv \mu E \quad . \tag{10.49}$$

μ is the *mobility*. In terms of it, the conductivity is

$$J = nq\langle v \rangle = \sigma E \quad \Rightarrow \quad \sigma = nq\mu \quad . \tag{10.50}$$

This relates Ohm's Law to microscopic material properties. The linear relationship between drift velocity and applied field holds only at sufficiently low fields, limited ultimately by the dielectric breakdown voltage of the material (which $\approx 5 \times 10^5$ V/cm for Si).

For undoped Si at room temperature, the electrons have a mobility of about 1350 cm^2/(V·s), and for GaAs it is 8500 cm^2/(V·s), which is why GaAs is used for high-speed devices. Note that this is still *much* slower than the propagation velocities that we found for electromagnetic waves. If Si is doped at 10^{17}/cm^3 the mobility falls to 800 cm^2/(V·s) because of the extra scattering from the dopant atoms, and at 10^{19}/cm^3 it is only 90. In a *High-Electron-Mobility Transistor* (*HEMT*) the doping material is confined to layers separated from where conduction takes place [Pavlidis, 1999]. This can be accomplished by using an alloy such as $Al_x Ga_{1-x} As$ that lets the band gap be tuned as a function of the composition x, ranging up to a few eV from GaAs to AlAs depending on the crystal orientation. These are examples of binary III–V semiconductors formed from elements in those columns of the periodic table; II–VI semiconductors such as CdSe are also useful, particularly for optoelectronics (Chapter 11).

10.3 JUNCTIONS, DIODES, AND TRANSISTORS

Fortified with the basic ideas of energy bands and the Fermi–Dirac distribution, we are now ready to tackle devices made out of semiconductors. These will rely on the properties

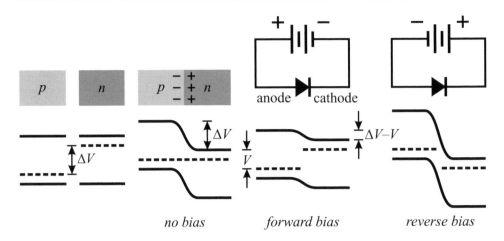

Figure 10.5. Biasing a p–n diode.

of junctions between materials. The Fermi energy (or, to be correct at $T \neq 0$ K, the chemical potential) in a material can be thought of as the height of an energy hill that the electrons are traveling on. If a drop of water is added to a bucket, its energy depends on how much water is already in the bucket. If two buckets filled with differing amounts of water are brought into contact and a partition between them is removed, then water will spill from one bucket to the other in order to equalize the energy difference. Similarly, when two semiconductors are brought into contact as shown in Figure 10.5, there is a transient current from the material with the higher Fermi level to the lower one until the Fermi levels are aligned, removing the energy gradient that is driving the current. A potential difference then appears between the bands that is equal to the potential difference ΔV between the two Fermi energies. This energy gradient is associated with a local electric field that is produced in the *transition* (or *depletion* or *space-charge*) region by the charge that moved between the materials. Once the bands have "bent" at the interface no average current will flow. Even though electrons and holes will recombine if they are given a chance because that lowers their energy, the electrons on the n side cannot climb up the potential hill to reach the p side. Similarly, holes behave oppositely from electrons, and so they cannot climb down the hill to reach the electrons.

If an electron is thermally excited from the valence band to the conduction band it leaves a hole behind, creating an *Electron–Hole Pair* (*EHP*). Normally these will quickly recombine, but if they form near the interface then the junction field will sweep the electron to the n side and the hole to the p side. This results in a *generation current*. In addition, there is a probability proportional to $e^{-\Delta E/kT} = e^{-q\Delta V/kT}$ for a carrier to be thermally excited over the energy barrier at the junction and then diffuse out, creating a *diffusion current*.

If a bias potential V is applied across the junction in a *p–n diode*, it will split the Fermi energies, reducing or increasing the size of the barrier depending on the polarity. The diffusion current will then be

$$I_{\text{diffusion}} = Ae^{-q(\Delta V - V)/kT} = Ae^{-q\Delta V/kT}e^{qV/kT} \quad , \quad (10.51)$$

where A is a constant that depends on device details including the junction geometry

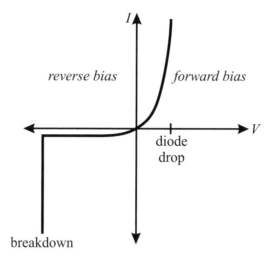

Figure 10.6. *I–V* curve for a *p–n* diode.

and the amount of doping. It can be found by recognizing the the generation current is independent of the bias voltage (until that becomes large enough to eliminate the band bending), and that at zero bias the two currents must cancel, so that their sum is

$$I = I_{\text{generation}} \left(e^{qV/kT} - 1 \right) \quad . \tag{10.52}$$

This characteristic *I–V* curve is shown in Figure 10.6. As the forward bias is increased, the current quickly increases and the diode turns on. The bias voltage appears as a *diode drop* potential in conduction across the diode; 0.6 V is a typical value. The diode blocks current in the opposite direction, letting through just the small generation current that is independent of bias but does depend on temperature (providing a useful thermometer).

This is a DC relationship, which will roll off at higher frequencies because of the capacitance associated with the charge induced at the junction. It also breaks down at higher fields through two important mechanisms. If the bands are bent far enough, the valence and conduction bands come so close at the junction that the wave function for carriers overlaps between them, creating a probability for them to *tunnel* between the bands. Because the voltage at which this *Zener breakdown* occurs can be selected by the doping, it is useful for providing voltage references. And if the field is strong is enough to accelerate a carrier so fast that it excites more carriers when it scatters, *avalance breakdown* occurs. The ensuing cascade makes is possible to detect very small numbers of electrons or photons.

Something similar happens at the interface between a semiconductor and a metal, shown in Figure 10.7. When the materials are brought together, a current must initially flow to equalize the Fermi levels. But because the metal cannot have an internal electric field the band bending occurs entirely on the semiconductor side. This is called a *Schottky barrier*, and it once again rectifies the current, which can be either a bug or a feature. It's a simpler way to fabricate a diode because all that's needed is a metallization, but it also means that any lead attached to a semiconductor becomes a diode. Creating linear *ohmic*

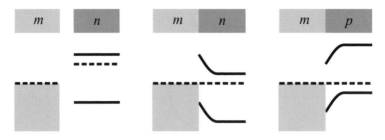

Figure 10.7. A Schottky barrier between a semiconductor and a metal.

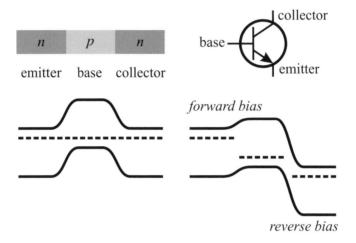

Figure 10.8. An n–p–n transistor.

contact requires extra effort, such as heavily doping the semiconductor at the interface to keep the transition region thin enough to permit tunnelling.

If one junction is good, two are even better. Consider the back-to-back diodes shown in Figure 10.8, forming a *bipolar transistor*. The center region is called the *base* and the sides are the *emitter* and *collector*. In the absence of any bias, current cannot flow into either the emitter or collector. But if the emitter–base junction is forward-biased and the collector–base junction is reverse-biased then a current I_{CE} can flow through the collector–emitter circuit. As before, the emitter–base junction will have an I–V curve of the form of equation (10.52), but now in addition to determining its own current flow V_{BE} will set that between the emitter and the collector if a voltage source is connected across them,

$$I_{CE} = I_S \left(e^{qV_{BE}/kT} - 1 \right)$$
$$= \beta I_{BE} \quad . \tag{10.53}$$

This is called the *Ebers–Moll model* of a transistor, with a *saturation current* I_S. Because a voltage determines a current this is a *transconductance* device, but since V_{BE} also produces a current I_{BE} it's simpler to understand the transistor as a current amplifier. What makes this device so useful is that a small current between the emitter and base

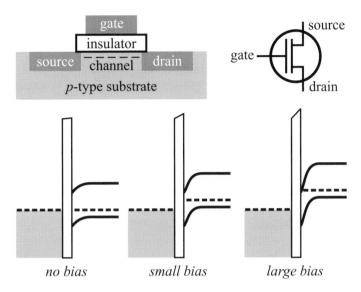

Figure 10.9. An NMOS FET.

can control a much larger current between the emitter and collector; the proportionality factor β is typically on the order of 100.

Bipolar transistors do have a significant liability: keeping them turned on draws a steady current through the base. This limits their use in applications for which power consumption or heat dissipation need to be minimized (i.e., almost all of them). Figure 10.9 show how a *Field-Effect Transistor* (*FET*) cures this by using an electric field rather than a current as the control input. A carrier source and drain are separated by a semiconducting channel, covered by a thin insulating layer and a metallic gate. One of silicon's great virtues is that it readily grows a tough SiO_2 oxide that is a good insulator, making this a Metal-Oxide-Semiconductor FET or *MOSFET*.

If the channel is p-type, and the source and drain are n-type, this is an *NMOS* transistor because electrons are the current carriers. Figure 10.9 plots the band structure along a vertical slice through the gate, oxide, and substrate. The overlap across the thin oxide aligns the Fermi levels of the gate and the substrate. As the gate is biased relative to the substrate the Fermi levels split, but the position of the bands at the interface is fixed by the material properties. This is accomplished once again by the formation of a field gradient in a transition region. As the substrate Fermi level gets closer to the conduction band at the surface, excess electrons begin to appear in the channel. The gate–oxide–substrate combination can be thought of as a capacitor with a semiconductor for one plate. Charge on the gate must be matched by image charge in the substrate, but because it is semiconducting this image charge also changes the conductivity. Unlike the continuous base–emitter current drawn by a bipolar transistor, a MOSFET is a voltage-controlled device that dissipates control current only when the gate voltage is changing and hence charging or discharging this capacitance.

Figure 10.10 plots the current I_{DS} between the drain and source as a function of the voltage V_{GS} between the gate and source, for a fixed voltage V_{DS} between the drain and source. An *enhancement mode* NMOS MOSFET is doped so that no current will flow

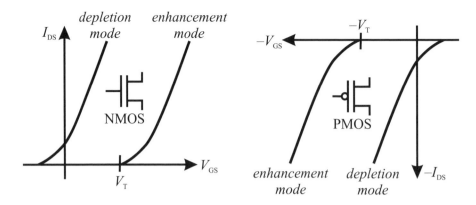

Figure 10.10. Threshold currents in MOSFETs, shown for a fixed V_{DS}.

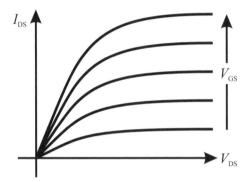

Figure 10.11. I–V curves for an NMOS FET as a function of the gate voltage.

for $V_{GS} = 0$. As V_{GS} is increased it reaches the threshold voltage V_T that brings the Fermi level between the valence and conduction bands so that electrons start to appear in the channel. Further increasing V_{GS} increases the number of carriers, reducing the channel resistance. A *depletion* mode device is doped so that the Fermi level starts out high enough for there to be some conduction carriers; a negative threshold voltage is needed to turn this kind of transistor off. In a *PMOS* MOSFET the channel is n-type and the source and drain are p-type, the current is carried by holes, and decreasing the gate voltage increases their concentration.

For small values of V_{DS} the current I_{DS} will be linear (ohmic), as shown in Figure 10.11. Increasing V_{GS} decreases the resistance, thereby increasing the slope. But as V_{DS} is increased the electrons in the channel are also pulled towards the source, eventually pinching off the channel and saturating the current.

10.4 LOGIC

TTL (*Transistor–Transistor Logic*) integrates bipolar transistors on a semiconducting substrate to implement logical functions. While historically significant, its use is limited by the static current drawn by a gate when it is turned on. MOSFETs are an attractive

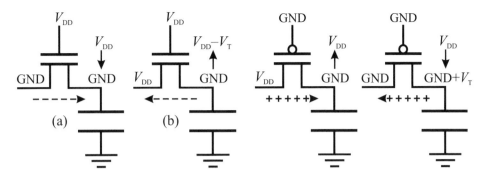

Figure 10.12. Charging and discharging capacitors through MOSFETs.

alternative, but the asymmetry shown in Figure 10.10 presents a serious obstacle to their use.

Consider the cases shown in Figure 10.12. An NMOS and a PMOS enhancement-mode FET are being used to drive another FET, represented by its gate capacitance. In case (a), an NMOS FET is turned on by applying the supply voltage to the gate. For MOSFETs this voltage is called V_{DD} (for historical reasons, the D is for drain; for bipolar transistors the supply is usually labeled V_{CC}, C for collector). Assume that the capacitor is initially charged to V_{DD} and that the input to the FET is grounded. Because electrons are the charge carriers in an NMOS FET, they must flow from the source at ground to the positive capacitor at the drain to discharge it. Since V_{GS} remains at $V_{DD} > V_T$ the FET stays on and the capacitor is fully discharged. Compare this to case (b), with the capacitor starting out grounded and V_{DD} applied to the input. This is a problem. For the capacitor to charge up, electrons must flow from it, so it is the source. But as its voltage rises, V_{GS} will eventually drop below V_T, shutting of the FET with $V_{DD} - V_T$ left on the capacitor instead of the desired V_{DD}. Because an NMOS FET can discharge a capacitor to ground, it can output a logical 0, but it can't output a logical 1 because it can't charge a capacitor up to the positive supply. Likewise, a PMOS FET can output a 1 but not a 0.

In integrated electronics, as in life, the solution to shared imperfections is a relationship based on complementary strengths. *CMOS (Complementary Metal Oxide Semiconductor)* logic uses pairs of MOSFETs, as shown in Figure 10.13 for the simplest circuit of all, an inverter. If the input is grounded, the PMOS transistor is on and the NMOS transistor is off, therefore the output is pulled up to V_{DD}, which the PMOS transistor can do well. If V_{DD} is input, the PMOS transistor turns off and the NMOS transistor turns on, bringing the output to ground which it can do well. We have inverted the input. For either state only one transistor is turned on, it is used in the mode in which it works best, and current is drawn only during state changes. In practice, it is important that the PMOS and NMOS threshold voltages be well matched, otherwise during transitions there may be a period when they are both turned on and a *crowbar current* will flow from V_{DD} to ground.

A gate with two inputs is shown in Figure 10.14. Now two NMOS transistors are connected in parallel to ground, and two PMOS transistors are connected in series to V_{DD}. If $A = B = $ GND then the output will be pulled up to V_{DD}, and in all other cases

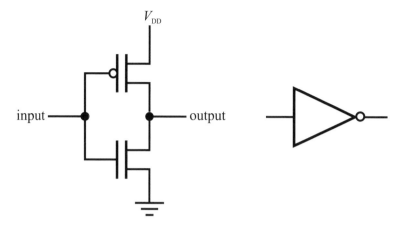

Figure 10.13. A CMOS NOT gate and its circuit symbol.

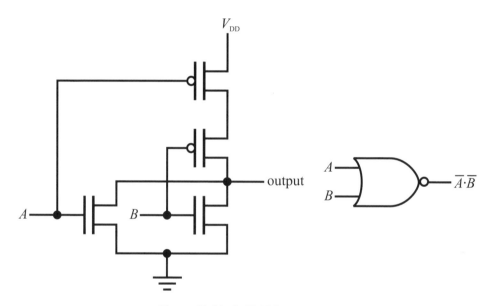

Figure 10.14. A CMOS NOR gate.

it will be pulled down to ground. This is the NOR (not-or) function. Similarly simple circuits can implement the other basic logical functions. From these two examples the essential principle of CMOS design should be clear: NMOS FETs are used only to pull outputs to ground, and PMOS FETs are used only to pull them to V_{DD}.

Because the NOR gate is a nonlinear function of its arguments (Table 10.1), it is possible to obtain any logical function by combining it with NOT gates [Hill & Peterson, 1993]. A different nonlinear gate such as AND could be used as a primitive instead, but it is not possible with a linear gate such as XOR (exclusive-or). An arbitrary logical function can in fact be realized in a *two-level* implementation using just a layer of NOT gates connected to a layer of NOR gates, so that the propagation delay of a signal through the circuit is fixed. This configuration is available packaged in a *Programmable Logic Array*

Table 10.1. *Linear* (XOR) *and nonlinear* (NOR) *logical functions.*

x	$1+x$	y	XOR(x,y)	XOR($1+x,y$)	NOR(x,y)	NOR($1+x,y$)
0	1	0	0	1	1	0
0	1	1	1	0	0	0
1	0	0	1	0	0	1
1	0	1	0	1	0	0

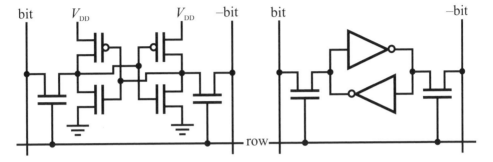

Figure 10.15. A CMOS SRAM cell.

(*PLA*) that can be used as a universal logical element. The technique most commonly used to reduce an arbitrary logical function to the smallest two-level implementation is the *Quine–McCluskey algorithm*, first developed by the philosopher W.V. Quine long before integrated circuits existed, in order to solve a puzzle in mathematical logic [Quine, 1952; McCluskey, 1956].

So far we've been discussing *combinatorial logic*, in which the output is determined by the instantaneous inputs. *Sequential logic* adds memory and a clock signal to drive transitions, so that the output can depend on the past as well as present values of the input. Clocks will be covered in Chapter 14; the last circuits to be considered here are *Random Access Memories* (*RAM*), starting with the *Static RAM* (*SRAM*) cell shown in Figure 10.15. The bit there is stored in a bistable configuration of two coupled inverters. If the input to one of the inverters is a logical 1 its output will be a 0, and this input to the other inverter will produce an output of 1 from it, agreeing with the input to the first inverter. The two inverters will also be in a stable configuration if the output of the first one is 1 and that of the second one is 0. To make this into a memory, transistors are connected between the outputs of the inverters and bit read/write lines. These pass transistors are turned on by row enable lines, letting a particular combination of row and bit lines address a unique bit. If sense amplifiers are connected to the bit lines, they can measure the state of the inverters and read out the bit, and if drive amplifiers are connected to the bit lines they can write a bit by forcing the inverters into a desired state. Two bit lines (the bit and its complement) are needed to make sure that both inverters end up in the desired state.

The basic SRAM cell requires six transistors. In 1966 Bob Dennard at IBM realized that it is possible to make a memory with just one transistor and one capacitor per bit, significantly increasing the bit density [Dennard, 1968]. In such a *Dynamic RAM* (*DRAM*) cell the bit is stored as charge on a capacitor, as shown in Figure 10.16. When

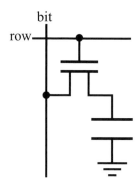

Figure 10.16. A CMOS DRAM cell.

the row and bit enable lines are turned on, charge can be written into the capacitor to store a bit, or an amplifier can detect the charge to read the bit. Unlike an SRAM cell, this read operation is destructive because the capacitor is used to charge up the bit line while it is being read, and even if a bit isn't read the charge will eventually leak away from the capacitor, therefore DRAM cells require complex refresh circuits. However, the space saving from having 1 transistor per bit much more than makes up for this extra complexity at the periphery of the memory. DRAM is also slower than SRAM because the bit line is passively driven by a capacitor rather than actively driven by an inverter. Because of this, less dense SRAM is used for fast cache memory, and denser DRAM is used for larger slower primary memories.

Both SRAM and DRAM are *volatile* memories that must be powered to maintain their data. Since power can run out long before the value of data does, particularly in mobile or embedded applications, *non-volatile* memories are needed. The most common solution is to add a *floating gate* to the MOSFET structure shown in Figure 10.9. This is an electrode between the gate and the channel, completely surrounded by the insulating oxide. Charge stored on the floating gate can be read through the image charge it induces in the substrate changing the conductivity of the channel, and because it is completely isolated the charge retention times can be very long (many years). In *EPROM* (*Erasable Programmable Read-Only Memory*), charge is deposited on the floating gate by using a write voltage large enough to excite high-energy "hot" electrons over the gate's barrier, and the entire memory is erased by exposing the die to ultraviolet light with enough energy to knock the electrons back out. Because of the tens of volts and ultraviolet light needed for writing and reading, dedicated programmers are used for changing EPROM. In *EEPROM* (*Electrically Erasable Programmable Read-Only Memory*), the oxide thickness is reduced from ∼100 nm to ∼10 nm, making it possible for electrons to tunnel onto and off of the floating gate [Fowler & Nordheim, 1928]. This requires an extra transistor per cell to control the charging, and can reduce the charge storage time and reliability of the device, but it permits in-circuit access to read and write arbitrary bits. *Flash memory* is a compromise that writes with hot electrons and erases with tunneling, permitting in-circuit programming, using just one transistor per cell at the expense of restricting erasure to memory sectors rather than individual bits.

Because of its relative ease of fabrication, low power consumption, and high packing density, CMOS dominates integrated circuit production. This has historically been accompanied by slower switching speeds because of the RC charging time associated with gate transitions. This is why higher-frequency applications have used higher-power bipolar TTL or *ECL* (*Emitter-Coupled Logic*) families, or higher-mobility materials such as GaAs. But the speed of CMOS has increased beyond 1 GHz because of beneficial features of device scaling. Once the materials become pure enough, and the channel drops well below 1 μm, an electron can travel through a transistor without scattering. Such *ballistic* or *hot electron* devices can operate at speeds far beyond what the bulk mobility would suggest. CMOS is still limited in its ability to source or sink current; for this reason *BiCMOS* processes marry the best of both worlds by using MOSFETs for on-chip logic and bipolar transistors for driving external signals.

10.5 LIMITS

In the 1960s, Gordon Moore of Intel noticed that the number of transistors on a chip, along with almost every other specification, was doubling every one and a half years. This exponential growth has come be known as *Moore's Law* [Moore, 1979]. It is not a law of nature; it is an observation about exceptional engineering efforts. The many decades over which it has applied have made it possible to foretell the future of *Very-Large-Scale Integrated circuits* (*VLSI*) with surprising prescience. On the one hand, seemingly insurmountable obstacles have been overcome each year to continue the scaling; on the other hand, sometime around 2020–2040 most devices parameters will simultaneously reach fundamental physical limits [Keyes, 1987]. Wires as we know them cannot be thinner than one atom, memories cannot have fewer than one electron, and to be viable the chip fab plants must cost something less than the GDP of the planet.

While these impending limits are a cause for concern about continued progress in improving the performance of electronics, the reality is that gate speeds or bit densities are already no longer the serious constraints they once were for many applications. Some of the most interesting challenges now lie beyond processing with detecting, communicating, and presenting electronic information. Nevertheless, because each decade of device improvement has led to corresponding new and generally unforeseen opportunities, it's still worth asking how to maintain and extend this pace.

The present battleground is the physics of *microfabrication* [Brodie & Muray, 1982]. Chips are currently produced using *lithographic* techniques to optically define device features. The diffraction limit has pushed the optical systems to shorter and shorter wavelengths, but this is still well above atomic sizes. True atomic-scale patterning has been accomplished using *Atomic Force Microscopes* (*AFMs*) [Cooper *et al.*, 1999] and *Scanning Tunneling Miscroscopes* (*STMs*) [Stroscio & Eigler, 1991], which piezoelectrically scan a sharp tip above a sample and follow the atomic topography by measuring either the cantilever deflection or the electron tunneling current. While these *scanning probe* systems are slow, lithographically-produced parallel arrays of tips promise to yield commercially useful writing speeds.

The billions of dollars that must be invested in the machinery to deposit, expose, etch, implant, dope, diffuse, dice, and test chips in a single fab line are quickly becoming the more serious scaling constraint. An alternative is to eliminate the fab line entirely and use table-top printing processes, which can attain nanometer features [Jackman *et al.*, 1998;

Ridley *et al.*, 1999]. A related limit that receives less attention but may become even more economically significant is the lowest cost per packaged part, which unlike most other specifications has remained relatively constant over the VLSI scaling era at around 10 cents. An alternative approach to bring this down below a cent for applications such as electronic tagging of commodity objects is to remotely interrogate natural materials [Fletcher *et al.*, 1997].

One of the reasons chip fabrication is so expensive is that as the size of chips grows while their minimum feature size shrinks, the impact of a single speck of dust is magnified. A very small defect can doom an entire part. The conventional response has been to use ultrapure materials in ultraclean rooms, but even so the yields of state-of-the-art chips are usually so bad that they are considered a sensitive trade secret (on the order of a few percent). A radically different approach is to design machines with the expectation that most of their parts will be faulty [Heath *et al.*, 1998]. Components can be hierarchically packaged in modules of greater and greater complexity, which can then be adaptively rewired based on self-testing. There is some empirical basis for this kind of partitioning, through *Rent's rule*, the observation that many engineered (and biological) systems have a power-law scaling relationship between the number of connections to a subsystem and the number of functional units in that subsystem, with an exponent typically between 1/2 and 1 [Landman & Russo, 1971; Vilkelis, 1982].

Thermodynamics presents profound limitations that are also of great short-term significance [Gershenfeld, 1996]. 10 W laptops run out of power before airplane trips end, the 100 W desktop computers in a building taken together can consume more power than air conditioning systems use or can handle, and it's a challenge to keep a 100 kW supercomputer from melting through the floor. As we saw in Section 4.5, the roots of information theory grew out of the study of the efficiency of steam engines, and are now returning to help optimize the thermodynamic performance of computing machines [Leff & Rex, 1990]. Rolf Landauer resolved a long-standing puzzle by showing that erasure is where computation necessarily incurs a thermodynamic cost, because the heat associated with the change in entropy that follows from resetting an unknown bit to a known state is $Q = T dS = kT \log 2$ [Landauer, 1961]. Charles Bennett went further to unexpectedly demonstrate that universal computation is possible without any erasure by reversibly rearranging inputs and outputs [Bennett, 1973]. Since $kT \log 2 \approx 10^{-21}$ J, it was originally thought that these limits were remote. More recently, it's been appreciated that the design guidance they provide is applicable at much higher energy scales. *Reversible logic* seeks to recover rather than dissipate the energy associated with bits being erased [Merkle, 1993; Younis & Knight, 1993], and *adiabatic logic* makes changes no faster than they are needed (Problem 10.5) [Athas *et al.*, 1994; Dickinson & Denker, 1995]. Both principles have been used in fabricating circuits that show promising reductions in power consumption.

Even more fundamental are limits associated with physical constants. One is the speed of light. Synchronous logic requires distributing the clock over an entire chip each cycle; aside from the charging energy this entails, it also limits the cycle time to the chip size divided by the speed of light. One response is to eliminate clock delivery by using *asynchronous logic*, in which gates assert their output when they receive valid inputs rather than a global clock signal [Birtwistle & Davis, 1995]. This can also be beneficial for reducing dissipation and wiring complexity, although the ultimate limit on clock speed comes from the quantum energy–time uncertainty relationship, which argues for using the maximum available energy in the minimum possible number of gates in order to minimize the communication time [Lloyd, 2000].

Another is the size of atoms. As feature sizes drop below 0.1 μm, continuum approximations can no longer be made. This shows up in *electromigration*, the motion of individual atoms due to the momentum transported by the electronic current, which leads to wiring failures that must be prevented through careful attention to the metallurgy and current density. The discreteness of current is turned from a bug into a feature in a *Single-Electron Transistor* (*SET*) [Likharev & Claeson, 1992; Grabert & Devoret, 1992]. An electron can tunnel across an insulator onto a conducting island only if states are available to it on both sides, creating a periodic modulation called the *Coulomb blockade* in the charging current due to the integer number of electrons allowed on the island. Among other applications, this can be used to create a memory cell that stores a single electron [Durrani *et al.*, 1999].

Once devices reach these limits, further advances are possible only by finding new degrees of freedom to represent and manipulate information. One option is to recognize that analog nonlinear systems can be used for far more than binary logic; some examples will be seen in Chapter 13. Another is to retain the notion of bits, but use quantum mechanics to describe their logical as well as physical states. The remarkable implications of this will be explored in Chapter 15.

10.6 SELECTED REFERENCES

[Ashcroft & Mermin, 1976] Ashcroft, N., & Mermin, N.D. (1976). *Solid State Physics*. New York: Holt, Rinehart and Winston.

A very readable introduction to solid state physics. The index deserves special attention.

[Sze, 1981] Sze, S.M. (1981). *Physics of Semiconductor Devices*. 2nd edn. New York: Wiley-Interscience.

The definitive device physics text.

[Streetman, 1990] Streetman, B. (1990). *Solid State Electronic Devices*. Englewood Cliffs: Prentice-Hall.

This is a more accessible introduction to device physics.

10.7 PROBLEMS

(10.1) (*a*) Derive equation (10.28) by taking the integral and limit of equation (10.27).

   (*b*) Show that equation (10.2) follows.

(10.2) What is the expected occupancy of a state at the conduction band edge for Ge, Si, and diamond at room temperature (300 K)?

(10.3) Consider Si doped with 10^{17} As atoms/cm^3.

   (*a*) What is the equilibrium hole concentration at 300 K?

   (*b*) How much does this move E_F relative to its intrinsic value?

(10.4) Design a *tristate* CMOS inverter by adding a control input to a conventional inverter that can force the output to a high impedance (disconnected) state. These are useful for allowing multiple gates to share a single wire.

(10.5) Let the output of a logic circuit be connected by a wire of resistance R to a load of capacitance C (i.e., the gate of the next FET). The load capacitor is initially discharged, then when the gate is turned on it is charged up to the supply voltage V. Assume that the output is turned on instantly, and take the supply voltage to be 5 V and the gate capacitance to be 10 fF.

(a) How much energy is stored in the capacitor?

(b) How much energy was dissipated in the wire?

(c) Approximately how much energy is dissipated in the wire if the supply voltage is linearly ramped from 0 to 5 V during a long time τ?

(d) How often must the capacitor be charged and discharged for it to draw 1 W from the power supply?

(e) If a large IC has 10^6 transistors, each dissipating this charging energy once every cycle of a 100 MHz clock, how much power would be consumed in this worst-case estimate?

(f) How many electrons are stored in the capacitor?

11 Generating, Detecting, and Modulating Light

In this chapter we will see the light associated with the intersection between the electronic and optical properties of materials, called *optoelectronics*, or, with a bit less electronics and a bit more religion, *photonics*.

We'll start by looking at how electronic energy can be converted into illumination (both visual and intellectual), then how light can be converted back to electronic energy and information, and close with methods for modifying it.

11.1 GENERATION

11.1.1 Incandescence

The simplest way to generate light is by heating something to produce *incandescence*. To find the states available to a thermal photon in a box of side length L, following the derivation of the electron density of states (from equation 10.36) we'll impose periodic boundary conditions on the radiation field so that the wave vector $\vec{k}$ is indexed by integers n_x, n_y, n_z

$$\vec{k} = \frac{2\pi}{L} \left(n_x \hat{x} + n_y \hat{y} + n_z \hat{z} \right)$$
$$k^2 = \left(\frac{2\pi}{L} \right)^2 r^2 \quad , \tag{11.1}$$

which in the limit of large box can be taken to define a continuous variable r. In terms of the frequency,

$$\frac{2\pi}{c} \nu = k = \frac{2\pi}{L} r \tag{11.2}$$

or

$$r = \frac{L}{c} \nu \quad . \tag{11.3}$$

The total number of states in a volume $V = L^3$ that have indices between r and $r + dr$ is then given in terms of the density of states per volume N by a spherical shell

$$V N(r) \, dr = 2 \cdot 4\pi r^2 \, dr$$
$$= 8\pi r^2 \, dr$$

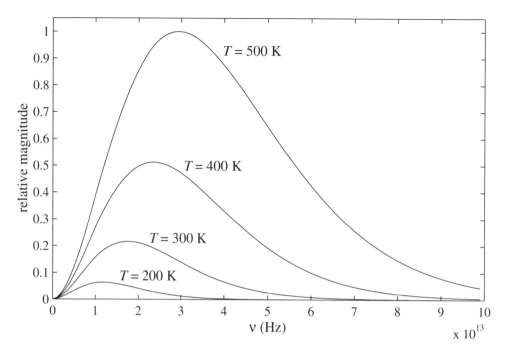

Figure 11.1. Planck's Law for thermal radiation.

$$= 8\pi \left(\frac{L}{c}\right)^2 \nu^2 \frac{L}{c} \, d\nu$$

$$= 8\pi \frac{V}{c^3} \nu^2 \, d\nu$$

$$N(\nu) \, d\nu = \frac{8\pi}{c^3} \nu^2 \, d\nu \quad . \tag{11.4}$$

The first factor of 2 comes from the two possible transverse photon polarizations (which will be discussed further in Section 11.3). The thermal photon energy per volume U in the box is then found by multiplying this density of states by the photon energy $h\nu$ and the Bose–Einstein distribution for the photon occupancy of the available sates,

$$U = h\nu \frac{8\pi}{c^3} \nu^2 \frac{1}{e^{h\nu/kT} - 1}$$

$$= \frac{8\pi h\nu^3}{c^3 \left(e^{h\nu/kT} - 1\right)} \quad . \tag{11.5}$$

This is *Planck's Law*, plotted in Figure 11.1.

The total energy per volume is found by integrating over the spectrum,

$$\int_0^\infty U(\nu) \, d\nu = \frac{8\pi h}{c^3} \int_0^\infty \frac{\nu^3}{e^{h\nu/kT} - 1} \, d\nu$$

$$= \frac{8\pi h}{c^3} \frac{1}{15} \left(\frac{kT\pi}{h}\right)^4 \quad . \tag{11.6}$$

This is done analytically in terms of the *Riemann zeta function*, arguably the most interesting integral in all of mathematics [Hardy & Wright, 1998].

Planck's Law applies to photons in thermal equilibrium, for example in a closed cavity with walls at a temperature T. If we open a hole in the cavity we'll disturb the distribution, but as long as the hole is not too large it can be used to sample the radiation. This idealization is called a *black-body* radiator, because any light entering the hole will have little chance of scattering out so it is an almost-ideal absorber and emitter. The total power per area R radiated from the hole is found by multiplying equation (11.6) by the speed of light c (to convert energy per volume to energy per time per area), dividing by 2 (because half the photons are headed towards the hole, and half away), and dividing by another factor of 2 (because the effective area of the opening must be scaled by the dot product of the surface normal with the uniformly-distributed photon orientations $\int_{-\pi/2}^{\pi/2} \cos\theta \, d\theta = 2$), giving

$$
\begin{aligned}
R &= \frac{c}{4} \int_0^\infty U(\nu) \, d\nu \\
&= \frac{\pi^2 k^4}{60 \hbar^3 c^2} T^4 \\
&\equiv \sigma T^4 \\
&= 5.67 \times 10^{-8} \, T^4 \, \frac{\text{W}}{\text{m}^2} \quad .
\end{aligned}
\tag{11.7}
$$

This is the *Stefan–Boltzmann* Law. For real surfaces it must be corrected for the *emissivity* deviating from the black-body idealization of unit efficiency, but for a wide range of materials it is a good approximation. The presence of Planck's constant h in the formula indicates its quantum origin. The inability to derive the correct form for thermal radiation during the latter part of the 19th century, at a time when physics was widely viewed as having been completed as a theory, was an irritation that lead to a revolution with the development of quantum mechanics by Einstein and others.

11.1.2 Luminescence: LEDs, Lasers, and Flat Panels

Light produced by quantum transitions rather than thermal means is called *luminescence*. If the excitation mechanism is an electrical current or voltage it is called *electroluminescence*, if the photons are produced by electron bombardment it is called *cathodoluminescence*, and if the excitation is by photons it is called *photoluminescence*. When the decay time is fast (on the order of the nanosecond time scales for direct electron–hole recombination) the radiation is called *fluorescence*, and if the decay is slow (seconds, minutes, even hours) it is called *phosphorescence* and the material is called a *phosphor* [McKittrick *et al.*, 1999].

We'll primarily be concerned here with the electroluminesence of semiconductor devices. The most important example of cathodoluminescence is the familiar *Cathode Ray Tube* (*CRT*), in which electrons emitted from a heated cathode are accelerated by an anode to strike a phosphor. A typical phosphor is ZnS doped with Cu as an *activator*; excited conduction electrons make a 530 nm (green) transition at a Cu ion. An increasingly significant application of photoluminescence is in *optical repeaters* for long-haul fiber links. These use silica fibers doped with erbium ions, which can be pumped by 980

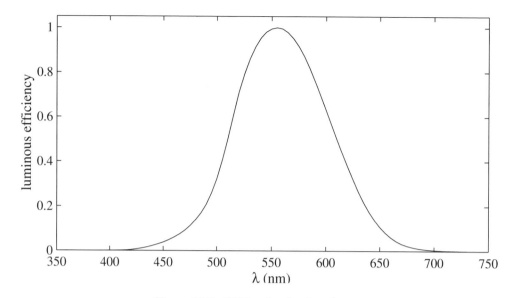

Figure 11.2. CIE luminosity function.

or 1480 nm light from a local laser into a metastable state with a transition that matches the 1.5 μm absorption minimum used for communications. When a signal photon arrives it stimulates emission from this excited state, providing gain in an all-optical system. This eliminates the losses, noise, and complexity of detecting the light, amplifying it electronically, and then generating it again [Delavaux & Nagel, 1995].

The apparent brightness of any light source is measured in *lumens*. According to the definition given in Chapter 2, one watt of 555 nm (green) light corrsponds to 683 lumens (lm). Unfortunately, for any other kind of light the meaning is less straightforward. In 1924 the *Commission Internationale de l'Eclairage* (*CIE*) used early perceptual studies to define a standard *luminosity function*, shown in Figure 11.2, that gives the relative sensitivity of the eye to wavelengths away from the 555 nm peak. The spectrum of a broadband light source must be weighted by this curve to determine its value in lumens. Even worse, later studies have shown that this curve underestimates the eye's sensitivity to short wavelengths, so lumen measurements are sometimes reported with more modern weightings. This is analogous to the ambiguity possible in specifying the reference level used for a decibel measurement, but now a function must be given.

Because of the eye's non-ideal response, an ideal source of white light over the visible range would produce about 200 lm per watt of power. A typical 75 W light bulb produces 1200 lm, giving 10–20 lm/W for incandescent sources. Fluorescent bulbs raise this to 50–100 lm/W by replacing electronic heating with electronic excitation of a mercury plamsa which releases ultraviolet photons that pump a phosphor coating. The most efficient lamps eliminate the down-conversion loss by directly using atomic transitions in a sodium vapor, approaching 200 lm/W [Hollister, 1987].

Semiconductors can display bulk electroluminescent effects through a range of mechanisms including carrier impact scattering, field emission around defects, and nanoscale quantum confinement in *porous silicon* [Fauchet, 1998]. More efficient, predictable, and

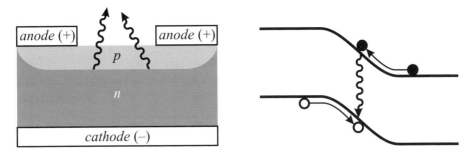

Figure 11.3. A Light-Emitting Diode.

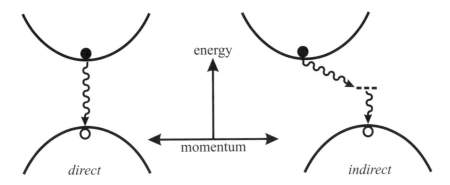

Figure 11.4. Semiconductor band gaps.

controllable is the *injection electroluminescence* associated with electrons and holes re-combining at a *p–n* junction. In a *Light-Emitting Diode (LED)*, shown in Figure 11.3, a photon is produced when a conduction electron falls into a valence hole. The junction is forward biased to drive excess carriers into the junction region, which unlike a conventional diode is wide and shallow to enhance the production and emission of light.

Some semiconductors, such as GaAs, have *direct band gaps* in which the conduction band minimum has the same crystal momentum as the valence band maximum. In others such as Si these are displaced, as shown in Figure 11.4. These *indirect band gap* materials make very inefficient emitters of light, because momentum conservation requires a phonon for electron–hole recombination. This reduces the recombination probability as well as releases energy to the lattice and broadens the radiation linewidth. For this reason, optoelectronics almost exclusively uses direct band gap semiconductors.

In $GaAs_{1-x}P_x$, as x is varied from 0 to 0.45, the band gap changes from 1.4 eV (IR) to 2 eV (red). This lets the color of an LED be selected by the composition. At higher concentrations the gap becomes indirect, but nitrogen impurity doping is used to introduce gap states that let concentrations up to $x = 1$ be used (2.2 eV, green). GaN has a larger direct band gap of 3.4 eV, making blue and even UV LEDs possible [Mukai *et al.*, 1999]. Through improvements in band-structure engineering and light collection, LEDs have become competitive beyond information displays as direct sources of illumination; AlInGaP LEDs have been produced with outputs over 10 lm at efficiencies over 20 lm/W [Fletcher *et al.*, 1993]. Less efficient but more versatile are *Organic Light-Emitting*

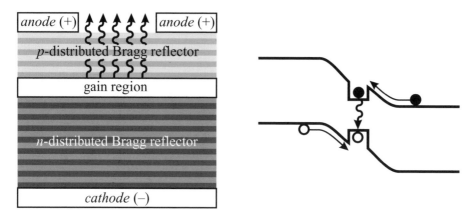

Figure 11.5. A Vertical-Cavity Surface-Emitting Laser.

Diodes (*OLEDs*), which can be produced as thin, flexible devices [Sheats *et al.*, 1996; Friend *et al.*, 1999].

A typical linewidth for an LED is 100 Å, which is too broad for fiber-optic links that rely on low dispersion and on wavelength-division multiplexing, and the lack of control over where the photons are emitted prevents their use in the diffraction-limited optics needed for storage applications. These are among the many reasons for the growing importance of *diode lasers*, which can provide linewidths below 1 Å with fundamental beam mode-shapes.

Lasers rely on *stimulated emission* [Corney, 1978]. If a system such as an atom or a electron–hole pair that can make a radiative transition is pumped into its excited state, then an incident photon at that frequency can stimulate a transition to the ground state with the emission of a photon, in the inverse process to the absorption of a photon driving a transition to the excited state. And because this is a resonant effect, the emitted photon matches the phase of the incident one. The result is two phase-coherent photons instead of one; if they can be kept around long enough there will be net optical gain with the mode shape determined by the mirrors defining the optical cavity.

Lasing requires maintaining a *population inversion* of excited states, and needs high-reflectivity mirrors that let photons pass through the gain medium many times. An elegant semiconductor solution is the *Vertical-Cavity Surface-Emitting Laser* (*VCSEL*), shown in Figure 11.5 [Lott *et al.*, 1993; Someya *et al.*, 1999].

This is still a p–n diode just like an LED, but the junction is now sandwiched between two mirrors. These are *Distributed Bragg Reflectors* (*DBRs*), periodic quarter-wavelength dielectric layers that scatter coherently at their interfaces, the opposite of an antireflection coating (Problem 8.6). The index of refraction can be controlled by varying the composition of $Al_xGa_{1-x}As$. Not only does this let the mirror also serve as a part of the semiconductor junction that can still be doped, it avoids the losses due to the conductivity of a metal mirror. The lower mirror reflectivity can be better than 99%; the upper one is intentionally slightly lower to couple some light out. The heart of the junction itself consists of undoped GaAs layers, called *quantum wells*. Because these have a smaller band gap, the carriers being injected across the junction by the forward bias

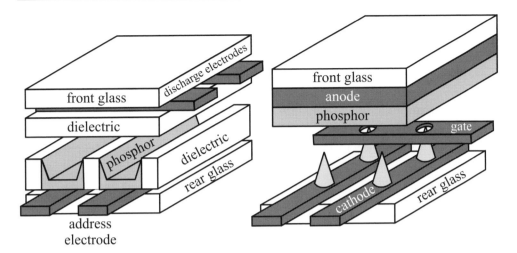

Figure 11.6. A Plasma Display Panel (left) and a Field Emission Display (right).

drop into these thin high–mobility layers where they can easily recombine. Since there is more than one interface, this is called a *heterojunction* device or a *heterostructure*.

The thickness of the gain region is chosen to support a single longitudinal mode, and the transverse mode structure is determined by the lateral shape. Because of the laser's gain, all of the light that emerges comes from this same cavity mode, providing the desired narrow linewidth and beam shape. VCSELs makes such good use of the carriers that, unlike early semiconductor lasers, just a few volts and mA are adequate for lasing at room temperature. The conversion efficiency from electricity to light can be over 50%. Because the light emerges from the top of the laser, these are easily integrated with other devices on a chip, and by coupling them into arrays it is possible to generate watts of output power.

Multiple light-emitting elements can be combined to form an *emissive display* (in Section 11.3 we'll look at alternatives that modulate rather than generate light). While many diodes can be fabricated on a single substrate, making a display this way would entail covering large areas with expensive silicon wafers. The challenge in developing displays is to find scalable technologies that can match the performance of the eye, which can resolve a spatial frequency of 60 cycles per degree [Campbell & Green, 1966], and compete with ambient light ranging from $100 \, cd/m^2$ indoors to $10\,000 \, cd/m^2$ in sunlight. Two approaches that build on familiar light sources have received particular attention (Figure 11.6): *Plasma Display Panels* (*PDPs*) and *Field Emission Displays* (*FEDs*).

A PDP comprises an enormous number of tiny fluorescent lights [Bitzer, 1999]. Dielectric channels are coated with a phosphor and filled with a combination of inert gases such as He, Ne, and Xe. In columns above the channels are pairs of clear electrodes, encapsulated in a protective dielectric layer. An AC voltage is applied to them that is just below the breakdown voltage needed to ionize the gas. Below the channels are row electrodes that are used to turn the discharge on and off, creating plasmas that excite the phosphor with ultraviolet light. Because it's not possible to exercise fine control over the UV intensity, PDPs cycle the plasma many times in microsecond pulses to vary the brightness. Even though the physical structure is (relatively) straightforward, this requires complex control strategies to prime the discharge because of the time scales associated with the surface charge in the channel and the ionization state of the gas [Rauf & Kushner, 1999]. PDPs are attractive because they can be fabricated over large areas, and can match the brightness of fluorescent lighting because they *are* fluorescent lighting, although their efficiency drops to lumens per watt because of the losses associated with the small channel size.

A more efficient alternative is an FED, which can be thought of as a huge number of tiny cathode ray tubes [Ghrayeb *et al.*, 1997]. Although CRTs are a mature technology, they are inefficient because of the thermionc electron emission, and bulky because of the electron deflection optics. An FED replaces the single cathode with a huge number of tiny sharp metal tips. An electric field is applied by column gate electrodes above the rows of tips. Because the field gradient is significantly increased around the tip, the local potential difference produced by just a few volts can exceed the work function of the tip, causing it to emit electrons by *field emission* [Phillips *et al.*, 1998]. These are then accelerated towards a phosphor by an anode, much like a CRT but now each pixel has its own emitters. Although FEDs are more difficult to manufacture than PDPs, the electronic control is straightforward, and the efficiency is increased over 10 lm/W because there are no thermal or down-conversion losses.

11.2 DETECTION

The fundamental processes that generate photons from electrons can be run in reverse, to convert photons to electrical signals. Simplest of all is a *photoconductor*. This mechanism was discovered in the early days of studying semiconductors, when a curious oscillation developed in the conductivity of a sample that was finally explained by the shadow of a ceiling fan rotating above it. Photons can excite carriers across the conduction band, reducing the resistance of the material. The energy of the photons of interest must be larger than the gap energy; for visible light common photoconductors include CdS (2.4 eV, 0.52 μm) and CdSe (1.8 eV, 0.69 μm). For longer wavelengths lower-energy excitations must used; HgCdTe with a gap around 0.12 eV is used for detection of infrared light below 10 μm. Beyond that, it's possible to introduce gap states with dopants that provide lower-energy excitations. Hg in Ge is an acceptor that sites sit 0.09 eV (14 μm) above the valence band, and Cu acceptors in Ge are 0.04 eV (32 μm) above it. Valence band electrons that are excited into these acceptors leave holes behind that increase the conductivity. At these low energies, the number of carriers produced by thermal excitation becomes significant compared to those excited by weak optical signals, requiring that the detectors be cooled. This can be done with a Peltier cooler (Chapter 14), liquid nitrogen (77 K), or liquid helium (4.2 K).

The dominant noise mechanism in uncooled photoconductors is their $4kTR$ Johnson noise, which is intrinsic to this kind of detector because it works by measuring a resistance. A quieter alternative is to use an LED in reverse. An incoming photon can excite an electron–hole pair in the depletion region, which will then be swept apart by the junction field and measured as a current. Because carrier diffusion is a slow process, a faster variant is the *p–i–n photodiode*, shown in Figure 11.7. Thin *p* and *n* layers are sandwiched around a thicker insulating layer, expanding the depletion region to fill most of the diode. Now

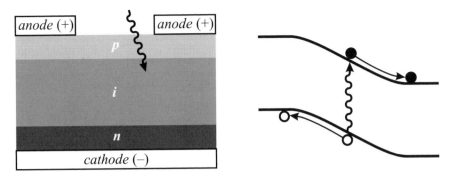

Figure 11.7. A p–i–n photodiode.

photocarriers are accelerated out by the junction field, and just have to diffuse through the thin doped layers. These devices can have response times approaching picoseconds.

If the power of the illumination of a photodiode is P, and it produces a current I, then the *quantum efficiency* η is defined to be

$$\eta = \frac{I/e}{P/h\nu} \quad . \tag{11.8}$$

The numerator divides the current by the charge to find the rate of electron–hole pair production, and the denominator divides the power by the photon energy to give the rate of photon arrival. The ratio is the probability that a photon will contribute to the current. The quantum efficiency falls off at long wavelengths because the photon energy is smaller than the gap energy, and at short wavelengths because the photon is absorbed before reaching the depletion region, producing an electron–hole pair that can recombine. At its peak, the quantum efficiency can approach unity; for Si this happens around 0.8 μm.

Photodiodes are ultimately limited by the shot noise of the photocurrent due to fluctuations in the rate of photon arrival. On top of this, there is noise from the *dark current* due to thermal carrier excitation, and Johnson noise associated with the load. While the shot noise limit is fundamental, the other two are instrumental and can be eliminated by *heterodyne* detection. This clever trick is the optical analog of electrical measurement techniques that will be covered in Chapter 13. The quantum transition probability to excite a transition is proportional to the intensity of the radiation, and hence the square of the electric field strength [Corney, 1978]. Heterodyne detection works by exploiting this nonlinearity. The idea is to add to the optical signal of interest, taken to have an electric field at the detector of $E_S e^{i\omega t}$, a much stronger local optical field $E_L e^{i[(\omega+\delta)t+\varphi]}$. The frequency shift δ is an intentional detuning, and φ is their relative phase. The total intensity is then

$$\begin{aligned}
|E|^2 &= |EE^*| \\
&= E_S^2 + E_L^2 + E_S E_L \left(e^{i[\omega t - (\omega+\delta)t - \varphi]} + e^{i[-\omega t + (\omega+\delta)t + \varphi]} \right) \\
&= E_S^2 + E_L^2 + 2E_S E_L \cos(\delta t + \varphi) \\
&\approx E_L^2 \left(1 + 2\sqrt{\frac{E_S}{E_L}} \cos(\delta t + \varphi) \right)
\end{aligned} \tag{11.9}$$

in the limit $E_L \gg E_S$. The magic happens in the product, where the fluctuations in the optical signal are scaled up by the much stronger local oscillator field, bringing them above the level of the photodiode's dark current and Johnson noise. The detected current will then just be proportional to equation (11.9), with the coefficient found from equation (11.8), so that in terms of the signal and local-oscillator powers

$$I = \frac{P_L \eta e}{h\nu} \left[1 + 2\sqrt{\frac{P_S}{P_L}} \cos(\delta t + \varphi) \right] \quad . \tag{11.10}$$

The magnitude of the current signal is then

$$S = \left\langle (I - \langle I \rangle)^2 \right\rangle$$

$$= \left(\frac{P_L \eta e}{h\nu} \right)^2 2 \frac{P_S}{P_L} \quad . \tag{11.11}$$

Because the detected current is shot-noise-limited (equation 3.33), the current noise magnitude is

$$N = 2e \langle I \rangle \Delta f$$

$$= 2e \frac{P_L \eta e}{h\nu} \Delta f \quad , \tag{11.12}$$

where Δf is the measurement bandwidth. This gives a quantum-limited SNR of

$$\frac{S}{N} = \frac{P_S \eta}{h\nu \Delta f} \quad . \tag{11.13}$$

For an SNR of 1, the photon arrival frequency equals the bandwidth of the detector.

Heterodyne detection does require a local light source matched to the signal. An alternative mechanism to improve sensitivity is used in an *Avalanche PhotoDiode (APD)*. When a *p–n* junction is illuminated, its *I–V* curve is shifted down by the photocurrent. Photodiodes are usually operated reverse-biased, where this current depends on the light intensity but is independent of the bias voltage. As the reverse bias is increased, avalanche breakdown is reached. Just short of that, a photocarrier can get enough energy from the junction field to excite another carrier by *impact ionization,* leading to a cascade that produces many electrons from one photon. This can lead to a gain of 100 or more in the current, although this does come at the expense of a slower response (because of the collisions) and more noise (because the thermal dark current also gets amplified).

An interesting thing happens as the bias becomes positive in Figure 11.8: the *IV* product changes sign. This means that the diode becomes a net exporter rather than importer of energy, generating power in a *photovoltaic* or *solar cell* [Chapin *et al.*, 1954]. For best efficiency, the load must be chosen to maximize the *IV* product in that quadrant. Then the efficiency of the solar cell is limited by the energy lost from missing those photons with an energy below the gap, and from thermalizing carriers produced by photons with an energy above the gap. For a single junction under solar illumination that results in a maximum efficiency of about 30%; strategies for raising the efficiency include stacking multiple junctions ranging from highest to lowest band gap, and using concentrators to collect photons from a larger area.

The converse to creating a display out of an array of LEDs is to use an array of photodetectors to record an image. The challenge is to integrate as many detectors as possible,

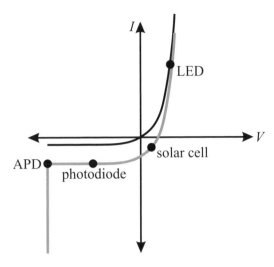

Figure 11.8. Biasing for a p–n junction.

while still managing to extract the signals from each device with acceptable fidelity. We've already seen one solution to this problem: a DRAM memory (Figure 10.16). A *CMOS imager* [Denyer *et al.*, 1995] adds a photodetector to each cell, most simply just a MOS capacitor (Figure 10.9). The charge produced by the photodetector is read out just as the charge on a DRAM cell capacitor is. But now, rather than trying to maximize the capacitance, it's important to maximize the collection area of the photodetector. While this cell design, called a *Passive Pixel Sensor* (*PPS*), does do that by having just one transistor per cell, it suffers from noise and delay associated with charging a long read line with a small capacitor. The analog to SRAM is an *Active Pixel Sensor* (*APS*), which adds one or more transistors for *transimpedance* conversion of current to voltage with gain (Chapter 13). This comes at the expense of losing collection area for the photodetector, but that can be ameliorated by fabricating microlenses above the pixels. The remaining bane of CMOS imagers is *Fixed Pattern Noise* (*FPN*), the systematic image errors that come from pixel-to-pixel sensitivity variations and cross-talk in the readout lines. This is dealt with by schemes for differential readout and background subtraction.

Charge-Coupled Devices (*CCDs*) take advantage of the ability to manipulate the surface band structure to move charge out directly [Boyle & Smith, 1971]. The interfacial band-bending in an MOS capacitor is used in a MOSFET to introduce carriers into the conduction band, but it also forms a potential well that can store carriers that arrive by other means (Figure 10.9). A CCD pixel accumulates photo-induced charge in this well. But instead of reading it out through a wire, the pixels are connected as shown in Figure 11.9. The depth of the well is a function of the gate voltage, which at the beginning of a cycle is set to retain charge below every third electrode. Then, the potential on that line is dropped while it is raised on the neighboring cells, creating a single larger well that fills up with the charge. At the end of the cycle the second well is lowered while the first one is rasied, shifting the charge over by one pixel. By repeating this cyclic pattern, the charge on each pixel is sequentially shifted out of the end of the row. The long scan lines are now actively driven from the periphery of the chip, with each pixel needing only to

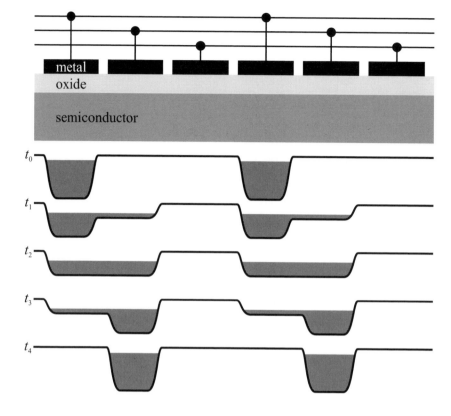

Figure 11.9. The operation of a CCD.

transfer charge to its neighbor. In practical CCDs there is an extra n-type layer on top of the p-type substrate, which forms a *buried channel* that moves the potential well down from the surface and its higher density of defect states. This CCD structure also finds application in analog memories and delay lines.

For low-light applications, CCDs are cooled to reduce the dark-current accumulation of thermally-induced charge, and they can be thinned to bring light in from the rear to completely avoid the losses associated with the wires and electrodes. With these enhancements, the quantum efficiency can approach 100% at matched wavelengths, and the noise introduced by the readout can be on the order of a single electron. The sensitivity is then determined solely by the collection time, permitting dimmer images to be recorded by accumulating charge for a longer time before scanning it out. Room-temperature CCDs read out at video rates cannot reach this sensitivity, but can still have readout noise of a few tens of electrons per pixel. A dominant contribution to this is the *reset noise* associated with resetting the readout circuit, which can be reduced by reading it out twice to perform a background subtraction.

Compared to CMOS imagers, CCDs offer good pixel density and noise performance, but they require higher power because of the charging currents associated with driving the readout cycle, and the device optimizations are not compatible with conventional CMOS design, requiring specialized fabs to make them and supporting chips to interface to them. CMOS imagers offer random pixel access and on-chip integration of related

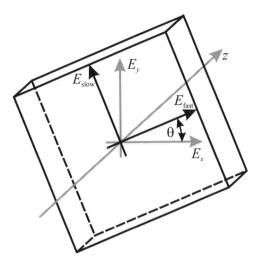

Figure 11.10. Axes in a birefringent crystal.

functions, historically with lower power, cost, and performance. But as this technology matures the performance difference is disappearing.

11.3 MODULATION

We've examined how to generate and detect light; this final section will look at passive and active mechanisms to modulate it.

11.3.1 Polarization

The eye cannot see polarization directly, but in many applications it does see the consequences of manipulating polarization states. This can be done with a *birefringent* material, one with an anisotropic ordering that causes its polarizibility ϵ and hence index of refraction n to depend on the orientation. The simplest of these are *uniaxial* materials that have a single optical axis with orthogonal fast and slow directions. For *calcite* (CaCO$_3$), these two refractive indices are $n_{slow} = 1.658$ and $n_{fast} = 1.486$, giving a birefringence difference of 0.172. *Quartz* (SiO$_2$) is less birefringent, with a difference of 0.009.

Just as ray matrices simplified the description of a series of optical elements in Section 8.2.1, the *Jones calculus* does the same for polarizing materials. If the transverse electric field components of a wave are $E_x e^{i\omega t}\hat{x} + E_y e^{i\omega t}\hat{y}$, we'll write the complex coefficients as a two-component vector (Figure 11.10)

$$\vec{E} = \left(\begin{array}{c} E_x \\ E_y \end{array} \right) \quad . \tag{11.14}$$

If E_y/E_x is pure real the wave is said to be *linearly polarized* because the components move back and forth in phase; if $E_y/E_x = i$ or a multiple of it, the wave is *circularly polarized* because the vector rotates around a circle; and between these cases it is *elliptically polarized*.

The field components relative to the axes of a birefringent material are found with a rotation matrix

$$
\begin{pmatrix} E_{\text{slow}} \\ E_{\text{fast}} \end{pmatrix} = \underbrace{\begin{pmatrix} \cos\theta & \sin\theta \\ -\sin\theta & \cos\theta \end{pmatrix}}_{\equiv \mathbf{R}(\theta)} \begin{pmatrix} E_x \\ E_y \end{pmatrix} \quad . \tag{11.15}
$$

Since the wave vector is $k = 2\pi/\lambda = n\omega/c$, the field components after propagating through a thickness of d will pick up a phase shift of e^{ikd} along each axis

$$
\begin{pmatrix} E_{\text{slow}} \\ E_{\text{fast}} \end{pmatrix}' = \begin{pmatrix} e^{-in_{\text{slow}}\omega d/c} & 0 \\ 0 & e^{-in_{\text{fast}}\omega d/c} \end{pmatrix} \begin{pmatrix} E_{\text{slow}} \\ E_{\text{fast}} \end{pmatrix} \quad . \tag{11.16}
$$

This can be written more symmetrically in terms of the sum

$$
\sigma = (n_{\text{slow}} + n_{\text{fast}})\frac{\omega d}{2c} \tag{11.17}
$$

and the difference

$$
\delta = (n_{\text{slow}} - n_{\text{fast}})\frac{\omega d}{2c} \tag{11.18}
$$

as

$$
\begin{pmatrix} E_{\text{slow}} \\ E_{\text{fast}} \end{pmatrix}' = e^{-i\sigma} \underbrace{\begin{pmatrix} e^{-i\delta} & 0 \\ 0 & e^{i\delta} \end{pmatrix}}_{\equiv \mathbf{B}(d)} \begin{pmatrix} E_{\text{slow}} \\ E_{\text{fast}} \end{pmatrix} \quad . \tag{11.19}
$$

The phase prefactor $e^{-i\sigma}$ can be left out unless the light will later be recombined with a reference beam. For a wave polarized along the laboratory axes, the change after passing through a birefringent material is found by rotating to the optical axes, applying the birefringence matrix, and then rotating back:

$$
\begin{pmatrix} E_x \\ E_y \end{pmatrix}' = \mathbf{R}(-\theta)\mathbf{B}(d)\mathbf{R}(\theta) \begin{pmatrix} E_x \\ E_y \end{pmatrix} \quad . \tag{11.20}
$$

A *dichroic* material has absorption coefficients that depend on polarization; there are both linearly- and circularly-polarized dichroics. If a linear dichroic material completely absorbs one component while passing the other, it is a *linear polarizer* with a Jones matrix

$$
\mathbf{L} = \begin{pmatrix} 1 & 0 \\ 0 & 0 \end{pmatrix} \tag{11.21}
$$

(ignoring the phase prefactor). Edwin Land developed synthetic polarizers using *hera-pathite*, which forms dichroic crystals that were discovered through the rather unusual laboratory accident of dropping iodine into the urine of dogs fed quinine [Land, 1951]. More stable polarizers are made from stretched sheets of *PolyVinyl Alcohol* (*PVA*) reacted with iodine.

In some magnetic materials, left- and right-circularly-polarized waves travel at different

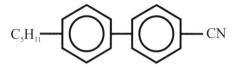

Figure 11.11. The liquid crystal 5CB.

speeds. A circularly-polarized basis is related to the linear one by

$$\begin{pmatrix} E_+ \\ E_- \end{pmatrix} = \underbrace{\frac{1}{2} \begin{pmatrix} 1 & i \\ 1 & -i \end{pmatrix}}_{\mathbf{C}} \begin{pmatrix} E_x \\ E_y \end{pmatrix} \quad . \tag{11.22}$$

Analogous to birefringence, after passing through a magnetic material these components become

$$\begin{pmatrix} E_+ \\ E_- \end{pmatrix}' = \begin{pmatrix} e^{i\theta_F} & 0 \\ 0 & e^{-i\theta_F} \end{pmatrix} \begin{pmatrix} E_+ \\ E_- \end{pmatrix} \quad , \tag{11.23}$$

where $\theta_F = (n_- - n_+)\omega d/2c$ is the *Faraday rotation* angle. In ferrite materials, $\theta_F = VBd$, where d is the thickness, B is an applied DC magnetic field, and V is the *Verdet constant*. For *Yttrium Iron Garnet (YIG)*, $Y_3Fe_5O_{12}$, $V = 0.1\ °/G\cdot cm$

In the linear basis, Faraday rotation is

$$\begin{pmatrix} E_x \\ E_y \end{pmatrix}' = \mathbf{C}^{-1} \begin{pmatrix} e^{i\theta_F} & 0 \\ 0 & e^{-i\theta_F} \end{pmatrix} \mathbf{C} \begin{pmatrix} E_x \\ E_y \end{pmatrix}$$

$$= \begin{pmatrix} \cos\theta_F & \sin\theta_F \\ -\sin\theta_F & \cos\theta_F \end{pmatrix} \begin{pmatrix} E_x \\ E_y \end{pmatrix}$$

$$= \mathbf{R}(\theta_F) \begin{pmatrix} E_x \\ E_y \end{pmatrix} \quad . \tag{11.24}$$

This is simply a rotation by the Faraday angle (hence the name). A magnetic material that rotates polarization by 45° and that is between linear polarizers rotated relative to each other by that angle is called a *Faraday isolator*: linearly polarized light can pass in one direction but not the other. This violation of reversibility is possible because magnetic interactions change sign under time reversal. Faraday isolators are used for preventing light from coupling back into lasers, which is important for their mode structure and stability, and for selecting a lasing direction in a symmetrical ring laser.

11.3.2 Liquid Crystals

The most visible application of polarization is in *Liquid Crystal Displays* displays (*LCDs*). Liquid crystals are fluids with order intermediate between the long-range periodicity of crystals and the short-range correlations of ordinary liquids.

An example of a liquid crystal molecule is *5CB* (4-pentyl-4'-cyanobiphenyl), shown in Figure 11.11. The hexagons are *benzene rings*, with the circles showing the resonance between the two equivalent ways to alternate single and double bonds between the carbon atoms on the vertices. The long axis of this anisotropic molecule is called the *director*.

5CB forms a *nematic* liquid crystal, in which the positions of the molecules are random,

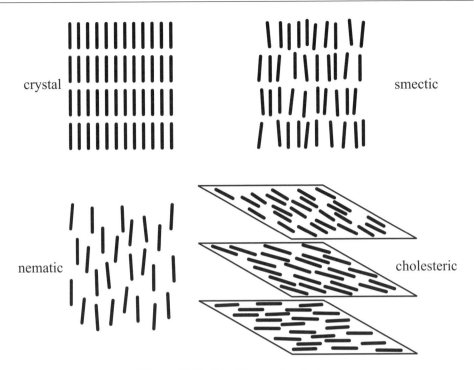

Figure 11.12. Liquid crystal ordering.

but they all point in the same direction. A *smectic* liquid crystal shares this long-range orientational order, and in addition the molecules are layered in planes. And in a *cholesteric* the positions are random, but the directors are aligned and twist in a helix (Figure 11.12).

The dipole moment of an anisotropic liquid crystal can be used to electrically switch its orientation, in the scheme shown in Figure 11.13 [Schadt & Helfrich, 1971]. The liquid crystal is contained between two glass plates with a spacing on the order of 10 μm. The inner surfaces have a thin coating of the polymer *polyimide*, which has been rubbed with a cloth in one direction to produce nanometer-scale grooves. This rubbing is one of a number of "black magic" procedures in LCD production, which are poorly understood but essential operations that are more likely to be considered proprietary trade secrets than research topics.

The director will line up with the grooves because there would be a bending energy associated with crossing them. In a *Twisted Nematic* (*TN*) display the plates are rotated by 90°. This induces a net rotation, like a cholesteric, but it is the result of boundary conditions applied to a nematic. Because there are two possible rotation directions, a small amount of cholesteric is in fact added to break that symmetry. The glass plates have clear electrodes deposited on them, usually *Indium Tin Oxide* (*ITO*). Because the plates are so close together, when a few volts are applied across them the electrostatic dipole energy becomes more significant than the liquid crystal orientational energy, and the molecules rotate to align with the field.

Because of the anisotropy, liquid crystals are also birefringent. When the cell is in the twisted state, it can be considered to be a stack of infinitesimally-thick rotated birefringent plates. If the state of the light coming into the cell is $\vec{E}_0$, then after passing through the

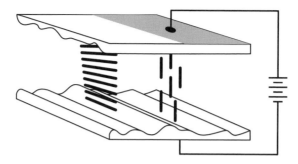

Figure 11.13. A twisted-nematic liquid crystal display.

first of these layers it is

$$\vec{E_1} = \mathbf{R}^{-1}(\theta)\mathbf{B}(d)\mathbf{R}(\theta)\ \vec{E_0} \quad , \tag{11.25}$$

where d is the layer thickness and θ is the angle change over the layer. After two layers,

$$\begin{aligned}\vec{E_2} &= \mathbf{R}^{-2}(\theta)\mathbf{B}(d)\mathbf{R}^{2}(\theta)\ \mathbf{R}^{-1}(\theta)\mathbf{B}(d)\mathbf{R}(\theta)\ \vec{E_0} \\ &= \mathbf{R}^{-2}(\theta)[\mathbf{B}(d)\mathbf{R}(\theta)]^{2}\vec{E_0} \quad ,\end{aligned} \tag{11.26}$$

and after N layers

$$\vec{E_N} = \mathbf{R}^{-N}(\theta)[\mathbf{B}(d)\mathbf{R}(\theta)]^{N}\vec{E_0} \quad . \tag{11.27}$$

If both θ and d are small, this reduces to [Chandrasekhar, 1992]

$$\mathbf{E}_N = \mathbf{R}(N\theta)\mathbf{B}(Nd)\ \vec{E_0} \quad . \tag{11.28}$$

In this *adiabatic* limit the light rotates with the pitch of the liquid crystal, also picking up the phase shift of the unrotated cell's thickness. If crossed polarizing filters are put before and after the cell, aligned with the direction of the polyimide texture, then when no voltage is applied the polarized light exiting the first filter will be rotated to pass through the second. But when the molecules align with an applied voltage, they no longer rotate the light and the second filter blocks the transmission. This provides a switchable light valve based on moving molecules rather than macroscopic materials.

TN displays are addressed with row and column electrodes that rely on each pixel's nonlinear response to the field to isolate the part of the drive waveform intended for it. This limits the size of the display because the on–off voltage ratio becomes too small as the number of pixels is increased [Alt & Pleshko, 1974], reducing the contrast and increasing the switching time. For this reason, twisted nematics are used in LCDs for applications such as watches and control panels, but not in larger computer screens. One way to increase the resolution is by decreasing the voltage range over which the cell switches, which is done in a *Super-Twisted Nematic* (*STN*) by using a twist angle of 270° instead of 90°. The larger index of refraction change also leads to a chromatic aberattion, giving an objectionable color difference between the off and on states. This is eliminated in a *Double Super-Twisted Nematic* (*DSTN*) display by adding a second index-compensating film or LCD layer. Note that the same acronym is used in a *Dual Scan Twised Nematic*, which splits the display into subpanels that are addressed separately.

DSTN displays can reach hundreds but not thousands of pixels. For that, it's necessary

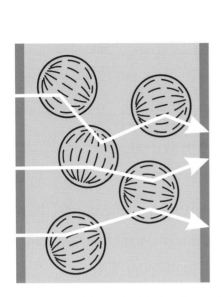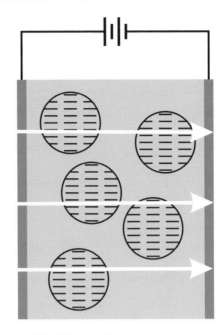

Figure 11.14. A polymer-dispersed liquid crystal panel.

to use some kind of active switch. A *Thin Film Transistor* (*TFT*) LCD does this with the same addressing scheme as a DRAM memory (Figure 10.15), where the capacitor becomes the pixel electrodes [Fischer *et al.*, 1972; Brody, 1996]. This brings the contrast up from around 10:1 to 100:1, and the switching time down from about 100 ms to 10 ms. The transistors have been made using *amorphous silicon* (*a-Si*) with a *silicon nitride* (Si_3N_4) gate deposited on the glass, which has a mobility on the order of 1 $cm^2/(V \cdot s)$, and increasingly with *polycrystalline silicon* (*p-Si*) because its mobility of ~ 100 $cm^2/(V \cdot s)$ is close enough to that of single-crystal silicon (over 1000 $cm^2/(V \cdot s)$) for some of the supporting electronics to be integrated in the same process.

Manufacturing TFT panels requires lithographic fabrication over large areas, bringing down the yield (and increasing the cost) of acceptable panels because defects are so easy to see. Another limition of TFT panels is their power consumption: after passing through the polarizing filters, the liquid crystal, the electrodes and drive transistors, and the color filters, less than 10% of the light makes it out.

One approach to reducing the cost is to take advantage of existing CMOS processes to make small displays that are used with external optics. This is done in a *Liquid Crystal On Silicon* (*LCOS*) display by putting the liquid crystal on top of a CMOS wafer and using it in a reflection mode. A benefit of this approach is that the pixel spacing can become comparable to the wavelength of light, letting the display control color and optical elements by using diffractive structures [Alvelda & Lewis, 1998].

In the other direction, *Polymer-Dispersed Liquid Crystals* (*PDLCs*) are used to cover large areas, such as electronically-controllable windows [Fergason, 1985]. The idea is shown in Figure 11.14. The liquid crystal is contained in small voids in a polymer matrix. With no field applied, the directors line up based on the local asymmetry in their environment, giving an random distribution of orientations. This causes light to scatter

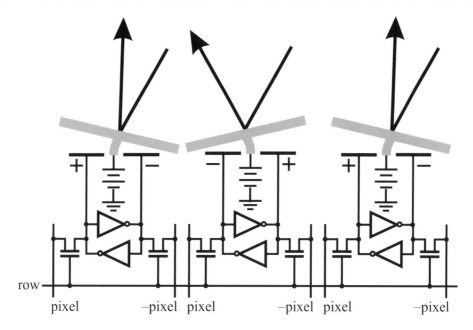

Figure 11.15. A Digital Micromirror Device.

many times, allowing it to be separated out in an optical system, or obscuring what is behind a window. When a voltage is applied to the electrodes, the dipole orientational energy once again dominates and the directors align. If the polymer is index-matched to the liquid crystal then light can pass straight through, switching the panel from opaque to clear.

11.3.3 Smoke and Mirrors

Small particles and shiny surfaces were among the first means used to modulate light; this final section will look at some of the new ways they are being reinvented to address serious limitations in more (recently) conventional displays.

Video projection is growing in importance, initially for presentations to groups, increasingly as a replacement to film in theaters, and ultimately as a way to illuminate smart spaces [Underkoffler *et al.*, 1999]. In the last section we saw that about 10% of the light incident on a liquid crystal panel makes it through; all of the rest is dissipated internally. This represents a significant heat load in a display that is required to produce thousands of lumens, which is a particularly serious issue for the long-term stability of optical materials. Another problem with liquid crystals for projection applications is the display area lost to addressing and TFTs, which can be apparent when the pixels are magnified many times. And for video applications at 60 frames per second, the 17 ms switching time per frame is close to the time scales required to establish the molecular alignment, leading to blurring artifacts.

Figure 11.15 shows an alternative that is easy to understand but hard to implement, a *Digital Micromirror Device* (*DMD*). This starts with the layout of an SRAM memory, but then fabricates above it electrodes on either side of the inverters, and a mirror on a

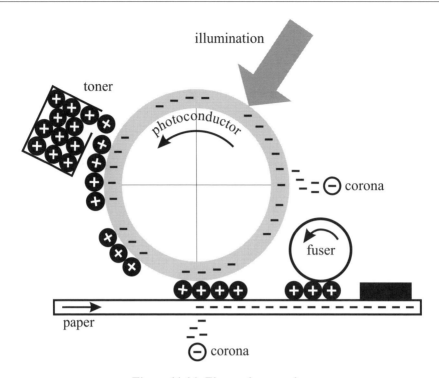

Figure 11.16. Electrophotography.

deformable support that can be electrically biased [van Kessel *et al.*, 1998]. Depending on the bit stored in the cell below it, the mirror is tilted to one side or the other. This can be used to deflect incident illumination into or out of the exit optical path. But unlike an LCD, very little energy is absorbed by the mirrors, the mirrors can fill the surface area of the chip, and they switch in microseconds rather than milliseconds. Because of the difficulty in controlling the magnitude of the bending force, the mirrors are driven between stops in either direction, with grayscale variation coming from modulation of the switching waveform. Such a structure is an example of a *Micro-Electro-Mechanical System* (*MEMS*), extending CMOS fabrication techniques to selectively etch supporting layers to yield free-standing mechanical structures that bridge between the mechanical and electronic worlds [Rodgers *et al.*, 1997]. Beyond the sophistication of the extra lithographic steps required to build them, MEMS encounter a host of forces that are not issues in larger machines. For DMDs, one of the biggest problems was simply preventing the mirrors from sticking to the substrate because of weak inter-atomic forces and capillary adhesion from moisture [Hornbeck, 1998].

The paper you're holding is one of the most interesting alternatives to a mirror for deflecting light. Its constituent fibers are translucent; the white color comes from photons bouncing many times and then diffusing back out. This lets it convert incident light from almost any direction into uniform background illumination, with contrast coming from absorption in the ink. The same mechanism occurs with the emulsion of fat globules in milk or water droplets in a cloud; it is related to the phenomena of *weak localization* in which coherent scatterers become trapped in random media [Yoo *et al.*, 1989; Hastings *et al.*, 1994].

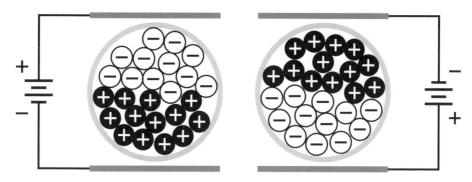

Figure 11.17. Microencapsulated electrophoretic electronic ink.

The ubiquitous connection between a computer and a piece of paper is based on Chester Carlson's invention of *electrophotography* in 1938. The essential elements he used then appear today in laser printers and copiers. The printing cycles starts with a charge source, typically a *corona* discharge from a wire held at many kilovolts. This ionizes the air around it, attracting the positive ions and repelling the negative eletrons. These electrons accumulate on the surface of an insulating or semiconducting photo-conductor. Materials used include selenium, amorphous silicon, and increasingly organic photoconductors because of their chemical and mechanical flexibility.

After it is charged, the photoconductor is illuminated with the desired image. This is done by focusing light reflected from a scanned document, or a linear array of light-emitting diodes, or by switching on and off a rastered laser beam [Starkweather, 1980]. The result is photo–induced carriers, with the positive ones being attracted to the surface electrons to neutralize their charge, leaving negative charge in the complement to the illumination. Then charged *toner* is applied, adhering to the photoconductor in these charged areas. These are pigmented thermoplastic particles, with sizes on the order of 10 μm. Their charge, opposite to that on the photoconductor, is developed through *triboelectricity*, the charge transfer that occurs between two objects rubbed together because of differences in their electron affinity. Finally, a piece of paper is brought in contact, itself charged to pull the toner off of the photoconductor. The final step is to use heat and pressure to fuse the toner to the paper, and to reset the photoconductor for the next pass. This can all happen very quickly, at speeds approaching 1000 pages per minute.

A piece of paper is an ideal display medium: it is thin, flexible, and non–volatile, and it offers high resolution, and high contrast. Its one liability is that what is printed cannot be changed. This is being remedied with the development of *electronic inks* that retain the contrast mechanism of printing, but also provide electronic addressability. This can be done with *microencapsulated electrophoresis*, shown in Figure 11.17 [Comiskey *et al.*, 1998].

The synthesis starts with a solution of toner particles, with a color difference matched by a difference in their surface charge. These are then dispersed in a second liquid to form an emulsion of toner-containing droplets, on the order of 100 μm. Finally an *interfacial polymerization* step mixes in a binary system that grows a clear shell at the droplet–solution interface. This is the process used to encapsulate ink into shells that burst under pressure in carbonless copy paper, but the introduction of surface charge on the particles permits them to be moved relative to each other because of their differential motion in an electric field.

The resulting contrast, resolution, and packaging are competitive with conventional printing because the mechanism is so similar, but now the image can be changed after it is put down. This could be done in a simple printer that needs just an electrode array to reuse a sheet of paper, or by integrating the drive electronics on the substrate as part of the printing process [Ridley *et al.*, 1999]. While this technology is just beginning the technological scaling that more mature display technologies have been through, it promises to merge information display with the inks and paints used in everyday objects.

11.4 SELECTED REFERENCES

[Sze, 1998] Sze, S.M. (ed). (1998). *Modern Semiconductor Device Physics*. New York: Wiley-Interscience.

  Advances in optoelectronic (and many other kinds of semiconductor) devices.

[O'Mara, 1993] O'Mara, William C. (1993). *Liquid Crystal Flat Panel Displays: Manufacturing Science & Technology*. New York: Van Nostrand Reinhold.

  Everything you need to know to start your own LCD production facility.

[Pai & Springett, 1993] Pai, D.M., & Springett, B.E. (1993). Physics of Electrophotography. *Reviews of Modern Physics*, **65**, 163–211.

[Williams, 1993] Williams, Edgar M. (1993). *The Physics and Technology of Xerographic Processes*. Malabar, FL: Krieger.

  The remarkable sophistication of, and insight into, the familiar copier.

11.5 PROBLEMS

(11.1) (*a*) How many watts of power are contained in the light from a 1000 lumen video projector?

  (*b*) What spatial resolution is needed for the printing of a page in a book to match the eye's limit?

(11.2) (*a*) What is the peak wavelength for black-body radiation from a person? From the cosmic background radiation at 2.74 K?

  (*b*) Approximately how hot is a material if it is "red-hot"?

  (*c*) Estimate the total power thermally radiated by a person.

(11.3) (*a*) Find a thickness and an orientation for a birefringent material that rotates a linearly polarized wave by 90°. What is that thickness for calcite with visible light ($\lambda \sim 600$ nm)?

  (*b*) Find a thickness and an orientation that converts linearly polarized light to circularly polarized light, and evaluate the thickness for calcite.

(11.4) Consider two linear polarizers oriented along the same direction, and a birefringent material placed between them. What is the transmitted intensity as a function of the orientation of the birefringent material relative to the axis of the polarizers?

12 Magnetic Storage

Most of the world's bits are stored by orienting magnetic spins. The evolution of these magnetic storage devices is a good lesson in mature technology. For many years, confident and sensible predictions have shown why alternatives, such as optical storage, will soon replace magnetic media, but each year evolutionary innovations bring magnetic storage ever closer to fundamental physical limits and lead to revolutionary new applications. In the densest disks the bit spacing is on the order of 1 μm, reaching the diffraction limit for optical storage. This requires flying the recording head that close to the platter, which is comparable to the mean free path of an air molecule and hence in a regime where the air must be described by the discrete particles of kinetic theory rather than the continuum partial differential equations of hydrodynamics. Many gigabytes fit in drives that are just a few inches big, with costs that have dropped from thousands to hundreds of dollars [Grochowski *et al.*, 1993]. These improvements are the result of a combination of accumulated experience with this system, sophisticated study of the underlying mechanisms, and some luck in how nature responds to such aggressive scaling [Mallinson, 1996]. Surely a time will come when we will stop using spinning platters of what is essentially rust for information storage, but that time remains further off than was once thought [Thompson & Best, 2000].

Magnetism is a surprisingly complex and poorly-understood subject. This chapter therefore starts with a review of the basic phenomenology of magnetic materials, and an introduction to the mechanisms that cause it. These are then applied to explain magnetic and magneto–optical recording, and the ubiquitous but less familiar application in magnetic tags.

12.1 MAGNETISM

In Chapter 5 we saw that the energy density in a field is

$$U \equiv \frac{1}{2}(\vec{E} \cdot \vec{D} + \vec{B} \cdot \vec{H}) \quad \left(\frac{\text{J}}{\text{m}^3}\right) \quad , \tag{12.1}$$

and that in a magnetic material

$$\vec{B} = \mu\vec{H} = \mu_0\mu_r\vec{H} = \mu_0(1 + \chi_m)\vec{H} = \mu_0(\vec{H} + \vec{M}) \quad , \tag{12.2}$$

where $\vec{B}$ is the magnetic flux density, $\vec{H}$ is the magnetic field strength, and $\vec{M}$ is the magnetization. The magnetization is equal to the magnetic moment $\vec{m}$ of the material

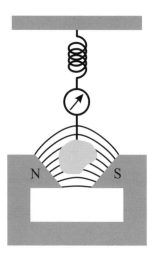

Figure 12.1. Measurement of permeability by an inhomogeneous field.

per volume:

$$\vec{M} = \frac{\vec{m}}{V} \quad . \tag{12.3}$$

Unlike most engineering practice, in the study of magnetic media CGS units are most commonly used because the magnitudes of the quantities are more appropriate. MKS (SI) magnetic fields are measured in amps per meter; the conversion to the CGS (EM) unit, the *oersted*, is

$$H: \; 1 \, \frac{\text{A}}{\text{m}} = \frac{4\pi}{10^3} \, \text{Oe} \quad . \tag{12.4}$$

The SI flux density is measured in tesla; the conversion to *gauss* is

$$B: \; 1 \, \text{T} = 10^4 \, \text{G} \quad . \tag{12.5}$$

Now consider what happens if a material is brought into a magnetic field that has a gradient in the z direction. This might be created by a magnet with tapered pole pieces, shown in Figure 12.1. If the material has a volume V, assumed to be small compared to the length scale of the gradient, then the change in energy after it is brought into the field is equal to the field energy stored in the material minus the energy that was there beforehand:

$$\begin{aligned}
\Delta E &= \frac{1}{2} \int_{\text{material}} \vec{B} \cdot \vec{H} \, dV - \frac{1}{2} \int_{\text{no material}} \vec{B} \cdot \vec{H} \, dV \\
&= \frac{1}{2} V \mu_0 \mu_r H^2 - \frac{1}{2} V \mu_0 H^2 \\
&= \frac{1}{2} V \mu_0 (\mu_r - 1) H^2 \\
&= \frac{1}{2} V \mu_0 \chi_m H^2 \quad . \tag{12.6}
\end{aligned}$$

There will be a force on the material, measured by the scale in Figure 12.1, that is equal to the gradient of this energy

$$F = -\frac{d\Delta E}{dz}$$

$$= -V\mu_0\chi_m H \frac{dH}{dz} \quad . \tag{12.7}$$

The force will be proportional to the magnetic susceptibility χ_m, which is equal to the relative permeability μ_r minus 1. This technique, proposed by Faraday, provides a simple way to measure the permeability of a material. It leads to the following unexpected experimental result: some materials (*diamagnetic, superconducting*) move up the gradient towards the weaker field, and some (*paramagnetic, ferromagnetic, ferrimagnetic*) move down it towards the stronger field. Diamagnetic materials have a small negative susceptibility ($\mu_r = 0.99996$ for Au), paramagnetic materials have a small positive susceptibility ($\mu_r = 1.00002$ for Al), and ferromagnetic and ferrimagnetic materials have a huge positive susceptibility ($\mu_r \sim 10^4$ for steel). In a superconductor, the *Meissner effect* requires that there be no flux lines in the material. This implies that $\vec{B} = 0$ and so

$$\vec{H} = -\vec{M} \Rightarrow \chi_m = \frac{M}{H} = -1 \quad . \tag{12.8}$$

This susceptibility is many orders of magnitude larger than that for a normal diamagnetic material; this strong repulsion can be used for magnetic levitation of bearings and vehicles [Nakashima, 1998].

Why do materials have such different opinions about how to behave in a magnetic field? We now turn to the microscopic origin of magnetic phenomena. Ferromagnetism is the most important mechanism for magnetic storage, but it will be instructive to relate all of them.

12.1.1   Diamagnetism

Lenz's Law states that a time-varying magnetic field induces a current in a loop that acts to oppose the field; diamagnetism comes from this effect operating on the electrons in an atom. Although this is a quantum system, a simple model due to Langevin is in good quantitative agreement with experimental measurements. Viewed semiclassically, the magnetic moment of an electron orbiting a nucleus is

$$m = IA = \frac{qv}{2\pi r}\pi r^2 = \frac{qvr}{2} \quad . \tag{12.9}$$

A time-varying field threading this loop gives rise to an induced potential around the loop

$$V = -\frac{d\Phi}{dt} = -\frac{d(BA)}{dt} = -\mu_0\frac{d(HA)}{dt} \quad , \tag{12.10}$$

taking the magnetic field direction to be normal to the loop. This accelerates the electron by

$$a = \frac{dv}{dt} = \frac{F}{m_e} = \frac{qV}{2\pi r m_e} = -\mu_0\frac{qr}{2m_e}\frac{dH}{dt} \quad . \tag{12.11}$$

Integrating both sides as the field is ramped up from 0 to H in a time T gives the total change in velocity

$$\int_0^T \frac{dv}{dt}\,dt = \int_0^T -\mu_0 \frac{qr}{2m_e}\frac{dH}{dt}\,dt$$

$$\Delta v = -\mu_0 \frac{qrH}{2m_e} \quad , \tag{12.12}$$

which in turn gives the change in the moment

$$\Delta m = \frac{q\Delta v r}{2} = -\mu_0 \frac{q^2 r^2 H}{4m_e} \quad . \tag{12.13}$$

The magnetization caused by this induced moment is

$$M = \frac{m}{V} = -\mu_0 \frac{q^2 Z r^2 H}{4m_e V} \quad , \tag{12.14}$$

where V is the volume of the atom, and the factor of Z has been added to account for multiple electrons in the atom. The susceptibility is then

$$\chi_m = \frac{M}{H} = -\mu_0 \frac{q^2 Z r^2}{4m_e V} \quad . \tag{12.15}$$

Even though this estimate has ignored both thermodynamics and quantum mechanics, it gives numbers that are in line with observed values for diamagnetic materials (Problem 12.1), and shows why diamagnetism is not strongly temperature dependent.

12.1.2 Paramagnetism

The effective current loop used in the preceeding calculation is not fixed in space; under an applied field it can change its orientation as well as speed up or slow down. For the simplest quantum mechanical case of non-interacting spin-1/2 magnetic moments, this corresponds to flipping between states parallel and antiparallel to the field (Chapter 15). If the magnetic moment is m, the energy of the two states is $\pm mB$. If the density of these moments is n, then the magnetization is found from the expected value of the spin orientation

$$M = nm\langle s \rangle$$

$$= nm\frac{\sum_{s=-1,1} s e^{-E_s/kT}}{\sum_{s=-1,1} e^{-E_s/kT}}$$

$$= nm\frac{e^{mB/kT} - e^{-mB/kT}}{e^{mB/kT} + e^{-mB/kT}} \quad . \tag{12.16}$$

mB is usually much smaller than kT, so the exponentials can be expanded as $1 \pm mB/kT$, giving

$$M = \frac{nm^2 B}{kT}$$

$$= \frac{nm^2 \mu_0 H}{kT} \tag{12.17}$$

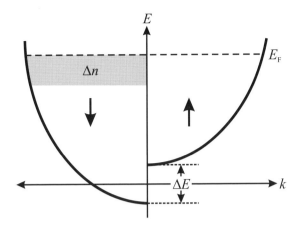

Figure 12.2. Spin band splitting in a field.

or

$$\chi_m = \frac{M}{H}$$
$$= \frac{nm^2\mu_0}{kT}$$
$$\equiv \frac{C}{T} \quad . \tag{12.18}$$

This is *Curie's Law* and the constant C is (can you guess?) the *Curie constant*.

Curie's Law might be expected to apply to the conduction electrons in a metal, which have a spin magnetic moment of

$$\mu_B = \frac{e\hbar}{2m_e} = 9.274 \times 10^{-24} \frac{J}{T} \quad , \tag{12.19}$$

but the susceptibility of most metals is found to be relatively independent of temperature rather than inversely proportional to it. Pauli solved this mystery by pointing out that the derivation of Curie's Law used a canonical partition function, which is appropriate only in the high-temperature limit. Otherwise, the Fermi–Dirac distribution must be used.

The band diagrams in Chapter 10 were drawn as a function of momentum, with each momentum state containing a spin-down and a spin-up state. Figure 12.2 replots a band diagram with all the spin-down states on the left and the spin-up states on the right; each spin state is now associated with $+k$ and $-k$ momentum states. An applied field will split the spin energies. Because the Fermi energies must remain equal, some electron spins flip (assuming that there are available states for them to go into). The number of electrons transferred is approximately equal to the energy split times the density of states at the Fermi energy $n(E_F)$. The two populations will equalize when half the difference is transferred, but there are two momentum states for each spin state:

$$\Delta n = \Delta E\, n(E_F) = B\mu_B n(E_F) \quad . \tag{12.20}$$

The magnetization is the induced moment per volume

$$M = \frac{m}{V} = \mu_B\, \Delta n = \mu_B^2 B n(E_F) \quad , \tag{12.21}$$

therefore the susceptibility is

$$\chi_{\mathrm{m}} = \frac{M}{H} = \mu_0 \mu_B^2 n(E_F) \quad .$$

(12.22)

This is the spin paramagnetism of a metal. It is positive, roughly temperature–independent, and it will vanish if the density of states vanishes at the Fermi energy because the valence band is full. A paramagnetic material still has the diamagnetic magnetization from the electron orbits, but if the paramagnetic magnetization is large enough it will dominate. This is one example of how materials can be paramagnetic (partially-filled conduction band), diamagnetic (filled valence band), or have little susceptibility (diamagnetism cancels paramagnetism).

Other than obeying Fermi–Dirac statistics, Pauli paramagnetism assumes that the spins are independent. One sign of the failure of this approximation is that in many materials Curie's Law is empirically found to need an offset

$$\chi_{\mathrm{m}} = \frac{C}{T - T_C} \quad .$$

(12.23)

This is the *Curie–Weiss* Law. T_C is the *Curie temperature*, and can be quite large: 1043 K in iron, for example. If it is defined in terms of the Curie constant as $T_C = \lambda C$, the susceptibility can be rewritten as

$$\begin{aligned} \chi_{\mathrm{m}} &= \frac{M}{H} \\ &= \frac{C}{T - \lambda C} \\ HC &= MT - M\lambda C \\ \frac{C}{T} &= \frac{M}{H + \lambda M} \quad . \end{aligned}$$

(12.24)

This recovers the original form of Curie's Law, if we assume that the spins see a local field λM added to the applied field H. The offset λM is called the *molecular field*, and to understand it we must understand the origin of ferromagnetism and its relatives.

12.1.3 Ferro-, Antiferro-, and Ferri-magnetism

Diamagnetism and paramagnetism can arise from a range of mechanisms, all relatively weak. Ferromagnetic materials behave very differently in an applied field: the response is large and *hysteretic*. This dependence of the present state of the sample on its past history provides the memory mechanism needed for magnetic storage. The magnetization of a ferromagnet could be measured by the apparatus in Figure 12.1, or its modern cousin the *Vibrating Sample Magnetometer* (*VSM*) that vibrates the sample in a fixed applied field and listens in nearby pickup coils to the signal due to the moving magnetization. If the induced magnetization is plotted as a function of the applied field, the result will look something like Figure 12.3. It is still true by definition that $B = \mu_0(H + M)$, but now the simple ratio $\mu = B/H$ must be replaced with the *differential permeability* $\mu = dB/dH$.

As the applied field is ramped up, the magnetization grows until it reaches a saturation value M_S that is independent of the field. When the field is brought back to zero

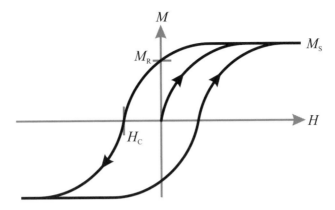

Figure 12.3. A hysteresis loop.

a *remnant magnetization* M_R remains, and if the field is further decreased to the trm-coercivityCoercivity $-H_C$ the remnant magnetization will be removed. This is called a *hysteresis loop*; it can also be plotted as B vs H.

Such persistent magnetic ordering cannot be explained by magnetic forces, which are typically much smaller than kT. They can actually be ignored; to understand this we must turn to the much larger electrostatic interactions (Problem 12.2), and quantum mechanics. Let's start with the simplest possible model for a solid, two nuclei and two electrons. Although this is really just a hydrogen molecule, it's enough to introduce the origin of ferromagnetism.

Call the nuclei a and b, and the electrons 1 and 2. If we isolate one of the atoms, say electron 1 around nucleus a, and assume that the nucleus is so massive that it doesn't move, the Hamiltonian is just

$$\mathcal{H}_a = \frac{p_1^2}{2m} - \frac{e^2}{r_{1a}} \quad , \tag{12.25}$$

where r_{1a} is the distance between electron 1 and nucleus a. Let $\varphi_a(\vec{x}_1)$ be the lowest-energy eigenfunction of this Hamiltonian.

Now bring the atoms closer together. The single-electron eigenfunctions will no longer be eigenfunctions of the joint Hamiltonian, but we can use them as a basis to build those up. These joint wave functions $\psi(1,2)$ must satisfy two essential constraints. Because quantum particles are *indistinguishable*, expectation values must not change if the particles are interchanged, so

$$|\psi(1,2)|^2 = |\psi(2,1)|^2 \quad . \tag{12.26}$$

And because electrons are fermions, the wave function must be antisymmetric, changing sign if the electrons are swapped

$$\psi(1,2) = -\psi(2,1) \quad . \tag{12.27}$$

ψ must describe both the spin and spatial degrees of freedom of the electrons. Chapter 15 will work out the form of the spin states for two electrons; these will be either symmetric or antisymmetric under particle interchange. This means that the corresponding spatial wave functions must be either antisymmetric or symmetric, to preserve the overall antisymmetry. We can construct such states from our single-electron eigenfunctions by combining them as

$$\psi(\vec{x}_1, \vec{x}_2) = \varphi_a(\vec{x}_1)\varphi_b(\vec{x}_2) \pm \varphi_a(\vec{x}_2)\varphi_b(\vec{x}_1) \quad . \tag{12.28}$$

The plus sign goes with the antisymmetric spin state, and the minus sign with the symmetric spin state. These will no longer be eigenfunctions of the joint Hamiltonian, but will be a good approximation as the atoms begin to come together, and will be part of a complete basis set to expand an arbitrary solution.

The Hamiltonian for the joint system becomes

$$\mathcal{H} = \underbrace{\frac{p_1^2}{2m} + \frac{p_2^2}{2m} - \frac{e^2}{r_{1a}} - \frac{e^2}{r_{2b}}}_{\mathcal{H}_0} + \underbrace{\frac{e^2}{r_{ab}} + \frac{e^2}{r_{12}} - \frac{e^2}{r_{1b}} - \frac{e^2}{r_{2a}}}_{\mathcal{H}_{\text{int}}} \quad . \tag{12.29}$$

$\mathcal{H}_0$ is the sum of the Hamiltonians for the individual atoms, having the single-electron eigenstates, and $\mathcal{H}_{\text{int}}$ is the interaction Hamiltonian that arises from bringing the atoms together. The energy associated with the interaction is $E_{\text{int}} = \langle \psi | \mathcal{H}_{\text{int}} | \psi \rangle$. If we evaluate this for our basis wave function,

$$
\begin{aligned}
E_{\text{int}} &= \langle \psi | \mathcal{H}_{\text{int}} | \psi \rangle \\
&= \int \int \psi^* \left(\frac{e^2}{r_{ab}} + \frac{e^2}{r_{12}} - \frac{e^2}{r_{1b}} - \frac{e^2}{r_{2a}} \right) \psi \, d\vec{x}_1 d\vec{x}_2 \\
&= 2 \int \int |\varphi_a(\vec{x}_1)|^2 |\varphi_b(\vec{x}_2)|^2 \left(\frac{e^2}{r_{ab}} + \frac{e^2}{r_{12}} - \frac{e^2}{r_{1b}} - \frac{e^2}{r_{2a}} \right) d\vec{x}_1 d\vec{x}_2 \\
&\quad \pm 2 \int \int \varphi_a^*(\vec{x}_1)\varphi_b^*(\vec{x}_2)\varphi_a(\vec{x}_2)\varphi_b(\vec{x}_1) \left(\frac{e^2}{r_{ab}} + \frac{e^2}{r_{12}} - \frac{e^2}{r_{1b}} - \frac{e^2}{r_{2a}} \right) d\vec{x}_1 d\vec{x}_2 \\
&= E_{\text{overlap}} \pm E_{\text{exchange}} \quad . \tag{12.30}
\end{aligned}
$$

The first term is the *overlap* integral, and the second is the *exchange* integral. Here is the essential point: the preferred spin orientation will be the one that minimizes the contribution from the exchange integral. The relative orientation of the electron spins determines the symmetry of the spin wave function. It in turn constrains the spatial wave function to be either symmetric or antisymmetric, determining the sign of the exchange integral. This integral is a function of electrostatic forces, setting an energy scale much larger than the magnetic forces associated with the spin ordering. This is how electrostatic interactions lead to stable magnetic ordering.

The overlap integral is really a manifestation of the Pauli exclusion principle. The electrons can't be in the same state, leading to an effective force between them. Although its origin lies in the foundations of symmetry in quantum mechanics, its consequence is a very real interaction. For spin 1/2, in Chapter 15 we will see that the dot product $\vec{S}_1 \cdot \vec{S}_2$ of two spins $\vec{S}_1$ and $\vec{S}_2$ can have eigenvalues of $-3/4$ for the antisymmetric spin state or $+1/4$ for the symmetric state. Through the exchange integral, these spin states are associated with overall energies $E_{\text{antisymmetric}}$ and $E_{\text{symmetric}}$. This relationship can be described by an effective spin Hamiltonian

$$\mathcal{H}_{\text{spin}} = \frac{1}{4}(E_{\text{antisymmetric}} + 3E_{\text{symmetric}}) - (E_{\text{antisymmetric}} - E_{\text{symmetric}})\vec{S}_1 \cdot \vec{S}_2 \quad , \tag{12.31}$$

verified by plugging in $\vec{S}_1 \cdot \vec{S}_2 = -3/4, +1/4$. Dropping the constant that does not depend on the spins, calling the prefactor J, and generalizing to more than two spins gives the *Heisenberg Hamiltonian*

$$\mathcal{H}_{\text{spin}} = -\sum_{i,j} J_{ij} \vec{S}_i \cdot \vec{S}_j \quad . \tag{12.32}$$

This interaction is called $\mathcal{J}$ *coupling*. If J is positive, as it is for the metals Fe, Co, and Ni, then the spins will want to point in the same direction, giving *ferromagnetic* ordering.

In an *antiferromagnet* such as Mn or Cr the exchange energy is negative, therefore neighboring spins alternate orientation and there is no net moment even though there is long-range magnetic order. A *ferrimagnet* is a ceramic oxide that has a spontaneous moment but is a good insulator. The moment arises because it has an antiferromagnetic coupling, but there are interpenetrating spin-up and spin-down lattices that have different moments that do not cancel. Most common ferrimagnets are made from materials containing iron oxides, called *ferrites*. Because they do not conduct, they do not screen electric fields or have eddy current heating, and so they are useful for a range of microwave applications as well as guiding flux in coils. One example is the microwave equivalent of optical Faraday rotation, which is used in a "magic T" to steer microwave signals in different directions depending on whether they arrive at the input or the output port. This apparent violation of reversibility is possible because magnetic interactions break time reversal invariance, since the sign of time appears in the velocity in the basic $\vec{v} \times \vec{B}$ law. Cables are often wrapped around ferrites, such as the beads on computer monitor cables, to add inductance to filter out unwanted high-frequency components.

Equation (12.32) can include terms coupling non-adjacent spins. The exchange interaction between overlapping wave functions in equation (12.30) is called *direct exchange*; it's also possible for an exchange interaction to pass through many intervening particles. This is called *indirect exchange*. An important example occurs in NMR, where bonding electrons mediate an exchange interaction between atomic nuclei [Ernst *et al.*, 1994], and indirect exchange is the origin of the strength of rare-earth magnets [Buschow, 1991]. Although a great deal is known about the behavior of the Heisenberg Hamiltonian, quantitatively calculating the J_{ij}'s and its solution from first principles remain dauntingly open problems because of the challenge of handling these many-body effects beyond the independent electron approximation [Mattis, 1988].

At high temperatures, ferromagnets become paramagnets when thermal excitations become more significant than the exchange energy. This is observed to be a sharp transition, with the saturation magnetization vanishing at the Curie temperature T_C (Figure 12.4). Likewise, antiferromagnets become paramagnets above the *Néel temperature*. As a ferromagnet is lowered below its Curie temperature, the saturation magnetization reaches a limiting value when all of the spins in the material are aligned.

The remnant magnetization is what's left of the saturation magnetization after the applied field is taken away. If it is large, the material is said to be *hard* and is useful as a permanent magnet. If it is small, the material is said to be *soft*. Microscopically, these materials differ in their local anisotropy. Because the energy stored in a magnetic field is

$$E = \frac{1}{2\mu} \int B^2 \, dV \quad , \tag{12.33}$$

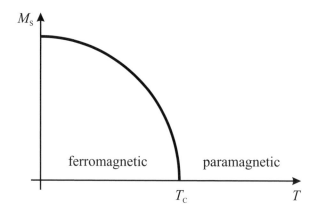

Figure 12.4. The temperature dependence of the saturation magnetization.

the energy will be minimized if the magnetic flux density is greatest in regions of high permeability. Therefore, materials with high permeability "pull in" the field from around them; that is why this property is called permeability. It can be used to guide the field in a recording head or transformer core. Since there is dissipation associated with a hysteresis loop (Problem 12.5), these applications use a soft magnetic material with a narrow hysteresis loop. This is particularly true of *metallic glasses* that are formed by rapidly quenching metallic alloys to prevent the growth of crystalline ordering. Transformer core materials are also often laminated from thin sheets; this reduces eddy current heating if the laminations are perdendicular to the direction electrons are accelerated by the field.

Because of the exchange energy, in a ferromagnet the atomic spins are locally aligned forming *domains*. However, the domain size is usually much smaller than the sample size because of competing factors such as thermal fluctuations which can reorient domains, and because of the energy stored in the fields produced by the domains. To see the latter effect, consider what happens to a sample that is initially homogeneously magnetized (Figure 12.5). There is a large energy that is stored in the external return flux, which can be reduced by splitting the spins into two opposite domains so that the return flux does not have to travel as far. This can be further reduced by splitting into four domains, and so forth. The process does not continue indefinitely, because at the boundary between domains the spins require a few hundred lattice spacings to change direction in what is called a *Bloch wall*. Through the dot product, the exchange energy is proportional to the cosine of relative spin orientations θ. In the limit of a small misalignment this can be expanded as $1 - \theta^2$. Expanding the wall over multiple spins incurs a linear increase in energy from the number of spins, but saves a quadratic amount of energy by reducing the relative angles. This spreads the wall out, up to a size limited by favorable global spin alignment. The final domain size is a result of the tradeoff among all of these mechanisms; characteristic sizes are 1–100 μm. A *magnetic bubble* is a small domain that is just a single loop of a Bloch wall; magnetic bubbles were once of interest for non-volatile memories, but were limited by the speed at which they could be moved.

Each trip around a hysteresis loop starts with most of the spins pointing in the same direction; because of temperature some will point in other directions. As an external field is swept, domains with spins pointing in the opposite directions will be seeded and grow,

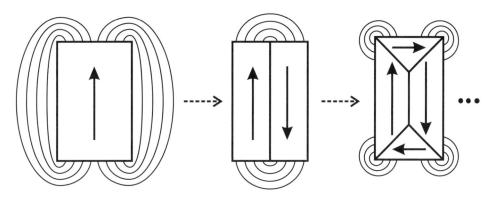

Figure 12.5. Magnetic domains reduce return flux energy.

eventually taking over the entire sample. This process takes time; moving the field quickly has little effect. Therefore the shape of a hysteresis loop will also depend on how fast the field is changed, reaching a limit in the case of slow field changes. If the field does not reach the saturation magnetization then some domains will be left with opposite spins. This is why slowly reducing an oscillating magnetic field will demagnetize a sample. If a hysteresis loop is examined in detail, the continuous curve is actually made of discrete steps called *Barkhausen steps*. These arise from discontinuous jumps in the size of the magnetic domains; within a single step the change in the magnetization is reversible.

12.2 MAGNETIC RECORDING

Magnetic recording started with Valdemar Poulsen crudely recording speech on a steel wire in 1898 by using a electromagnet hung from a trolley running along the wire. Since then the sophistication of the media and recording systems has progressed profoundly, but the basic principle is unchanged: write the message as magnetization in a suitable media, and then later detect that magnetization. Because thermal fluctuations can move domain walls, all magnetic media will eventually erase themselves if left at room temperature; like most digital media they are not suitable for very long-term archival storage unless there is regular error correction.

12.2.1 Magnetic Media

Most magnetic media consist of ferrite particles in a binder. To give the sharpest and most stable hysteresis curve, the particles are chosen to be just small enough (~ 1 μm) so that they cannot support a transverse domain wall. The oldest material used, and still one of the most common, is the phase γ-Fe_2O_3 (*gamma ferric oxide*). It has a coercivity of 300 Oe and a Curie temperature of 600 °C, although it undergoes a phase transition at 400 °C. *Chromium dioxide* (CrO_2) became popular for analog recording because of its higher coercivity of 450 Oe, but this comes at the price of a Curie temperature reduced to just 128 °C. One of the highest coercivities occurs in $BaFe_{12}O_{19}$ at ~ 6000 Oe, making it useful for magnetic stripes on credit cards. This is actually too large to be useful for recording; Co–Ti is added to reduce the coercivity down to ∼1000 Oe. Other materials can be added to improve the mechanical properties of the medium, such as ceramic particles that help protect it from hard disk head crashes (discovered accidently from using ceramic materials to grind iron oxide powder in a ball mill).

Magnetic tape is made by dispersing the ferrite in a solvent and binder and spreading it on a substrate, typically a *polyester* such as *mylar*, that is ∼1 mil thick (0.001 inch, 25.4 μm). A strong field is applied to orient the particles along the tape axis, the solvent is dried by heating, and then the tape is compressed and polished between rollers. Floppy disks are made in a similar manner except that the particles are randomly aligned, leading to a smaller remnant magnetization on the order of 1000 G instead of 1500 G. About 2000 square miles of recording media were coated in 1990.

The most sophisticated hard disks replace this process by the vacuum deposition of thin magnetic films such as CoCr or CoNi. A film of 500 Å can have a coercivity of 1000 Oe, and when deposited on a glass or diamond-turned Al substrate it can be flat and smooth enough to permit extremely close head–platter distances. Thin films also have the advantage of hysteresis curves that are almost rectangular, so that the transition between orientations is very sharp. Further improvements to the media are coming from lithographic patterning to eliminate the interaction energy between adjacent bits, and storing the bits with a vertical domain orientation to pack them together more closely [Bertram *et al.*, 1998; Todorovic *et al.*, 1999]. Such refinements have brought magnetic recording over a density of 10 Gbit/in^2, challenging the diffraction limit of optical storage.

12.2.2 Magnetic Recording

The most common recording heads contain an inductor wound around a loop of a magnetically soft permeable material such as *permalloy* ($Ni_{78}Fe_{22}$), which has a permeability over ∼10^5. The large permeability guides the field to a gap that produces a fringing field that is used for reading and writing (Figure 12.6). In a *laminated head* sheets of permalloy approximately 1 mil thick are stacked and pressed; this helps confine the flux within the head because there is a cost for it to cross between laminations, and it reduces eddy current losses. The gap is polished and then filled with a spacer such as glass.

If an analog signal to be recorded was applied directly to the write head, the recording would be dreadful because of the hysteresis of the media. This can be cured by adding a high-frequency *bias* signal, typically in the range of 100–400 kHz and with an amplitude ∼10 times that of the desired signal. The bias takes the media quickly around the hysteresis loop. With no write signal, this just swings between the saturation magnetizations. However, when the write signal is added, one side of the cycle is slightly less magnetized than the other, and it was experimentally discovered that this difference is surprisingly linear in the write signal as long as it is not too large. The high-frequency bias is removed when the recording is read because it is far out of the bandwidth of the read electronics.

Permalloy is a mechanically soft material and so permalloy heads suffer from wear and poor dimensional control. Much more durable heads, such as those needed for video recording, are made from a ferrite with a layer of SiO_2 grown at an interface to provide the gap. The most precise heads are made by thin film deposition on a substrate of a permalloy layer, a layer of copper coils, a top permalloy layer, and then a SiO_2 overlayer; these are used in very high density computer disks.

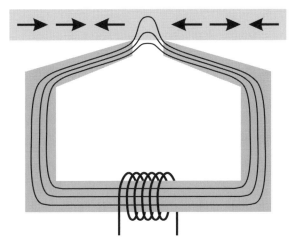

Figure 12.6. An inductive recording head.

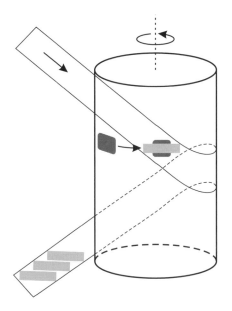

Figure 12.7. Tracks written by helical scan magnetic recording.

Videotape and digital audio tape systems require high frequencies, $\sim$10 MHz. Even if the read and write electronics could operate at these frequencies, the corresponding wavelength of the recording on the tape would be too short for the spins to follow. A solution to this problem, developed by the Ampex Corporation around 1956, is to move the head relative to the tape so that the relative speed between the head and the tape can be much higher. This is now done with *helical scanning*, shown in Figure 12.7.

The heads that have been discussed so far use inductive coils for picking up the signal. As the size of the area that is written decreases it becomes necessary to use more and more turns in the coil to maintain an adequate signal strength, but this increases the inductance, slowing down the response of the head. Also, for a constant speed drive the signal strength will vary from the edge to the center of the platter because the relative velocity between the platter and the head changes. An alternative is *magnetoresistive* heads, which use a material that has a resistance that depends on an applied magnetic field. A common example is permalloy, which changes its resistance by a few percent in the fields used for recording. Aside from having no inductance, magnetoresistive heads have the great advantage that the response does not depend on the relative velocity of the head and the substrate, and so it is expected that they will become dominant for high-performance applications.

There has been a great deal of interest in *giant magnetoresistance* materials, such as multilayer or granular structures of NiFeCo/Cu, which can have magnetoresistances of tens of percent [Baibich *et al.*, 1988]. These operate by using the field to modify spin-dependent electron transport properties [Parkin, 1994]. Even larger magnetoresistance is seen in materials related to $La_{0.7}Ca_{0.3}MnO_3$ [Ramirez *et al.*, 1997]. This change can be 100% or more, and has come to be called *colossal magnetoresistance* (of course).

12.2.3 Recording Systems

Commercial magnetic storage systems range from pocket tape recorders and PCMCIA hard drives up to terabyte mainframe storage systems; some typical parameters for examples across this spectrum are listed below. Perhaps the most aggressive scaling has been in PC disk drives, which continue to get cheaper, smaller, denser, and more reliable. Because of this, they are being assembled into *RAID* (*Redundant Arrays of Inexpensive Disks*) systems to provide much greater capacity and fault tolerance at prices well below traditional large storage systems. As with all other aspects of magnetic storage, it is hard to beat mature solutions refined by big markets.

Audiotape (cassette)

- Frequency range: 20 Hz–20 kHz (CrO_2 tape)
- Bias frequency: 100 kHz
- SNR: 80 dB (Dolby S)
- Tape speed: 1-7/8 inches per second
- Shortest wavelength: 2 μm

Videotape (VHS-SP)

- Tape speed: 1-5/16 inches per second
- Tape width: 0.5 in
- Track pitch: 58 μm
- Track angle from horizontal: 6°
- Drum diameter: 2.45 in
- Drum rotation rate: 1800 revolutions per minute
- Luminance modulation: 3.4–4.4 MHz (FM)
- SNR: 42 dB
- Relative head speed: 220 inches per second
- Shortest wavelength: 1 μm

Floppy Disk (3.5 in HD)

- Formatted capacity: 1.44 MB
- Data transfer rate: 500 kbit/s
- Bit density: 17434 bits per inch
- Track density: 135 tracks per inch
- Rotation rate: 300 revolutions per minute

Hard Disk (IBM Ultrastar 72ZX)

- Disk size: 3.5 in
- Number of disks: 11
- Number of heads: 22
- Capacity: 73.4 GBytes
- Bit density: 352 000 bits per inch
- Track density: 20 000 tracks per inch
- Rotation rate: 10 000 revolutions per minute
- Transfer rate: 160 MByte per second
- Media transfer rate: 473 Mbit per second
- Storage density: 7 Gbit per square inch

12.3 SELECTED REFERENCES

[Hummel, 1993] Hummel, Rolf E. (1993). *Electronic Properties of Materials*. 2nd edn. Berlin: Springer-Verlag.

An applied introduction to the magnetic (as well as many other kinds of) properties of materials.

[Mattis, 1988] Mattis, Daniel C. (1988). *The Theory of Magnetism I: Statics and Dynamics*. New York: Springer-Verlag.

The quantum mechanical theory of magnetism.

[Mee & Daniel, 1996] Mee, C. Denis, & Daniel, Eric D. (eds). (1996). *Magnetic Storage Handbook*. 2nd edn. New York: McGraw-Hill.

[Mallinson, 1993] Mallinson, John C. (1993). *The Foundations of Magnetic Recording*. 2nd edn. Boston: Academic Press.

These two books cover details of practical magnetic storage systems.

12.4 PROBLEMS

(12.1) (*a*) Estimate the diamagnetic susceptibility of a typical solid.

(*b*) Using this, estimate the field strength needed to levitate a frog, assuming a gradient that drops to zero across the frog. Express your answer in teslas.

(12.2) Estimate the size of the direct magnetic interaction energy between two adjacent free electrons in a solid, and compare this to size of their electrostatic interaction energy. Remember that the field of a magnetic dipole $\vec{m}$ is

$$\vec{B} = \frac{\mu_0}{4\pi} \left[\frac{3\hat{x}(\hat{x} \cdot \vec{m}) - \vec{m}}{|\vec{x}|^3} \right] \quad . \tag{12.34}$$

(12.3) Using the equation for the energy in a magnetic field, describe why:

(*a*) A permanent magnet is attracted to an unmagnetized ferromagnet.

(*b*) The opposite poles of permanent magnets attract each other.

(12.4) Estimate the saturation magnetization for iron at 0 K.

(12.5) (*a*) Show that the area enclosed in a hysteresis loop in the (B,H) plane is equal to the energy dissipated in going around the loop.

(*b*) Estimate the power dissipated if 1 kg of iron is cycled through a hysteresis loop at 60 Hz; the coercivity of iron is 4×10^3 A/m.

(12.6) Approximately what current would be required in a straght wire to be able to erase a γ-Fe_2O_3 recording at a distance of 1 cm?

(12.7) Assuming digital recording with a bit size equal to the shortest wavelength that is recorded in the medium, how long would a videotape need to be to store 1 Gbyte? 1 Tbyte?

13 Measurement and Coding

As the number of electrons or photons used to represent a bit becomes small enough to be counted without taking one's shoes off, the means to measure them must become correspondingly sophisticated. Weak signals must be separated from strong backgrounds, using devices that may present a range of constraints on how they can and cannot be used. The only certainty is that mistakes will be made; to be useful, a system must be able to anticipate, detect, and correct its errors. And all this must of course be done at the lowest cost, highest speed, greatest density, ...

This chapter will study a collection of techniques for addressing these problems, starting with the low-level instrumentation that measures a signal, turning to the mid-level modulation used to detect it, and closing with the high-level coding that represents information in it. A striking example that both demonstrates and helped develop these ideas is communication with deep-space probes. As they've traveled further and further out into the solar system the rate at which they can send data back to the Earth has remained roughly constant, because the decreasing signal strength has been matched by increasing communications efficiency due to using bigger antennas, with more sensitive electronics, and better compression and error correction [Posner & Stevens, 1984].

These important topics might appear to be mundane matters of engineering detail, hardly worth considering in a book about physics. That's wrong at three levels. First, without these details all the clever physical insight in the world would not be able to influence anything, so they provide the context needed to understand how to develop mechanisms into working devices. Second, these details make or break practical systems, turning fundamental physical limits into engineering design constraints. And finally, there are in fact very deep connections between these ideas and the character of physical law. We'll see that as we come to understand both engineering and nature better and better, it makes less and less sense to distinguish between the physical laws governing a system and the information represented in it. Novel physical mechanisms such as quantum logic (Chapter 15) offer promising replacements and enhancements to the present practice described here.

13.1 INSTRUMENTATION

13.1.1 Amplifiers

Measuring a signal usually requires some combination of amplification and filtering. The workhorse for manipulating analog signals is the *operational amplifier* (*op-amp*), an (almost) ideal amplifier that is remarkably versatile. Op-amps are available with input noise floors down to $nV/\sqrt{Hz}$, and output power up to kilowatts, at costs ranging from pennies to hundreds of dollars.

The key insight that led to the development of op-amps is that, while it is difficult to build an amplifier with a specified gain, a differential amplifier that has an enormous gain can have its properties determined solely by a feedback network. Furthermore, since the input–output relationship is determined by passive components in the feedback network, such an amplifier can also be very linear even though its transistors or vacuum tubes are not [Black, 1934].

An op-amp has two inputs; the output is the difference between the signal at the positive side and the signal at the negative side, multiplied by a gain of $\sim 10^6$. The exact value of the gain is not a reliable parameter, but consider the circuits shown in Figure 13.1. The op-amp will drive the output so that its non-inverting input is at the same potential as the inverting input. In these cases the non-inverting connection is grounded, therefore the inverting lead acts as a *virtual ground*: it isn't actually connected to ground, but it behaves like one as long as the op-amp is able to drive its output so that the inverting input matches the grounded non-inverting input.

Most op-amps draw so little input current that it is a good approximation to assume that no current flows into the inputs. Requiring that the total current coming into and going out of the inverting node of the first circuit in Figure 13.1 adds up to zero gives the relationship

$$\frac{V_{in} - 0}{R_{in}} + \frac{V_{out} - 0}{R_{out}} = 0 \Rightarrow V_{out} = -\frac{R_{out}}{R_{in}} V_{in} \quad . \tag{13.1}$$

The output, which is inverted relative to the input, is given simply by the ratio of the two resistors. Related configurations accept current inputs or provide current outputs (Problem 13.1), and replacing one or the other of the resistors with a capacitor gives an integrator or a differentiator (Figure 13.1). Note that in a practical integration circuit a large resistor is usually added in parallel with the feedback capacitance, otherwise any small offset voltage error in the op-amp will be integrated up and eventually drive the output to the power supply rails (limits).

Op-amp integrators and differentiators can be used as low- or high-pass filters, and even to solve differential equations in an *analog computer* (although analog computers usually solve equations written just in terms of integrals because differentiation can increase the noise in the result). They were very important up to the 1950s for solving differential equations, and although they've been almost entirely replaced by digital computers they are still useful when fast, cheap, and continuous solutions are needed.

Balancing currents at the inverting nodes shows that the circuits in Figure 13.2 sum or difference their inputs. A differential amplifier is particularly useful for instrumentation because it can be used to measure a small difference in two signals that have a large common component, such as the same external interference. Because the performance is limited by how close the resistor values are, carefully matched pairs of resistors are available for differential amplifiers. Another limitation is the *Common Mode Rejection Ratio* (*CMRR*) of the op-amp, the ratio of the response to the difference in the input signals divided by the common value of the signals. This can easily be over 100 dB.

Common op-amps are internally *compensated* with a single-pole filter [Gershenfeld,

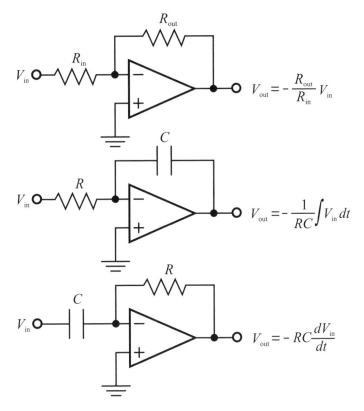

Figure 13.1. Op-amp amplifier, integrator, and differentiator.

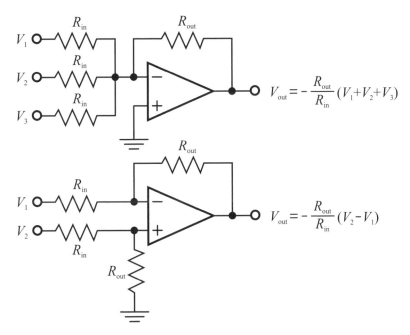

Figure 13.2. Summing and differential amplifiers.

1999a] to ensure stability [Millman & Grabel, 1987]. Their gain as a function of frequency
is

$$G(\omega) = \frac{G_{ol}}{1 + i\frac{\omega}{\omega_{ol}}} \quad , \tag{13.2}$$

where G_{ol} is the open-loop DC gain without an external feedback circuit, and ω_{ol} is
where the open-loop filter rolls off. The frequency ω_1 where the gain is reduced to unity
is easily found to be

$$1 = \left| \frac{G_{ol}}{1 + i\frac{\omega_1}{\omega_{ol}}} \right|$$

$$= \frac{G_{ol}}{\sqrt{1 + \omega_1^2/\omega_{ol}^2}}$$

$$\omega_1^2 = (G_{ol}^2 - 1)\omega_{ol}^2$$

$$\omega_1 \approx G_{ol}\omega_{ol} \tag{13.3}$$

(since $G_{ol} \gg 1$). This is why ω_1 is called the *gain–bandwidth product*. It determines the
highest frequency that an op-amp can operate at; if the frequency response is reduced
then higher gain is possible (Problem 13.2).

The *input impedance* and *output impedance* of an amplifier are other important
specifications. These are the effective impedances seen by a device driving, or being
driven by, the amplifier. The input impedance should be as large as possible so that the
amplifier does not load its source; in an FET op-amp it can be $\sim 10^{12}$ Ω, while in a bipolar
op-amp it can be as small as $\sim 10^9$ Ω The output impedance should be as low as possible,
otherwise the output voltage will depend on how much current is being drawn. Typical
values range from ohms to kilo-ohms.

A differential amplifier has two practical constraints: its CMRR depends on how well
the resistors are matched, and the input impedance is set by the input resistors. For
very high output-impedance sources, the current drawn by these input resistors can be
unacceptable. These problems can be fixed by using an *instrumentation amplifier*, shown
in Figure 13.3. The inputs go directly into buffer amplifiers so that the input impedance is
just the (large) amplifier input impedance. The outputs are connected in a clever divider
circuit that amplifies the difference between the signals but not their common mode, and
this goes to a unity gain differential amplifier that can have precision trimmed on-chip
resistors. Balancing currents at the inverting pins gives

$$\frac{V_{out-} - V_-}{R_1} + \frac{V_+ - V_-}{R_2} = 0 \qquad \frac{V_{out+} - V_+}{R_1} + \frac{V_- - V_+}{R_2} = 0 \tag{13.4}$$

or

$$V_{out-} = \frac{R_1}{R_2}(V_- - V_+) + V_- \qquad V_{out+} = \frac{R_1}{R_2}(V_+ - V_-) + V_+ \quad . \tag{13.5}$$

The important change here is that the difference between the inputs is being amplified
by R_1/R_2, while the individual signals which can contain common mode noise are passed
through without gain. The output from the differential amp is then

$$V_{out} = V_{out+} - V_{out-} = \left(1 + \frac{2R_1}{R_2}\right)(V_+ - V_-) \quad . \tag{13.6}$$

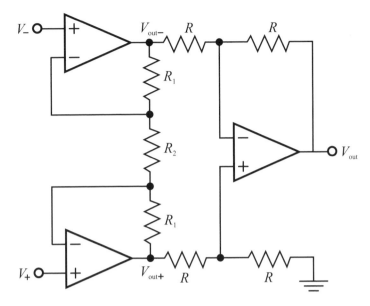

Figure 13.3. An instrumentation amplifier.

The differential amp has a much easier job than before, removing the smaller common mode noise from the larger amplified difference signal from the first stage.

13.1.2 Grounding, Shielding, and Leads

An amplifier is only as good as its leads. While this reasonable observation has led to the unreasonable marketing of rather pathological cables to gullible audiophiles, it is true that small changes in wiring can have a very large impact on a system's performance (both good and bad). The goal is to make sure that as much of a signal of interest gets to its destination, and as little as possible of everything else. The polite term for this area is *electromagnetic compatibility*, asking, say, how to ground your battleship so that its electronics can withstand a nuclear electromagnetic pulse [Hunt & Fisher, 1990; Mitchell & George, 1998].

Although the principles for good wiring practice can appear to be closer to black magic than engineering design, they are really just an exercise in applying Maxwell's equations. Consider the series of circuits shown in Figure 13.4. In (a), a source is directly connected to a single-ended amplifier, introducing two serious problems. First, any other fluctuating voltages around the signal lead can capacitatively couple into it, producing interference. Second, the source and amplifier are grounded in different locations. Current must flow through the pathway connecting the grounds, and so any resistance there will lead to a change in the relative potentials. Even worse, this difference will depend on the load, and on everything else using the ground. This is called a *ground loop*; thick conducting braid is a favorite tool for combating it by reducing the resistance between ground locations. Well-designed systems go further to maintain separate ground circuits for each function, with plenty of capacitance added to each as filters: one ground for digital logic with its high-frequency noise, one for motors with their large current surges, a quiet one for

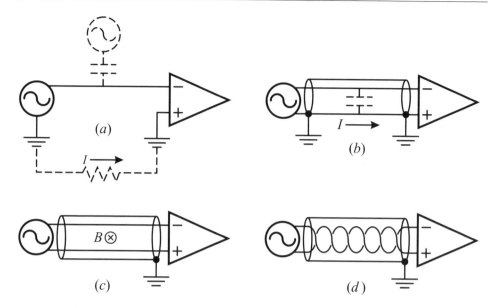

Figure 13.4. Grounding woes: (*a*) ground loop and capacitative pickup, (*b*) cross-talk and improper shield grounding, (*c*) magnetic pickup, and (*d*) shielded twisted pair.

sensors requiring little current but good voltage stability, and so forth. These join only at a single *ground mecca* node.

Circuit (*b*) cures the capacitative pickup by surrounding the wire with a conducting shield, establishing an equipotential around it. Related tricks are building a conducting box around a sensitive circuit to provide electrostatic protection, and winding leads coming into and out of a circuit around a toroidal transformer core to provide inductive filtering of high-frequency noise. A cable shield comes at the cost of introducing a large capacitance from the source to the shield; for typical coaxial cable this can be tens of picofarads per foot, resulting in significant signal loss. That can be cured by using a unity-gain amplifier to drive the outer shield with the potential of the inner conductor. As long as the amplifier is fast enough, the shield will track the signal, effectively removing the cable capacitance. Special *followers* are available for this purpose, because if the amplifier is not fast enough, or cannot source enough current, then the dynamics of the cable-shield system can swamp the signal of interest. Circuit (*b*) also grounds the shield at both ends. This is effective if a heavy shield is used, so that the resistance of the connection is very small, but otherwise it brings the ground loop even closer to the signal lead.

In (*c*), both ends of the signal source are connected to a differential amplifier, and the cable shield is tied at one end. Not only is the shield not used as a continuous circuit, we don't want it to be available as one: its job is just to maintain the equipotential around the signal leads. And because the signals now arrive differentially, the amplifier can remove any common-mode interference that remains. That unfortunately does not help with another important noise source, time-varying magnetic flux linking the circuit, frequently coming from power lines. Even a high-permeability shield can't keep all the flux from threading between the conductors, and the induced potential appears as a voltage difference rather than a common mode shift.

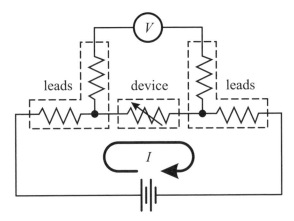

Figure 13.5. A four-terminal measurement.

The straight conductors are replaced in (*d*) with *shielded twisted pair*. The loops do two things: they reduce the cross-sectional area for flux pickup, and the direction of the induced current alternates between loops, approximately averaging it out. This is why shielded twisted pair, grounded at one end, is used most often for low-level signals. It ceases to be useful when the signal wavelength becomes on the order of the conductor spacing, but that cutoff can extend up to microwave frequencies.

If the measurement apparatus need only respond to a signal, a high input-impedance amplifier can be used that does not load the source. But if the apparatus is also responsible for providing current to excite the measurement, there can be a substantial voltage drop across the connecting leads that will vary as the load changes. This problem is cured by making a *four-terminal measurement*, shown in Figure 13.5. Each lead on the device under test, here taken to be a variable resistor, has two connections. One goes to a voltage or current source that drives the current through the leads and the device. And the others are used to measure the voltage drop across the device. The resistance in the current loop does not matter, because the current is the same everywhere. And the resistance in the voltage loop does not matter, because the voltmeter draws essentially no current. This is why precision reference resistors have four terminals, even though they appear in two apparently identical pairs.

When all these techniques fail, it's still possible to give up on electromagnetic shielding entirely and couple optically. For long runs, information can be sent in optical fibers, and many kinds of sensing are possible with all-optical devices (Chapter 14). Even within an electronic circuit, *optoisolators* pair an LED with a photodiode in a single package to provide a logical connection without an electrical one. These are used, for example, in the *Musical Instrument Digital Interface* (*MIDI*) specification to prevent ground loops in audio equipment [Lehrman & Tully, 1993], and in medical instruments to prevent ground loops in people.

13.1.3 Bridges

Many sensors, such as strain gauges and magnetoresistive heads, require detecting small impedance changes. While it's almost always preferable to measure small changes in small

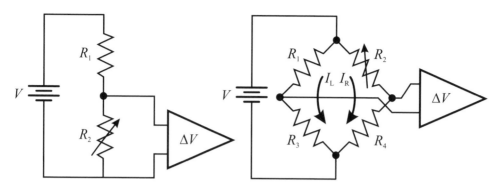

Figure 13.6. Measuring a small resistance change with a bridge.

signals rather than find small changes in the value of a large signal, it's rarely possible to arbitrarily change the baseline impedance of these devices. *Bridge* circuits provide a solution to this problem.

If a voltage V is dropped over two resistors in series, a fixed one with resistance R_1 and a variable one R_2 (Figure 13.6), the measured voltage across the variable resistor will be

$$\Delta V = \frac{V R_2}{R_1 + R_2} \quad . \tag{13.7}$$

If the two resistors differ by the desired sensor signal, $R_1 = R$ and $R_2 = R + \delta$, then

$$\Delta V = \frac{V R}{2R + \delta} \approx \frac{V}{2}\left(1 - \frac{\delta}{2R}\right) \quad . \tag{13.8}$$

A small resistance change leads to a small change in a large voltage. If instead the resistances are arranged in a *Wheatstone bridge*, the voltage difference between the arms of the bridge is

$$V = I_L(R_1 + R_3) = I_R(R_2 + R_4)$$
$$\Delta V = I_R R_4 - I_L R_3$$
$$= V\left(\frac{R_4}{R_2 + R_4} - \frac{R_3}{R_1 + R_3}\right) \quad . \tag{13.9}$$

Now if $R_1 = R_3 = R_4 = R$ and $R_2 = R + \delta$ then

$$\Delta V = V\left(\frac{R}{2R + \delta} - \frac{R}{2R}\right) \tag{13.10}$$
$$= V\left(\frac{1}{2 + \delta/R} - \frac{1}{2}\right)$$
$$\approx -\frac{V}{2}\frac{\delta}{2R} \quad .$$

The small voltage change can now be measured directly without a large offset. This same analysis applies to complex impedances for variable capacitors and inductors.

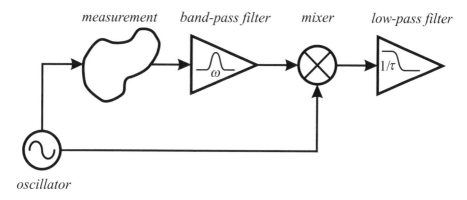

Figure 13.7. A lock-in amplifier.

13.2 MODULATION AND DETECTION

So far we've worried a great deal about the integrity of the signals we're trying to measure, but not at all about their design. Since this can usually be selected in an engineered system, the next group of techniques seek representations of information that help with *signal separation* (distinguishing between noise and the signal), and with satisfying constraints such as limited bandwidth.

13.2.1 Synchronous Detection

If the quantity of interest can periodically be modulated by the measurement apparatus, *synchronous detection* with a *lock-in* amplifier can find a weak signal buried in much larger noise (Figure 13.7). For a bridge circuit the modulation could be done by replacing the DC voltage source with an AC one; for an optical measurement the modulation might periodically vary the intensity of the light source. For this to work the noise must not depend on the excitation. A lock-in can reduce amplifier Johnson noise that is present independent of the input, but not photodetector shot noise that turns on and off with the light. Problem 13.3 looks at typical numbers for this kind of noise reduction.

In a lock-in an oscillator generates a periodic excitation $\sin(\omega t)$ that drives the measurement, resulting in a signal $A(t)\sin(\omega t) + \eta(t)$ that includes the desired response $A(t)$ along with unwanted noise $\eta(t)$. Since multiplication in the time domain is equal to convolution in the frequency domain, the detected signal is convolved around the oscillator (the positive-frequency components are shown in Figure 13.8). An immediate advantage of this is that the subsequent amplification can happen at the oscillator's frequency rather than near DC, away from the amplifier's $1/f$ noise. The front end also includes a bandpass filter centered on the oscillator that is broad enough to include the bandwidth of $A(t)$, but that rejects the remaining out-of-band noise in $\eta(t)$.

Next, the output from the filter is multiplied by the same oscillator signal to demodulate

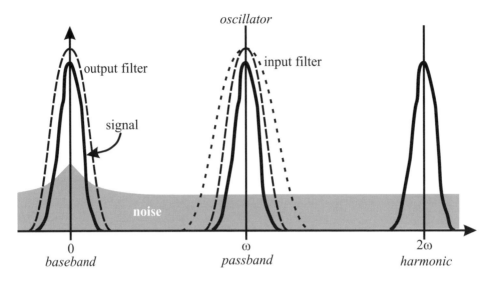

Figure 13.8. Lock-in amplification in the frequency domain (not to scale).

it, generating sum and difference terms:

$$[A(t)\sin(\omega t) + \eta(t)]\sin(\omega t) = A(t)\sin^2(\omega t) + \eta(t)\sin(\omega t)$$
$$= \frac{1}{2}A(t)\cos(\omega t - \omega t) - \frac{1}{2}A(t)\cos(\omega t + \omega t)$$
$$+ \eta(t)\sin(\omega t)$$
$$= \frac{1}{2}A(t) - \frac{1}{2}A(t)\cos(2\omega t) + \eta(t)\sin(\omega t) \quad . \ (13.11)$$

This is called *homodyne* detection; if a different signal source is used in the mixer it is *hetereodyne* detection. Heterodyne detection is used, for example, to down-convert a radio signal to an *IF* (*Intermediate Frequency*) stage for further amplification before final demodulation.

The final step in a lock-in is to pass the demodulated output through a low-pass filter to separate out the measurement component near DC from the modulated noise and sum signals, leaving just $A(t)/2$. In the time domain, the low-pass filter response is found by convolving the input with its impulse response, which performs a weighted average of the signal

$$\langle[A(t)\sin(\omega t) + \eta(t)]\sin(\omega t)\rangle = \langle A(t)\sin^2(\omega t)\rangle + \langle\eta(t)\sin(\omega t)\rangle$$
$$\approx \frac{1}{2}A(t) \quad . \tag{13.12}$$

This assumes that the noise is uncorrelated with the oscillator; the actual value of their overlap will depend on the duration over which the average is taken.

The lock-in has projected out the component of its input with the phase and frequency of the excitation. The noise rejection will depend on the output filter time constant, which can be quite long for a measurement near DC. It's instructive to view the output filter from before the mixer, where it appears to be a band-pass filter centered around the oscillator. But, unlike a conventional band-pass filter, we can make this one as narrow

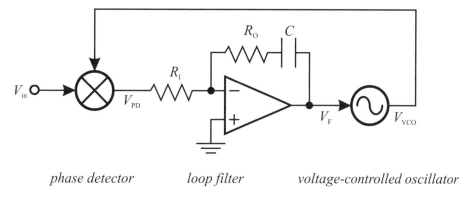

phase detector   *loop filter*   *voltage-controlled oscillator*

Figure 13.9. A Phase-Locked Loop.

as we want by increasing the output filter time constant, and if the oscillator drifts the effective band-pass filter will automatically track it. In theory the input band-pass filter is not even needed at all, but in practice too much noise can lead to nonlinearities and cross-talk in the input stage that do get detected as a signal.

If there is any delay in the measurement there will be a phase shift that turns A into a complex quantity. To distinuish between changes in amplitude and phase, it's necessary to follow the input amplifier by two mixers and low-pass filters, one using $\sin(\omega t)$ to find the real component, and the other $\cos(\omega t)$ for the imaginary component. This is called *quadrature* detection, and the resulting values I and Q (for *In-phase* and *Quadrature*). The analog multiplication can be performed by a *Gilbert cell*, based on varying the current flowing through a differential amplifier [Gilbert, 1975]. The multiplier (or *mixer*) can also be replaced by a switch toggling between the signal and its inverse. This is the same as demodulating with a square wave; although there will be some noise pickup on the harmonics, it's much easier to make a nearly-ideal switch than a multiplier.

If the signal is digitized, the detection algorithm can instead be implemented in software in a digital signal processor. This provides algorithmic flexibility along with greater demands for power, complexity, and cost than a comparable analog circuit, although a periodic signal can be synchronously *undersampled* at less than its period as long as there is good phase stability [Smith, 1999].

13.2.2 Phase Detection and Encoding

If a lock-in measures components I and Q, the phase angle of the signal is given by $\tan^{-1}(I/Q)$. While the phase can be determined this way, a *Phase-Locked Loop* (*PLL*) is a handy cousin of the lock-in that eliminates the need for an inverse trigonometric function and can be used as a signal source as well as a signal analyzer [Wolaver, 1991]. An example is shown in Figure 13.9, with a mixer and *Voltage-Controlled Oscillator* (*VCO*) connected in a feedback loop around an active filter.

The multiplier is called a phase detector here. If $V_{\text{in}} = \cos(\omega t + \theta_{\text{in}})$, and $V_{\text{VCO}} = \sin(\omega t + \theta_{\text{VCO}})$, then if $\theta_{\text{in}} \approx \theta_{\text{VCO}}$ their product will be

$$2\sin(\omega t + \theta_{\text{VCO}})\cos(\omega t + \theta_{\text{in}}) = \sin(\theta_{\text{VCO}} - \theta_{\text{in}}) + \sin(2\omega t + \theta_{\text{VCO}} + \theta_{\text{in}})$$

$$\approx \theta_{\text{VCO}} - \theta_{\text{in}} \qquad (13.13)$$

(the sum signal is removed by the filter). The output is a DC value proportional to the phase difference

$$V_{PD} = K_{PD}(\theta_{VCO} - \theta_{in}) \tag{13.14}$$

with a coefficient K_{PD} that can include gain from the multiplier. As with a lock-in, this can also be implemented with a switch instead of a multiplier.

If the signal and VCO instead have a small frequency difference, then

$$2\sin\left[(\omega_{in} + \delta\omega)t\right]\cos(\omega_{in}t) = \sin(\delta\omega\ t) + \sin(2\omega_{in}t + \delta\omega\ t)$$
$$\approx \delta\omega\ t \tag{13.15}$$

the result is a slow ramp with a slope given by the frequency error.

Next comes the loop filter. Balancing currents into the non–inverting node (Problem 13.1),

$$\frac{dV_F}{dt} = -\frac{R_O}{R_I}\frac{dV_{PD}}{dt} - \frac{V_{PD}}{R_IC} \quad . \tag{13.16}$$

This is followed by the VCO, which has an instantaneous frequency $V_{VCO} = \cos(\omega_{VCO}t)$. Since we want to compare this to the input, their difference defines the time-dependent phase

$$\sin(\omega_{VCO}t) = \sin(\omega_{in}t + \theta_{VCO}(t)) \quad . \tag{13.17}$$

Since the frequency is the time derivative of the argument,

$$\omega_{VCO} = \frac{d\omega_{VCO}t}{dt} = \omega_{in} + \frac{d\theta_{VCO}}{dt} \quad . \tag{13.18}$$

The VCO puts out a frequency that is proportional to the input voltage, with a constant offset

$$\omega_{VCO} = K_{VCO}V_F + \omega_0 \quad . \tag{13.19}$$

Therefore

$$\frac{d\theta_{VCO}}{dt} = K_{VCO}V_F + \omega_0 - \omega_{in} \quad . \tag{13.20}$$

Now if the input frequency and phase are constant, the derivative of equation (13.14) will be

$$\frac{dV_{PD}}{dt} = K_{PD}\frac{d\theta_{VCO}}{dt} \quad , \tag{13.21}$$

so that

$$\frac{1}{K_{PD}}\frac{dV_{PD}}{dt} = K_{VCO}V_F + \omega_0 - \omega_{in} \quad , \tag{13.22}$$

or taking the second derivative,

$$\frac{d^2V_{PD}}{dt^2} = K_{PD}K_{VCO}\frac{dV_F}{dt} \quad . \tag{13.23}$$

Plugging this into equation (13.16) gives

$$\frac{1}{K_{PD}K_{VCO}}\frac{d^2V_{PD}}{dt^2} + \frac{R_2}{R_1}\frac{dV_{PD}}{dt} + \frac{1}{R_1C}V_{PD} = 0 \quad . \tag{13.24}$$

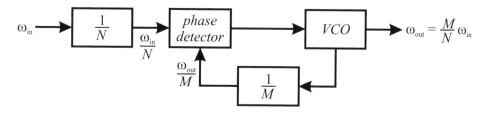

Figure 13.10. A digital PLL frequency synthesizer.

The phase detector output satisfies the equation of motion for a simple harmonic oscillator. The mass is set by the gain, the damping by the resistance ratio, and the restoring force depend on the feedback capacitance. These can be chosen to critically damp the PLL so that it locks onto the signal as quickly as possible. Because it can track changes in frequency as well as phase this is really a PFLL, but that doesn't have quite the same ring to it.

One of the most important applications of a PLL is in generating and recovering timing clocks. Consider the digital variant shown in Figure 13.10. A reference oscillator at ω_{in} goes to a counter which divides the frequency down by a divisor N. This alone could be used to synthesize other frequencies, but the resolution would be very poor for small values of N, requiring a very high input frequency. But here it is compared in the phase detector to the VCO output divided down in a second counter by a ratio M. The phase detector will move the VCO's frequency until these are equal, so that $\omega_{in}/N = \omega_{out}/M$, or $\omega_{out} = M\omega_{in}/N$. Now the frequency is determined by the ratio of M/N, giving much higher and more uniform frequency resolution for a given reference frequency.

On the recovery side, a PLL can lock the phase of a receiver onto a carrier sent by a remote transmitter. Once they share a phase reference, it's possible to use phase as well as frequency and amplitude to store information. Some possible modulation schemes are shown in Figure 13.11. The first, *On–Off Keying (OOK)*, simply turns the carrier amplitude on and off. This is the digital version of *Amplitude Modulation (AM)*. It works, but there's no way to distinguish between the off state and interference that blocks reception of the on state. Better is *Binary Phase-Shift Keying (BPSK)*. Once the PLL is locked, it's possible to keep the carrier amplitude constant and switch just its sign. Now the logical states are independent of the signal strength; BPSK receivers can intentionally clip the input and use a digital PLL to eliminate the amplitude information. This provides much more reliable reception of weak and fluctuating signals. As with a lock-in amplifier, it's possible to add a second demodulation channel with the VCO output phase-shifted by $90°$ to separately determine the I and Q components. Now *Quadrature Phase-Shift Keying (QPSK)* can be done, encoding information in four states based on the signs of the I and Q components. This send two bits instead of one per transmitted symbol (*baud*), but it's possible to do better still. The spacing of the states in the (I,Q) plane need only be as large as the expected channel noise. In *Quadrature Amplitude Modulation (QAM)* the amplitude information is used to squeeze in many more symbols; a V.32 modem sends 9600 bits per second in a 2400 Hz phone channel by using a constellation of 16 QAM states. By considering a string of symbols to be a vector in a higher-dimensional space it's possible to be even more efficient in arranging these. The underlying question of how to best pack spheres in a high-dimensional space is

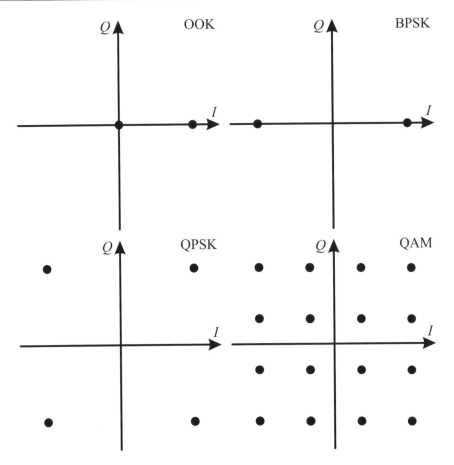

Figure 13.11. Modulation schemes.

surprisingly deep [Conway & Sloane, 1993]. Finally, because the PLL can track changes in frequency as well as phase, information can also be sent that way in *Frequency-Shift Keying (FSK)*. This can be less efficient in using available spectrum, and requires the transmitter and receiver to be designed to operate over a range of frequencies, but the analog version is familiar in *FM (Frequency Modulation)* radios.

13.2.3 Spread Spectrum

In a lock-in amplifier, or an AM radio link, the measurement or message is multiplied by a narrowband carrier signal. Before demodulation the signal retains its original bandwidth, now centered on the carrier (Figure 13.8). This means that the system is susceptible to interference at that frequency. Any background noise (or even intentional jamming) that is near to the carrier will be detected as a valid signal. Even worse, if many different messages are transmitted at adjacent frequencies in a given band, the edges of their distributions will overlap and lead to interference among them. These problems put a premium on reducing the bandwidth of the transmitted signal, using very stable oscillators and narrow filters. *Spread-spectrum* communications systems instead use as much bandwidth as possible

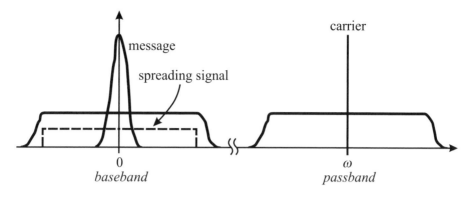

Figure 13.12. Spread-spectrum modulation.

for each signal. Although this might appear to be a perverse response, it leads to a number of significant benefits.

A *direct-sequence* spread-spectrum system multiplies the original message by a broad-band spreading signal that is generated by a *pseudo-random* noise source, one that appears to be random but is generated from a deterministic algorithm. This convolves the spectra so that the message fills the bandwidth of the spreading signal, which can then be modu-lated by another communications carrier to mix it up to the available transmission band (the positive-frequency component is shown in Figure 13.12). A variant is *frequency-hopping* spread spectrum, which uses the pseudo-random sequence to select the carrier frequency. The signal before modulation is said to be in the *baseband*, and after modu-lation it is in the *passband*. In the receiver, the high-frequency carrier is first removed by demodulation, then the message is retrieved from the spreaded signal by using a synchronized replica of the pseudo-random noise generator. We will henceforth ignore the (relatively) straightforward steps of modulation and demodulation by the carrier and consider just the spreading and de-spreading.

This is like a lock-in with a very noisy oscillator. The spreading process significantly increases the information content in the transmitted message and thereby reduces the overlap between the signal and interfering noise. For noise to be picked up in the receiver, it must now match the exact sequence of the pseudo-random generator rather than just a carrier's frequency and phase. Anything else will be rejected in the detector, up to a limit of how long we are willing to average the signal. This means that:

- Noise in the channel will be much less likely to be accepted as a valid signal.
- Different spreaded messages using the same bandwidth will interfere incoherently and so contribute only a small broadband component to the noise floor of the link. This kind of channel sharing, called *Code Division Multiple Access* (*CDMA*), degrades with load more gracefully than the alternatives of *TDMA* (*Time Divi-sion Multiple Access*, in which systems take turns using the channel), *FDMA* (*Frequency Division Multiple Access*, assigning them to different frequencies), and *CSMA* (*Carrier Sense Multiple Access*, where they take turns based on listening for channel activity).
- The chance of accidental or intentional reception by an unintended receiver is significantly reduced since without a synchronized replica of the spreading signal

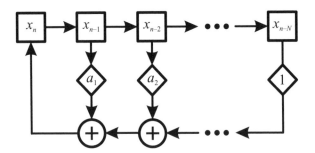

Figure 13.13. A linear feedback shift register.

the message will appear to be random noise. Note that a determined eavesdropper can still decode this; for true security stronger cryptographic techniques must be used.

- A jamming signal must operate over a much larger bandwidth in order to be effective.

- Because there is energy everywhere in the spectrum, the peak power density is reduced, often an important regulatory consideration.

- The resolution in time that the arrival of a message can be measured to is approximately equal to the inverse of its bandwidth because of the *frequency–time uncertainty*, therefore increasing the bandwidth improves timing measurements.

For these reasons spread-spectrum links are more robust and make better use of the available communications bandwidth than the alternatives we've considered, and hence are increasingly common in new designs ranging from sensitive laboratory equipment to communications modems to the GPS satellite positioning system. Their noise rejection is measured by the *coding gain*, which is the ratio in decibels of the energy per bit required to obtain a given *Bit Error Rate* (*BER*) with and without coding, for a fixed noise power (Problem 13.4).

Any spread-spectrum system must provide the transmitter and receiver with identical synchronized copies of an ideal pseudo-random noise source. The earliest patent for an implementation was granted during World War II to the actress Hedy Lamarr (Hedy Markey) and the composer George Antheil (#2,292,387, *Secret Communication System*, 1942) based on storing the noise sequence in a piano-roll mechanism. For those applications without ready access to a piano, a *linear feedback shift register* (*LFSR*) can be used instead. An *order N* LFSR satisfies the recursion relation

$$x_n = \sum_{i=1}^{N} a_i x_{n-i} \quad (\text{mod } 2) \quad , \tag{13.25}$$

shown in Figure 13.13. x and a are binary variables, and the mod 2 operator gives the remainder after dividing by 2. The frequency at which the register gets updates is called the *chip rate*.

The last bit is fed back to the first after passing through as many adders as there are register stages, introducing a propagation delay that can be significant. This is corrected in the equivalent *Galois* or *modular* configuration, shown in Figure 13.14. This puts the

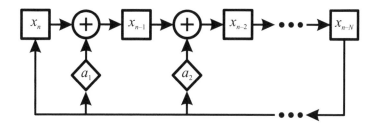

Figure 13.14. The Galois configuration for a shift register.

Table 13.1. *For a maximal LFSR* $x_n = \sum_{i=1}^{N} a_i x_{n-i}$ (mod 2), *lag* i *values for which* $a_i = 1$ *for the given order* N *(all of the other* a_i's *are 0)*.

N	i	N	i	N	i
2	1, 2	13	1, 3, 4, 13	24	1, 2, 7, 24
3	1, 3	14	1, 6, 10, 14	25	3, 25
4	1, 4	15	1, 15	26	1, 2, 6, 26
5	2, 5	16	1, 3, 12, 16	27	1, 2, 5, 27
6	1, 6	17	3, 17	28	3, 28
7	3, 7	18	7, 18	29	2, 29
8	2, 3, 4, 8	19	1, 2, 5, 19	30	1, 2, 23, 30
9	4, 9	20	3, 20	31	3, 31
10	3, 10	21	2, 21	32	1, 2, 22, 32
11	2, 11	22	1, 22	33	13, 33
12	1, 4, 6, 12	23	5, 23	34	1, 2, 27, 34

adders between the register stages, so that the addition is accumulated into the bits as they advance through it.

The sequence of bits generated by an LFSR will repeat after the register returns to the same state; the maximum possible sequence length is $2^N - 1$ (the all-0's state is not allowed because the system would get stuck there). Values for the taps a_i that give such *maximal* sequences can be found by asking that the z-transform of the recursion relation not have smaller polynomial factors, in much the same way as prime numbers are found [Gershenfeld, 1999a]. The resulting sequences satisfy many tests of randomness [Simon *et al.*, 1994], including a power spectrum that is flat up to the recurrence time, an autocorrelation function that equals 1 for a delay of 0 and the inverse of the sequence length otherwise, and the same number of 0's and 1's (plus 1, because the all-0's state has been omitted). Table 13.1 gives the coefficients for maximal LFSRs for a number of register lengths.

The hard part in any spread-spectrum implementation is synchronizing the transmitting and receiving shift registers. This comprises two parts: *acquisition* (setting the registers to the correct bit sequence) and then *tracking* (following drifts between the local clocks). The most straightforward, and widely-used, solution is the brute-force one of incrementally shifting the receiving register and cross-correlating with the incoming signal until an overlap is found. This slow process limits the speed of signal recovery.

An approximate alternative is to add the message into the transmitting LFSR

$$x_n = m_n + \sum_{i=1}^{M} a_i x_{n-i} \qquad (13.26)$$

and then add this to the output of a receiving register fed with the same sequence

$$r_n = x_n + \sum_{i=1}^{M} a_i x_{n-i}$$

$$= m_n + \sum_{i=1}^{M} a_i x_{n-i} + \sum_{i=1}^{M} a_i x_{n-i}$$

$$= m_n \qquad (13.27)$$

(remember that $x + x = 0 \mod 2$). This *self-synchronizing* configuration automatically recovers the message, but because the message enters into the register the noise is no longer guaranteed to be optimal, and the receiver is more susceptible to errors and artifacts.

An interesting alternative starts with the recognition that a PLL performs the desired acquisition and tracking for a periodic signal. This can be extended to pseudo-random sequences by replacing the LFSR with an *Analog Feedback Shift Register* (*AFSR*), which is a real-valued map

$$x_n = \frac{1}{2} \left[1 - \cos \left(\pi \sum_{i=1}^{N} a_i x_{n-i} \right) \right] \qquad (13.28)$$

with the same taps a_i as the corresponding LFSR. These two functions agree for digital values, but the analog freedom of the AFSR lets it lock onto a pseudo-random sequence coupled into it [Gershenfeld & Grinstein, 1995].

13.2.4 Digitization

After a signal is amplified, filtered, and demodulated, the final step is usually to digitize it in an *Analog-to-Digital Converter* (*ADC* or *A/D*). These usually start with a *sample-and-hold* circuit to store the voltage on a capacitor to keep it steady during the conversion, followed by one of a number of strategies for turning the analog voltage into a digital number. *Flash* A/Ds are the fastest of all, having as many analog comparators as possible output states (e.g., $2^8 = 256$ comparators for an 8-bit A/D). The conversion occurs in a single step, but it is difficult to precisely trim that many comparators, and even harder to scale this approach up to many bits. A *successive approximation* A/D uses a tree of comparisons to simplify the circuit, at the expense of a slower conversion, by first checking to see if the voltage is above or below the middle of the range, then testing whether it is in the upper or lower quarter of that half, and so forth.

In a *dual-slope* A/D, the input voltage is used to charge a capacitor, then the time required to discharge it is measured. This eliminates the need for many precise comparators, and also can reject some noise becomes the result depends only on the average charging rate. The number of bits is fixed by the timing resolution. A *delta-sigma* A/D also converts the voltage into the time domain, but in a way that permits the resolution

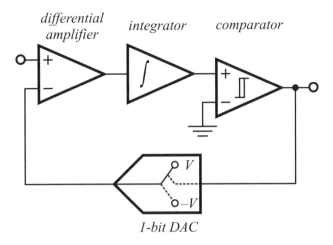

Figure 13.15. A delta-sigma ADC.

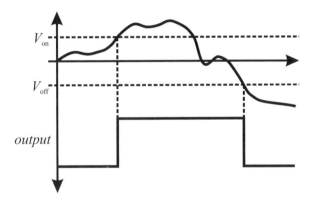

Figure 13.16. A Schmitt trigger.

to be dynamically determinted as needed (13.15). The input goes first to a differential amplifier, then to an analog integrator. After that a comparator outputs a digital 1 or a 0 if the integrator output is positive or negative. It is configured as a *Schmitt trigger* (Figure 13.16), which has some hysteresis in its response so that it doesn't rattle around as the input crosses the threshold voltage. This controls a switch between the voltage rails that goes to the negative differential amplifier input.

Consider what happens if the input is grounded. The integrator will be charged up by the supply voltage until it hits the comparator's threshold to turn on, flipping the output bit. Then, the switch will move to the opposite rail, driving the integrator down until it hits the lower comparator threshold. The output bit will cycle between 1 and 0, with a frequency set by the integrator time constant. If the input is non-zero, there will be an asymmetry between the charging and discharging times, with the relative rates set by where the input is in the voltage range. The fraction of time the resulting bit string is a 1 or a 0, which can be determined by a digital filter, gives the input voltage. The beauty of this approach is that more resolution can be obtained simply by changing the coefficients of the digital filter to have a longer time constant.

Something similar is possible with any A/D as long as it is noisy enough. The digitization process introduces errors into the signal, which can be approximated by a Gaussian noise source with a magnitude equal to the least significant bit. If in fact there is noise of that magnitude then repeated readings can be used to improve the estimate below the bit resolution. For this reason, high-performance converters intentionally add that much noise to the signal before digitization. This process is called *oversampling* because the conversions happen faster than what's required according to the Nyquist sampling theorem.

Similar devices operate in the opposite direction in a *Digital-to-Analog Converter* (*DAC* or *D/A*). A resistor ladder can be used to convert a set of bits to a voltage, but as with a flash A/D this requires precision trimming of the components and does not scale to high resolution. Here too a delta-sigma approach lets temporal resolution be used to obtain voltage resolution, using the same circuit as Figure 13.15 but now with digital logic. The difference is taken between the input and the output, summed into an accumulator, and used to trigger a comparator. This now controls a switch between the analog output rails, producing a waveform with the correct average voltage. An analog filter smooths this to produce the desired resolution; the frequency response is determined by the clock speed for updating the loop. Because that's much easier to increase than the precision of component values, delta-sigma converters dominate for high-performance applications such as digital audio.

13.3 CODING

Machines, like people, make mistakes, can talk too much, and have secrets. This final section takes a peek at some of the many techniques to reduce redundancy (*compression*), anticipate errors (*channel coding*) and fix them (*error correction*), and protect information (*cryptography*). These will all be phrased in terms of communicating digital messages through a channel, but the same ideas apply to anything that accepts inputs and provides outputs, such as a processor or a memory.

13.3.1 Compression

Our first step is *compression*. If there is redundancy in a message so that something is repeated over and over and over and over and over and over and over and over and over and over and over and over and over and over and over and over and over, it's more efficient to eliminate the redundancy by saying (and over)[15]. This is a simple example of a *run-length code* that replaces repeating blocks with a description of their length.

Better still is to recognize that common messages should require fewer bits to send than uncommon ones. This is accomplished by *Huffman* coding. The idea is shown in Figure 13.17, which shows the relative probabilities of vowels in the King James Bible. If the letters are simply encoded as bits then some bit strings will occur more often than others because of the unequal letter probabilities. In an optimally encoded string, 1's and 0's and all possible combinations are equally likely. Huffman coding starts by grouping the symbols with the smallest probabilities to define a new effective symbol, and proceeding in this manner trying to balance the probabilities in the branches. Reading back from the

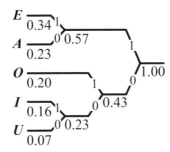

Figure 13.17. Huffman encoding of vowels.

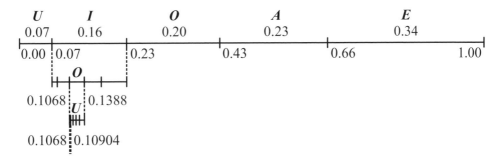

Figure 13.18. Arithmetic encoding of *IOU*.

right to decode a string, a variable number of bits is used depending on the frequency of the letter. A run-length Huffman code is used in the CCITT fax standard.

The success of Huffman compression depends on how well the probabilities in the tree can be matched up. For asymptotically long strings it will attain the Shannon entropy bound, but for shorter strings it won't. A more recent approach, *arithmetic* compression, comes much closer to this limit (Figure 13.18). The unit interval is divided up into segments with lengths corresponding to the relative probabilities of the symbols. A single number will be used to encode a string. It is constrained to be in the interval associated with its first symbol. Then that interval is subdivided in the same fractions to find the subinterval fixed by the second symbol. That interval is divided again according to the third symbol, and so forth. Now the whole Bible could by written by one number. That alone is of course not the compression, because the number is a very long one. But the compression comes in when the intervals are written as fixed-precision binary fractions rather than infinite-precision real numbers. Then the average number of bits used per symbol will reflect their relative probabilities, up to that precision.

Arithmetic compression still requires advance knowledge of the source's probabilities. This may not be possible because of non-stationarity or unfamiliarity. *Universal* compressors attempt to attain the Shannon bound for arbitrary sources by adaptively building a descriptions of them. The *Lempel–Ziv–Welch (LZW)* scheme [Welch, 1984] is shown in Figure 13.19; variants are used in most computer file compression utilities and modem compression standards. The encoder starts with a dictionary containing the possible symbols, in this case 0 and 1. It then works through the string, adding new entries to the dictionary as they are encountered, and transmitting the address of the known prefix to

$$\underline{1}\ \underline{0}\ \underline{1}\ \underline{0}\ \underline{1}\ \underline{0}\ \underline{1}\ \underline{0}\ 1$$

prefix	next character	transmitted entry	dictionary
1 ⟍ 0		1	0 1 10
0 ⟍ 1		0	0 1 10 01
1 ⟍ 0		*	0 1 10 01
10 ⟍ 1		10	0 1 10 01 101
1 ⟍ 0		*	0 1 10 01 101
10 ⟍ 1		*	0 1 10 01 101
101 ⟍ 0		101	0 1 10 01 101 1010
0 ⟍ 1		*	0 1 10 01 101 1010
01 ← 0		01	0 1 10 01 101 1010 010

Figure 13.19. Lempel–Ziv–Welch encoding of a periodic string.

the decoder which can follow the reverse algorithm to reconstruct both the dictionary and the string. If there are N address bits used for dictionary addresses it's possible to store 2^N strings, which can be much longer than N. As the dictionary fills up, the encoder and decoder need to share an algorithm for pruning it, such as throwing out the least-used entry.

So far we've been covering *lossless* compression which can be inverted to find the input string. *Lossy* compression cannot. While this might appear to be a dereliction of engineering duty, if the goal is to transmit a movie rather than a bank balance then all that matters is that it look the same. To see why this is needed, consider NTSC analog video, which provides roughly 640×480 bits of screen resolution at 30 frames per second [Pritchard & Gibson, 1980]. If we allow ourselves eight bits each for red, green, and blue color values, sending an NTSC channel digitally requires

$$640 \text{ pixels} \times 480 \text{ pixels} \times 24 \ \frac{\text{bits}}{\text{pixel}} \times 30 \ \frac{\text{frames}}{\text{s}} = 221 \times 10^6 \ \frac{\text{bits}}{\text{s}} \quad . \tag{13.29}$$

A fast network would be saturated by a standard that dates back to 1941. The *MPEG* (*Moving Picture Experts Group*) standards reduce by a few orders of magnitude the bit rate needed to deliver acceptable video [Sikora, 1997]. They accomplish this by taking advantage of a number of perceptual tricks, which is why lossy coding departs from rigorous engineering design and becomes an art that depends on insight into the application.

The details of the fine structure in an image are usually not important; the exact arrangement of the blades of grass in a field cannot be perceived. *Vector quantization* takes advantage of this insight to expand a signal in basis vectors and then approximate it by using nearby templates [Clarke, 1999]. And the ear will mask frequencies around a strong signal, so these can be discarded [Schroeder *et al.*, 1979]. MPEG compression also does *predictive* coding to send just updates to what a model forecasts the signal will do. This is only as effective as the model; the most sophisticated video coders build in enough physics to be able to describe the objects in a scene rather than the pixel values associated with a particular view of them [Bove, 1998].

13.3.2 Error Correction

Once a message is communicated as efficiently as possible, the next job is to make sure that it is sent as reliably as necessary. This is done by undoing some of the compression, carefully adding back enough redundancy so that errors can be detected and corrected.

The simplest error detection is to add up (mod 2) all of the bits in a data word to find the *parity* and append this value to the string. The receiver can then use the parity bit to catch any single bit error because it will change the parity. This prevents erroneous data from being used, but does not remove the error. If each bit is sent three times, then a majority vote can be taken, not only catching but correcting single-bit errors in the triple. This unfortunately also triples the data rate. Majority voting really overcorrects: it can repair as many errors as there are encoded bits, which may be far more than what's needed.

A *block code* corrects fewer bits with less overhead. In an (n, k) block code, k data symbols are sent in a block of n coded symbols, introducing $n - k$ extra ones for error correction. For a *Hamming* code of order m, $n = 2^m - 1$ and $k = 2^m - 1 - m$. The construction starts with the $(2^m - 1 - m) \times (2^m - 1 - m)$ *generator matrix* $\mathbf{G}$, which for $m = 3$ is

$$\mathbf{G} = [\mathbf{P}^T \ \mathbf{I}] = \begin{bmatrix} 0 & 1 & 1 & 1 & 0 & 0 & 0 \\ 1 & 0 & 1 & 0 & 1 & 0 & 0 \\ 1 & 1 & 0 & 0 & 0 & 1 & 0 \\ 1 & 1 & 1 & 0 & 0 & 0 & 1 \end{bmatrix} \quad , \tag{13.30}$$

where $\mathbf{I}$ is the $m \times m$ identity matrix, and $\mathbf{P}$ has as its columns all possible m-element vectors with more than one non-zero element. A data vector $\vec{d}$ with $(2^m - 1 - m)$ components is associated with a $(2^m - 1)$-element codeword $\vec{c}$ by

$$\vec{c} = \mathbf{G}^T \vec{d} \quad . \tag{13.31}$$

This is received as $\vec{r} = \vec{c} + \vec{\eta}$, with possible errors $\vec{\eta}$. The received vector is then multiplied by the *parity check matrix*

$$\mathbf{H} = [\mathbf{I} \ \mathbf{P}] = \begin{bmatrix} 1 & 0 & 0 & 0 & 1 & 1 & 1 \\ 0 & 1 & 0 & 1 & 0 & 1 & 1 \\ 0 & 0 & 1 & 1 & 1 & 0 & 1 \end{bmatrix} \tag{13.32}$$

to find the *syndrome*

$$\begin{aligned} \vec{s} &= \mathbf{H} \, \vec{r} \\ &= \mathbf{H} \mathbf{G}^T \, \vec{d} + \mathbf{H} \, \vec{\eta} \\ &= [\mathbf{I} \ \mathbf{P}] \begin{bmatrix} \mathbf{P} \\ \mathbf{I} \end{bmatrix} \vec{d} + \mathbf{H} \, \vec{\eta} \\ &= \mathbf{P} + \mathbf{P} + \mathbf{H} \, \vec{\eta} \\ &= \mathbf{H} \, \vec{\eta} \quad . \end{aligned} \tag{13.33}$$

The last line follows because in binary arithmetic $1 + 1 = 0 + 0 = 0$. Since each column of the parity check matrix is unique, if there is a single bit error the offending element of $\vec{\eta}$ can be read off and corrected. Errors of more than one bit will also be recognized,

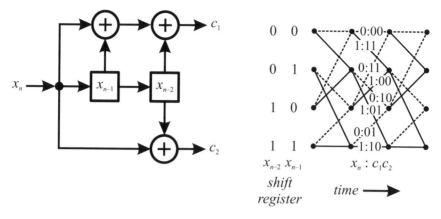

Figure 13.20. A convolutional coder, and the trellis that describes its output. The links are labelled by *data bit:code bits*, with dashed lines for links associated with 0's.

but because the syndrome is no longer unique they can't be corrected. This procedure works because all of the codewords differ by at least three bits (their *Hamming distance* is 3 or more), so that a vector within one bit can uniquely be identified.

Errors don't have to stick to blocks, and neither do coders. In a *convolutional code* memory is introduced so that the decoding depends on the history of what's received, helping fix errors by taking advantage of information that is not adjacent in time. The idea is shown in Figure 13.20. Data bits enter into a shift register, which is tapped and summed to obtain the code bits. This example has a rate of 1/2, turning each data bit into two code bits. There is not a design theory analogous to that for maximal LFSRs to find optimal tap sequences, but good values have been found experimentally [Larsen, 1973].

The action of a convolutional encoder can best be understood through the *trellis* shown in Figure 13.20. There are four possible shift register states, each of which can be followed by an input 0 or 1. All of the possible transitions at each time step are shown, labelled by the code bits associated with them.

When the decoder receives a string of code bits it can determine what was transmitted by finding the path through the trellis with the smallest Hamming distance from what was received. Because of the correlation created by the encoding, this can depend on the full history of the signal. Decoding might appear to be a daunting task: if there are N time steps and M code words, there are N^M sequences to check. But as the trellis makes clear, when two sequences join at a node then it's only necessary to keep track of the most likely one. Given a received string, decoding can progress by a forward pass through the trellis, evaluating the smallest error path arriving at each node, and then a reverse pass reading back from the final node with the smallest final error. This is called the *Viterbi* algorithm [Viterbi & Omura, 1979]. It drops the computational cost from N^M to order NM, quite a savings! Given that difference, it's not surprising that this insight recurs in probabilistics estimation [Gershenfeld, 1999a], and in statistical mechanics, where it is possible to design spin systems that have as their ground state a decoded sequence [Sourlas, 1989]. Problem 13.6 works through an example of Viterbi decoding.

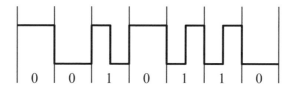

Figure 13.21. A Manchester code.

Table 13.2. *Encoding table for an RLL(2,7) code.*

Data	Code word
00	1000
01	0100
100	001000
101	100100
111	000100
1100	00001000
1101	00100100

13.3.3 Channel Coding

After compression and error correction comes an essential final step: *channel coding*. This is where errors are prevented by modifying the message to satisfy a channel's constraints. For example, if too many identical digits are written in a row to a magnetic disk then the read head will saturate in that direction, if transitions happen too infrequently then the system clock will lose synchronization, if the average number of 1's does not match the average number of 0's there will be a net magnetization of the readout, and if bit reversals happen too quickly it will not be possible to follow them. This is a *Run-Length Limited* (*RLL*) system.

A simple solution is to use a *Manchester* code, shown in Figure 13.21. This always flips the output at the beginning of each interval, and then flips it again in the middle for a 1. The logical bits are now represented by the timing of the transitions in the channel bits, with at least one transition per bit guaranteed. This is easy to understand and implement, but has the very great disadvantage of doubling the rate of the code. It is still used in applications for which the bit rate can vary significantly, such as credit card readers.

Much more efficient use can be made of the medium by explicitly building in its limits. Disk drives for many years used an RLL(2,7) code, which restricts the minimum distance between bits to 2 and the maximum to 7 by encoding the data according to a variable-length block lookup table (Table 13.2). Because of the importance of maximizing storage density still more efficient codes have superseded this; the most general way to understand them is through the language of *symbolic dynamics*, devising dynamical systems that transform symbols subject to a set of constraints [Lind & Marcus, 1995].

13.3.4 Cryptography

The preceeding techniques establish a reliable channel through imperfect devices, but this can be a bug rather than a feature if the receiver is not an intended one. Encoding information so that access can be controlled is the domain of *cryptography* [Simmons, 1992]. The essential insight behind modern cryptosystems is that there can be an asymmetry in the information required for encoding and decoding a message. In public key cryptography [Diffie & Hellman, 1976; Merkle, 1978] someone who wants to receive a secure transmission can openly publish a *public key* number that can be used by anyone to encrypt a message, but a secret *private key* is required to decrypt it. This relies on the existence of *one-way functions* which are easy to evaluate but very hard to invert.

The ubiquitous RSA scheme [Rivest *et al.*, 1978] relies on the difficulty of factoring. It starts by picking two large prime numbers p and q, with a product $n = pq$. Then two other integers e and d are selected for which

$$ed = 1 + (p-1)(q-1)r \tag{13.34}$$

for some integer r, i.e., d is the inverse of e mod $(p-1)(q-1)$. This combination is chosen because according to a version of *Fermat's Little Theorem* due to Euler [Koblitz, 1994]

$$m^{(p-1)(q-1)} = 1 \ (\text{mod } n) \tag{13.35}$$

for integer m that are not divisible by n. The number m can be formed from the bits of a message to be sent, and then encrypted with n and the public key e by

$$\mathcal{E}(m) = m^e \ (\text{mod } n) \quad . \tag{13.36}$$

This is easy to do, but hard to undo. But if the private key d is known, then it can be decrypted by another modular exponentiation

$$
\begin{aligned}
\mathcal{D}[\mathcal{E}(m)] &= \mathcal{D}[m^e \ (\text{mod } n)] \\
&= [m^e(\text{mod } n)]^d \ (\text{mod } n) \\
&= m^{ed} \ (\text{mod } n) \\
&= m^{1+(p-1)(q-1)r} \ (\text{mod } n) \\
&= m \left[m^{(p-1)(q-1)} \right]^r \ (\text{mod } n) \\
&= m \left[1 \ (\text{mod } n) \right]^r \ (\text{mod } n) \\
&= m \ (\text{mod } n) \quad .
\end{aligned}
\tag{13.37}
$$

Anyone with access to e can encrypt m, but only the holders of d can read it.

The security of this scheme rests on the presumed difficulty of finding prime factors, because if p and q could be found from n then equation (13.34) gives the secret key d. The best-known factoring algorithm is the *number field sieve* [Lenstra & Lenstra, Jr., 1993] which requires a number of steps on the order of $\mathcal{O}(e^{1.9(\log N)^{1/3}(\log \log(N))^{2/3}})$ to factor a number N. Because this is exponential in the number of digits in N, a linear increase in the key length imposes an exponential increase in the effort to find the factors. It is widely believed (but not proven) that it's not possible to factor in less than exponential time, unless you're fortunate enough to have a quantum computer (Chapter 15). We'll also see that quantum mechanics offers a way to distribute the private keys, which can't go through the public channel.

For some applications the presence of secret information itself must be kept secret, such as hidden IDs used to detect forgeries and copying. *Steganography*, the very old idea of hiding one kind of data in another kind of media, is becoming increasingly important as the range and value of new types of media and data proliferate [Johnson & Jajodia, 1998].

13.4 SELECTED REFERENCES

[Horowitz & Hill, 1993] Horowitz, Paul, & Hill, Winfield. (1993). *The Art of Electronics*. 2nd edn. New York: Cambridge University Press.

  A great electronics handbook, full of practical experience that is equally useful for beginners and experts.

[Wolaver, 1991] Wolaver, Dan H. (1991). *Phase-Locked Loop Circuit Design*. Englewood Cliffs: Prentice Hall.

  Fun with PLLs.

[Simon *et al.*, 1994] Simon, M.K., Omura, J.K., Scholtz, R.A., & Levitt, B.K. (1994). *Spread Spectrum Communications Handbook*. New York: McGraw-Hill.

[Dixon, 1984] Dixon, R.C. (1984). *Spread Spectrum Systems*. New York: John-Wiley & Sons.

  Most everything there is to know about spread spectrum.

[Sklar, 1988] Sklar, Bernard. (1988). *Digital Communications: Fundamentals and Applications*. Englewood Cliffs: Prentice Hall.

[Blahut, 1990] Blahut, Richard E. (1990). *Digital Transmission of Information*. Reading: Addison-Wesley.

  Good introductions to the theory of coding.

[Schroeder, 1990] Schroeder, M.R. (1990). *Number Theory in Science and Communication*. 2nd edn. New York: Springer-Verlag.

  A lovely introduction to the deep mathematical framework supporting coding theory.

13.5 PROBLEMS

(13.1) (*a*) Show that the circuits in Figures 13.1 and 13.2 differentiate, integrate, sum, and difference.

   (*b*) Design a non-inverting op-amp amplifier. Why are they used less commonly than inverting ones?

   (*c*) Design a transimpedance (voltage out proportional to current in) and a transconductance (current out proportional to voltage in) op-amp circuit.

   (*d*) Derive equation (13.16).

(13.2) If an op-amp with a gain–bandwidth product of 10 MHz and an open-loop DC gain of 100 dB is configured as an inverting amplifier, plot the magnitude and phase of the gain as a function of frequency as R_{out}/R_{in} is varied.

(13.3) A lock-in has an oscillator frequency of 100 kHz, a bandpass filter Q of 50 (remember that the Q or *quality factor* is the ratio of the center frequency to the width between the frequencies at which the power is reduced by a factor of 2), an input detector that has a flat response up to 1 MHz, and an output filter time constant of 1 s. For simplicity, assume that both filters are flat in their passbands and have sharp cutoffs. Estimate the amount of noise reduction at each stage for a signal corrupted by additive uncorrelated white noise.

(13.4) (*a*) For an order 4 maximal LFSR work out the bit sequence.

(*b*) If an LFSR has a chip rate of 1 GHz, how long must it be for the time between repeats to be the age of the universe?

(*c*) Assuming a flat noise power spectrum, what is the coding gain if the entire sequence is used to send one bit?

(13.5) What is the SNR due to quantization noise in an 8-bit A/D? 16-bit? How much must the former be averaged to match the latter?

(13.6) The message 00 10 01 11 00 (c_1, c_2) was received from a noisy channel. If it was sent by the convolutional encoder in Figure 13.20, what data were transmitted?

14 Transducers

Information technology requires technology, and it requires information. From the earliest computers (and people), external interfaces have been essential to their intelligence [Turing, 1950]. *Sensors* are used to convert a parameter of interest to a from that is more convenient to use. The most familiar sensors provide an electrical output, but many other kinds of mappings are possible. An optical sensor that modulates light in a fiber might be appropriate if interference, explosion, or speed is a concern. Sensors are sometimes called *transducers*, although transduction is more correctly used to refer to bidirectional mechanisms such as piezoelectricity that can convert between pressure and charge.

There are almost as many sensors as sensees. The enormous diversity of things that can be measured, and of people doing the measuring, is reflected in the range of available solutions. Because many of these mechanisms are complex, with embodiments that are considered to be trade secrets, and the safety of nuclear reactors or rocketships depends on them, the literature is a curious combination of profound insights and received wisdom that is repeated for reasons of liability rather than scientific merit. There is certainly a lot of room left for innovation.

We've already covered many important types of sensors. The purpose of this chapter is not to fill in the remaining ones – that would fill a shelf full of books. It is to introduce some of the physics that follows from relaxing assumptions we've been relying on, and show the applications to sensing that then follow. The first section will drop the independence of particle states used in deriving partition functions, which will lead to superconductivity and magnetic field sensors that can resolve individual flux quanta. Then comes eliminating the assumption that thermodynamic distributions are in equilibrium. Instead of just using a p-n junction as a thermometer by measuring the reverse-biased current, we'll see that because it is not in thermal equilibrium the presence of an electrical current through it lets it also act as a refrigerator. Finally, we'll question whether space and time themselves are independent or dependent variables, and find the relativistic corrections needed to make GPS work.

14.1 MANY-BODY EFFECTS

Because most devices operate somewhere above absolute zero, we've relied on statistical mechanics to find the occupancy of their available particle states. This has proceeded by finding the states for a single particle, and then maximizing the entropy of the distribution over those states subject to appropriate constraints. While a plausible approach, this ignores the possibility that the multi-particle states are very different from those available to single particles. The consequences of these much more challenging calculations are called variously *many-body*, *cooperative*, or *collective* effects. We'll see next that electrons in a metal provide a striking example of such behavior.

14.1.1 Superconductivity

In Chapter 10 electrons in a metal were found to have nearly-free plane-wave states, occupied up to a Fermi energy E_F. The negatively-charged electrons travel in a sea of positive ionic charge. A reasonable, and quite incorrect, assumption is that the electrons weakly repel each other. What actually happens is that the conduction electrons zipping along at the Fermi energy quickly pass the much heavier positive ions. An ion is attracted towards a passing electron, but by the time the ion begins moving the electron is long since gone. A second electron that passes by will be attracted by the positive wake left by the first electron, effectively being drawn towards the electron. What's actually happening is that the second electron is absorbing a lattice phonon excited by the first one.

To find the possible states of the two-electron system, let's look at those that have opposite momentum so that the combined system is at rest

$$\psi = \sum_k a_k e^{i\vec{k}\cdot\vec{x}_1} e^{-i\vec{k}\cdot\vec{x}_2} = \sum_k a_k e^{i\vec{k}\cdot(\vec{x}_1-\vec{x}_2)} \quad . \tag{14.1}$$

This must satisfy the two-electron Schrödinger equation

$$-\frac{\hbar^2}{2m}\left(\nabla_1^2 + \nabla_2^2\right)\psi + V(\vec{x}_1, \vec{x}_2)\psi = E\psi \quad , \tag{14.2}$$

where V is their effective interaction potential. Plugging in the wave function, multiplying both sides by $e^{i\vec{k}'\cdot(\vec{x}_1-\vec{x}_2)}$, and then integrating over all space and using the orthonormality of the eigenstates gives

$$(E - 2\epsilon_k)a_k = \sum_{k'} V_{kk'} a_{k'} \quad , \tag{14.3}$$

where ϵ_k are the single-particle eigenstate energies, and $V_{kk'}$ is the expectation value of the interaction potential between a pair of momentum states.

Because of Pauli exclusion with all of the core electrons, a_k will vanish for k below the Fermi level k_F (Chapter 10). And $V_{kk'}$ will be small above some cutoff $k > k_c$, because if an electron passes by an ion too quickly the ion won't move. A Nobel-prize-worthy approximation is to assume that the potential is zero below the Fermi energy E_F and also above a cutoff $E_F + E_c(k_c)$, and has a constant value $-V$ in between [Cooper, 1956]. Then the potential comes out of the sum

$$a_k = -V \frac{\sum_{k=k_F}^{k_c} a_{k'}}{E - 2\epsilon_k} \quad , \tag{14.4}$$

which can be simplified by summing both sides over k:

$$\frac{1}{V} = \sum_{k=k_F}^{k_c} \frac{1}{2\epsilon_k - E} \quad . \tag{14.5}$$

Because of the large number of states, the sum can be replaced by an integral weighted by the density of states N_F at the Fermi energy

$$\frac{1}{V} = N_F \int_{E_F}^{E_F + E_c} \frac{1}{2\epsilon - E} \, d\epsilon$$

$$= \frac{N_F}{2} \log \left(\frac{2E_F + 2E_c - E}{2E_F - E} \right)$$

$$E \left(1 - e^{-2/N_F V} \right) = 2E_F \left(1 - e^{-2/N_F V} \right) - 2E_c e^{-2/N_F V}$$

$$E \approx 2E_F - 2E_c e^{-2/N_F V} \quad . \tag{14.6}$$

This is a remarkable result: for any value of V, no matter how small, it is energetically favorable for the electrons to form bound pairs. These are called *Cooper pairs* in honor of Leon Cooper, who first did this calculation. At high temperatures thermal energies exceed the binding energy, but as a metal is cooled it becomes significant.

Even more remarkable is the symmetry implication. Because electrons are fermions their wave functions must be antisymmetric. But because a Cooper pair has two electrons, interchanging a pair of electrons changes the sign twice, leaving it unchanged. This means that Cooper pairs are bosons. And that means that an arbitrary number can be in the same state, and in particular they can all be in the ground state. When this happens to bosons it is called *Bose–Einstein condensation*, and it can even occur with entire atoms [Anderson *et al.*, 1995]. When it happens to Cooper pairs the result is *superconductivity*.

Superconductivity is explained by the *BCS theory*, named for its developers John Bardeen and Robert Schrieffer along with Leon Cooper [Bardeen *et al.*, 1957]. The difficulty in working out the details of the Bose–Einstein condensation of Cooper pairs is that each electron must wear two hats: it must be a member of a pair to act as a boson, and it must also act as a fermion to provide the exclusion that makes pairing possible. The resulting symmetry explains why superconductors carry *persistent currents* without loss. In a normal metal, resistance occurs through electrons losing energy in inelastic collisions to phonons. But in a superconductor, each electron is part of a collective state that reflects not just its pairing but also its contribution through exclusion to many other pairs. For one of these electrons to scatter it must destroy the symmetry of this much larger state, an enormously unfavorable event that happens exceedingly rarely.

Since the pairs are locally all in the same state, a collective wave function $\psi(\vec{x}, t) = \sqrt{n} \, e^{i\varphi}$ can be defined, where $|\psi|^2 = n$ is the pair density and φ is their phase. Ginzberg and Landau originally assumed this relationship [Ginzburg & Landau, 1950], and it was later justified as an approximation to the full BCS theory when the density and phase vary slowly compared to the pair size. In terms of ψ, the current associated with the pairs can be found from the expectation value for their velocity $\vec{v}$ times their charge $2e$

$$\vec{J} = \int \psi^* 2e\vec{v}\psi \, d\vec{x} \quad . \tag{14.7}$$

In the CGS units usually used for superconductivity, the momentum of a particle in an electromagnetic field has an extra piece

$$\vec{p} = m\vec{v} + q\vec{A}/c \tag{14.8}$$

due to the force associated with moving through the vector potential $\vec{A}$ [Goldstein, 1980], therefore plugging in the momentum operator shows that the velocity is found from

$$\vec{v} = -i\frac{\hbar}{m}\nabla - \frac{q}{mc}\vec{A} \quad . \tag{14.9}$$

The current is then

$$\vec{J} = \int \psi^* 2e\vec{v}\psi \; d\vec{x}$$

$$= \int \sqrt{n}\;\exp(-i\varphi)2e\left(-i\frac{\hbar}{m}\nabla - \frac{2e}{mc}\vec{A}\right)\sqrt{n}\;\exp(i\varphi)\;d\vec{x}$$

$$= \frac{2ne}{m}\left(\hbar\nabla\varphi - \frac{2e}{c}\vec{A}\right) \quad . \tag{14.10}$$

Taking the curl provides a relationship between $\vec{J}$ and $\vec{B}$

$$\nabla \times \vec{J} = -\frac{4ne^2}{mc}\nabla \times \vec{A} = -\frac{4ne^2}{mc}\vec{B} \tag{14.11}$$

called the *London equation*. Since $\nabla \times \vec{B} = (4\pi/c)\vec{J}$,

$$\nabla \times \nabla \times \vec{B} = \frac{4\pi}{c}\nabla \times \vec{J}$$

$$-\nabla^2\vec{B} = -\frac{4\pi}{c}\frac{4ne^2}{mc}\vec{B} \quad . \tag{14.12}$$

This differential equation shows that the magnetic field is exponentially screened at the surface, and hence that the supercurrent travels within the *London penetration depth* to cancel the field in the interior. This is the origin of the *Meissner effect* introduced in Chapter 12: superconductors expel magnetic fields.

Consider now a superconducting loop threaded by a flux Φ. In the interior where there are no currents, equation (14.10) becomes $\hbar c\nabla\varphi = 2e\vec{A}$. Integrating this around the loop,

$$\hbar c\nabla\varphi = 2e\vec{A}$$

$$\hbar c\oint \nabla\varphi \cdot d\vec{l} = 2e\oint \vec{A}\cdot d\vec{l}$$

$$\hbar c\Delta\varphi = 2e\int \nabla \times \vec{A}\cdot d\vec{A}$$

$$= 2e\int \vec{B}\cdot d\vec{A}$$

$$= 2e\Phi$$

$$\Delta\varphi = \frac{2e}{\hbar c}\Phi \quad . \tag{14.13}$$

But

$$\Delta\varphi = 2\pi m \tag{14.14}$$

must hold, where m is an integer, so that the phase is continuous around the loop. This means that the supercurrent in the loop varies as a function of the external field to enforce a *flux quantization* condition

$$\Phi = \frac{2\pi\hbar c}{2e}m = \frac{hc}{2e}m \equiv \Phi_0 m \quad . \tag{14.15}$$

14.1.2 SQUIDs

As with semiconductors, superconducting devices rely on junctions. Consider two super-conductors separated by an insulating layer so thin that Cooper pairs can tunnel through it. This geometry is called a *Josephson junction*, after Brian Josephson who as a graduate student predicted its unexpected behavior [Josephson, 1962].

Let ψ_1 and ψ_2 be the wave functions on either side of the junction, and r be the characteristic rate for tunneling. Then the rate of change of the wave functions due to tunneling is found from the time-dependent Schrödinger equation

$$i\hbar\frac{\partial\psi_1}{\partial t} = \hbar r\psi_2 \quad i\hbar\frac{\partial\psi_2}{\partial t} = \hbar r\psi_1 \quad . \tag{14.16}$$

Substituting $\psi_1 = \sqrt{n_1}\,\exp(i\varphi_1)$ and $\psi_2 = \sqrt{n_2}\,\exp(i\varphi_2)$ shows that

$$i\hbar\frac{1}{2\sqrt{n_1}}\frac{\partial n_1}{\partial t}\exp(i\varphi_1) + i\hbar\sqrt{n_1}\,\exp(i\varphi_1)\,i\frac{\partial\varphi_1}{\partial t} = \hbar r\sqrt{n_2}\,\exp(i\varphi_2)$$

$$\frac{\partial n_1}{\partial t} + 2in_1\frac{\partial\varphi_1}{\partial t} = -2ir\sqrt{n_1 n_2}\,\exp[i(\varphi_2 - \varphi_1)] \quad , \tag{14.17}$$

which has a real part

$$\frac{\partial n_1}{\partial t} = 2r\sqrt{n_1 n_2}\,\sin(\varphi_2 - \varphi_1) \quad . \tag{14.18}$$

The current density J across the junction is proportional to the rate of change of the pair density. If the superconductors are identical then $n_1 \approx n_2$, so combining all the constants into a coefficient J_0 gives

$$J = J_0\sin(\varphi_2 - \varphi_1) \quad . \tag{14.19}$$

This is quite unlike anything we've seen before: there is a current in the absence of an applied voltage, with a sinusoidal dependence on the quantum phase difference. This is called the *DC Josephson effect*.

Now turn on a potential V across the junction, adding an energy term to the Hamiltonian from the pair charge $-2e$

$$i\hbar\frac{\partial\psi_1}{\partial t} = \hbar r\psi_2 - eV\psi_1 \quad i\hbar\frac{\partial\psi_2}{\partial t} = \hbar r\psi_1 + eV\psi_2 \quad . \tag{14.20}$$

Once again plugging in the wave functions on each side,

$$\frac{\partial n_1}{\partial t} + 2in_1\frac{\partial\varphi_1}{\partial t} = -2ir\sqrt{n_1 n_2}\,\exp[i(\varphi_2 - \varphi_1)] + i\frac{2eV}{\hbar}n_1 \quad , \tag{14.21}$$

which has an imaginary part

$$\frac{\partial\varphi_1}{\partial t} = -r\sqrt{\frac{n_2}{n_1}}\cos(\varphi_2 - \varphi_1) + \frac{eV}{\hbar} \quad . \tag{14.22}$$

Taking $n_1 \approx n_2$ and subtracting the corresponding equation for $\partial\varphi_2/\partial t$ leaves

$$\frac{\partial(\varphi_2 - \varphi_1)}{\partial t} = -\frac{2eV}{\hbar} \quad . \tag{14.23}$$

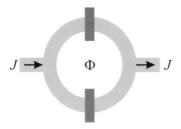

Figure 14.1. A SQUID. Current flows through two thin insulating barriers in a superconducting loop.

The current density through the junction is found by integrating this and plugging it into the real part of equation (14.21):

$$J = J_0 \sin \left(-\frac{2eV}{\hbar} t \right) \quad . \tag{14.24}$$

Now the current oscillates, with the proportionality between frequency and voltage given by the ratio of fundamental constants $e/\hbar$. This is the *AC Josephson effect*.

Finally, connect two Josephson junctions in a loop to from a *Superconducting Quantum Interference Device* (*SQUID*), as shown in Figure 14.1. From equation (14.13) the phase difference around the loop must be $2e\Phi/\hbar c$, and from equation (14.19) the total current is the sum over the two branches, each of which goes as the sine of the phase difference over the junctions, so the combined current is

$$J = J_0 \left[\sin \left(\varphi_0 + \frac{e}{\hbar c}\Phi \right) + \sin \left(\varphi_0 - \frac{e}{\hbar c}\Phi \right) \right]$$

$$= 2J_0 \sin(\varphi_0) \cos \left(\frac{e\Phi}{\hbar c} \right) \quad , \tag{14.25}$$

where φ_0 is an overall phase. Because of the interference between the two branches the current is a periodic function of the flux, which is why this geometry is called an interferometer. What makes it so interesting is the size of the period. A single *flux quantum* is just

$$\Phi_0 = \frac{2\pi\hbar c}{2e} = \frac{hc}{2e} = 2.07 \times 10^{-7} \text{ G} \cdot \text{cm}^2 \quad . \tag{14.26}$$

Measuring the current oscillations in a SQUID is not only the most sensitive way to detect magnetic fields, it can also be used to determine many other small quantities (such as voltage) by converting them to a magnetic field first (for voltage, with a current loop). Superconducting tunnel junctions can also be used as the basis for a high-speed, low-dissipation logic family [Likharev, 1999].

14.2 NON-EQUILIBRIUM THERMODYNAMICS

Many-body effects can profoundly change the states available to a system, as we saw in superconductivity, but these states are still populated in equilibrium with the maximum-entropy distribution (Section 3.4). Here we'll relax the assumption of thermal equilibrium, and find still more surprises in the behavior of electrons in a metal.

The entropy S is an *extensive* quantity that is a function of the size of a system, unlike *intensive* properties like temperature that have the same value in subsystems. It will in turn depend on other extensive variables, such as the total energy and number of particles, with densities $f_1, f_2, \dots$. Using the chain rule, the rate of change of the entropy is

$$\frac{\partial S}{\partial t} = \sum_i \frac{\partial S}{\partial f_i} \frac{\partial f_i}{\partial t}$$

$$\equiv \sum_i F_i \frac{\partial f_i}{\partial t} \quad . \tag{14.27}$$

This defines intensive thermodynamic forces F_i conjugate to the f_i's, which will vanish at the peak of the entropy. If there is a current density $\vec{J}_i$ transporting the quantity f_i (such as the local number of electrons), it will likewise lead to an entropy current

$$\vec{J}_S = \sum_i F_i \vec{J}_i \quad . \tag{14.28}$$

If f_i is conserved, then by Gauss' Theorem the local rate of change will be due to the divergence of the current

$$0 = \frac{\partial f_i}{\partial t} + \nabla \cdot \vec{J}_i \quad . \tag{14.29}$$

Because entropy can be created and destroyed, its rate of change is

$$\frac{dS}{dt} = \frac{\partial S}{\partial t} + \nabla \cdot \vec{J}_s$$

$$= \sum_i F_i \frac{\partial f_i}{\partial t} + \nabla \cdot \sum_i F_i \vec{J}_i$$

$$= \sum_i F_i \frac{\partial f_i}{\partial t} + \sum_i \vec{J}_i \cdot \nabla F_i + \sum_i F_i \underbrace{\nabla \cdot \vec{J}_i}_{-\frac{\partial f_i}{\partial t}}$$

$$= \sum_i \vec{J}_i \cdot \nabla F_i \quad . \tag{14.30}$$

Since $T dS = dQ$, the left hand side represents dissipation. The right hand side is the product of currents and forces that drive them; think of $P = IV$. It is experimentally found in many systems near equilibrium that these currents and forces are related linearly,

$$\vec{J}_i = \sum_j L_{ij} \nabla F_j \quad , \tag{14.31}$$

such as Ohm's Law for electrical conduction or Fourier's Law for thermal conduction. This equation says that the current moves in the direction that maximizes the entropy, with a flux proportional to the gradient, defining *linear non-equilibrium thermodynamics*.

The microscopic laws in a system put important constraints on the macroscopic L_{ij} coefficients. To see this, remember that

$$p(S) = \frac{e^{S/k_B}}{\int e^{S/k_B}} \tag{14.32}$$

(equation 3.46). Therefore the product of one of the extensive variables f_i and a thermodynamic force F_j is

$$
\begin{aligned}
\langle f_i F_j \rangle &= \int f_i F_j p(\vec{f}) \, d\vec{f} \\
&= \int f_i \frac{\partial S}{\partial f_j} p(\vec{f}) \, d\vec{f} \\
&= \frac{\displaystyle\int f_i \frac{\partial S}{\partial f_j} e^{S(\vec{f})/k_{\mathrm{B}}} \, d\vec{f}}{\displaystyle\int e^{S(\vec{f})/k_{\mathrm{B}}} \, d\vec{f}} \\
&= k_{\mathrm{B}} \frac{\displaystyle\int f_i \frac{\partial}{\partial f_j} e^{S(\vec{f})/k_{\mathrm{B}}} \, d\vec{f}}{\displaystyle\int e^{S(\vec{f})/k_{\mathrm{B}}} \, d\vec{f}} \\
&= -k_{\mathrm{B}} \frac{\displaystyle\int \frac{\partial f_i}{\partial f_j} e^{S(\vec{f})/k_{\mathrm{B}}} \, d\vec{f}}{\displaystyle\int e^{S(\vec{f})/k_{\mathrm{B}}} \, d\vec{f}} \qquad \text{(integrate by parts)} \\
&= -k_{\mathrm{B}} \delta_{ij} \quad .
\end{aligned}
\tag{14.33}
$$

For convenience, choose units for f and F that vanish in equilibrium. Then their deviation from equilibrium will relax with the same linear coefficients

$$
\frac{df_i}{dt} = \sum_j L_{ij} F_j \quad .
\tag{14.34}
$$

This is an experimentally-justified conjecture first made by [Onsager, 1931]. It implies that

$$
\begin{aligned}
\left\langle f_i \frac{df_j}{dt} \right\rangle &= \sum_k L_{jk} \langle f_i F_k \rangle \\
&= -k_{\mathrm{B}} L_{ji}
\end{aligned}
\tag{14.35}
$$

and

$$
\begin{aligned}
\left\langle \frac{df_i}{dt} f_j \right\rangle &= \sum_k L_{ik} \langle F_k F_j \rangle \\
&= -k_{\mathrm{B}} L_{ij} \quad .
\end{aligned}
\tag{14.36}
$$

The time derivatives are defined as the expectations of the change over an interval τ that is long compared to the time scale of microscopic dynamics but short compared to the time scale for macroscopic changes,

$$
\begin{aligned}
\left\langle f_i \frac{df_j}{dt} \right\rangle &= \frac{1}{\tau} \langle f_i(t) \, [f_j(t+\tau) - f_j(t)] \rangle \\
&= \frac{1}{\tau} \langle f_i(t) f_j(t+\tau) \rangle - \frac{1}{\tau} \langle f_i(t) f_j(t) \rangle
\end{aligned}
\tag{14.37}
$$

and

$$\left\langle \frac{df_i}{dt} f_j \right\rangle = \frac{1}{\tau} \left\langle [f_i(t+\tau) - f_i(t)] f_j(t) \right\rangle$$

$$= \frac{1}{\tau} \left\langle f_i(t) f_j(t+\tau) \right\rangle - \frac{1}{\tau} \left\langle f_i(t) f_j(t) \right\rangle \quad . \tag{14.38}$$

But most physical laws, other than magnetism, are unchanged if $t \to -t$, therefore

$$\left\langle f_i(t) f_j(t+\tau) \right\rangle = \left\langle f_i(-t) f_j(-t-\tau) \right\rangle$$

$$= \left\langle f_i(t+\tau) f_j(t) \right\rangle \quad . \tag{14.39}$$

This requires that

$$\left\langle \frac{df_i}{dt} f_j \right\rangle = \left\langle f_i \frac{df_j}{dt} \right\rangle \tag{14.40}$$

and hence

$$L_{ij} = L_{ji} \quad . \tag{14.41}$$

This is *Onsager's Reciprocity Theorem*. It provides a direct connection between apparently unrelated phenomena, as we'll see in the next section. If there are magnetic fields present, then B must be replaced with $-B$ along with changing the sign of t.

14.2.1 Thermoelectricity

Now let's apply this theory to electrons in a metal. Their energy U will vary with changes in both the entropy dS and the particle number dN

$$dU = TdS + (\mu + qV)dN \tag{14.42}$$

through the chemical potential μ (equation 10.5) and the electrical potential V. The sum $\mu + qV$ is called the *electrochemical potential*.

The entropy change is thus

$$dS = \frac{1}{T}dU - \frac{\mu + qV}{T}dN \quad , \tag{14.43}$$

and the currents are

$$\vec{J}_S = \frac{1}{T}\vec{J}_U - \frac{\mu + qV}{T}\vec{J}_N \quad , \tag{14.44}$$

or since $TdS = dQ$ the heat current J_Q is

$$\vec{J}_Q = \vec{J}_U - (\mu + qV)\vec{J}_N \quad . \tag{14.45}$$

Therefore from equation (14.30) the derivative of the entropy is

$$\frac{dS}{dt} = \vec{J}_U \cdot \nabla\left(\frac{1}{T}\right) - \vec{J}_N \cdot \nabla\left(\frac{\mu + qV}{T}\right)$$

$$= \vec{J}_U \cdot \nabla\left(\frac{1}{T}\right) - (\mu + qV)\vec{J}_N \cdot \nabla\left(\frac{1}{T}\right) - \frac{1}{T}\vec{J}_N \cdot \nabla(\mu + qV)$$

$$= \underbrace{\left[\vec{J}_U - (\mu + qV)\vec{J}_N\right]}_{\vec{J}_Q} \cdot \nabla\left(\frac{1}{T}\right) - \vec{J}_N \cdot \frac{1}{T}\nabla(\mu + qV) \quad . \tag{14.46}$$

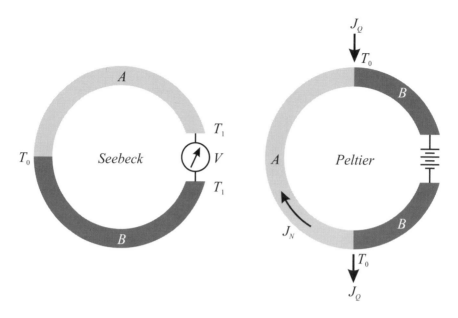

Figure 14.2. Thermoelectric effects.

Near equilibrium, these pairs of conjugate currents and forces will be related by linear coefficients

$$\vec{J}_Q = L_{QQ} \nabla \left(\frac{1}{T} \right) - L_{QN} \frac{1}{T} \nabla (\mu + qV)$$

$$\vec{J}_N = L_{NQ} \nabla \left(\frac{1}{T} \right) - L_{NN} \frac{1}{T} \nabla (\mu + qV) \quad . \tag{14.47}$$

The first represents the thermal current, and the second is the electrical current.

Crucially, these coefficients are material-dependent. Consider the two cases in Figure 14.2. In the first, two materials A and B meet at a junction at a temperature T_0, and are connected to a voltmeter at a temperature T_1. Since the voltmeter draws no electrical current, $J_N = 0$, so from equation (14.47)

$$0 = L_{NQ} \nabla \left(\frac{1}{T} \right) - L_{NN} \frac{1}{T} \nabla (\mu + qV)$$

$$-\nabla (\mu + qV) = -\frac{T L_{NQ}}{L_{NN}} \nabla \left(\frac{1}{T} \right)$$

$$= \frac{L_{NQ}}{T L_{NN}} \nabla T$$

$$\equiv S \nabla T \quad . \tag{14.48}$$

The gradient of the electrochemical potential is equal to the thermal gradient times the *thermopower* or *Seebeck coefficient* S. Integrating the thermal gradient along A and B will give the same result $T_1 - T_0$, but integrating the electrochemical gradient will give a different answer because the materials are different. This is called the *Seebeck effect*, and is used in *thermocouples* to measure temperature. There are many standard thermocouple pairs with known dependence of S on T. For example, a junction between copper and constantan (an alloy of 55% Cu with 45% Ni), called a *type T* thermocouple, produces 40.9 μV/°C at 25 °C.

There is one subtlety in the use of a thermocouple: there will be extra unintentional thermocouple junctions at the contacts between the leads and the voltmeter. Eliminating their influence requires either fixing their temperature (historically, with an ice bath) or measuring it (typically with a temperature-dependent resistor or diode, Chapter 10). While it might appear to be nonsensical to use a thermometer in order to make a thermometer work, among the many desirable features of thermocouples that lead to their routine use are their low cost, small size, and wide temperature range. These are all consequences of the fact that the sensor is just a junction between dissimilar metals. For many applications these benefits more than justify the addition of an extra electronic thermometer that does not need compensation at the instrument connections.

The second circuit in Figure 14.2 drives an electrical current around a loop between the two materials, which are held at the same temperature. Now $\nabla(1/T) = 0$, so the ratio of the currents is

$$\frac{J_Q}{J_N} \equiv \Pi = \frac{L_{QN}}{L_{NN}} \tag{14.49}$$

This defines the *Peltier coefficient* Π. To maintain the isothermal condition, a thermal current must flow into one junction and out of the other one if there is an electrical current. That means that the circuit acts as a heat pump, generating heat on one side and removing it on the other. Unlike ordinary ohmic heating this process is reversible, and unlike ohmic heating it can also be used to cool things.

Because the Onsager relations require that $L_{QN} = L_{NQ}$,

$$\Pi = ST \quad . \tag{14.50}$$

There's no *a priori* reason to expect a relationship between the coefficients in circuits acting as a thermometer and a refrigerator; it is a consequence of the time-reversal invariance of their governing equations.

Materials for practical *Peltier coolers* require a large thermopower, along with a high electrical conductivity σ and low thermal conductivity κ. These can be combined to define a dimensionless figure of merit

$$ZT \equiv \frac{S^2 \sigma}{\kappa} \quad . \tag{14.51}$$

The most common choice is doped semiconducting bismuth telluride, which has $ZT \sim 1$.

The junctions in a Peltier cooler are wired in series to increase the device resistance in order to reduce the ohmic heating losses, and they can be stacked in parallel to increase the temperature drop. They find application in cooling without moving parts everything from chips to submarines. One of the most important applications is cooling other sensors, such as infrared detectors, which have sensitivities that improve with decreasing temperature. The maximum temperature difference that can be achieved is set by the decrease of the thermopower at low tempeatures as the electron scattering rate that establishes the thermalization decreases. A temperature difference of about 100 °C is the limit for bismuth telluride with no heat load, with a maximum efficiency ratio of heat pumped to power supplied of ~0.25–0.5 due to internal losses.

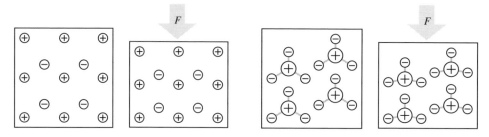

Figure 14.3. Charge locations in the unit cell for ferroelectric (left) and non-ferroelectric (right) piezoelectric materials.

14.2.2 Piezoelectricity

A second important example of a reciprocity relationship is between force and electric fields, called *piezoelectricity*. If an electric field $\vec{E}$ and a stress (force per area) $\vec{T}$ are applied to a material, the resulting polarization $\vec{P}$ is

$$\vec{P} = \mathbf{d}\vec{T} + \chi\vec{E} \quad , \tag{14.52}$$

where χ is the susceptibility matrix and $\mathbf{d}$ is the *piezoelectric* matrix. The corresponding strain $\vec{S}$ (relative displacement) is

$$\vec{S} = \mathbf{c}\vec{T} + \mathbf{d}\vec{E} \quad , \tag{14.53}$$

where $\mathbf{c}$ is the mechanical *compliance* matrix. Because of Onsager's Theorem the $\mathbf{d}$'s are the same. Once again, there's no *a priori* reason to expect the electrical field produced in response to a mechanical force to have the same coefficient as the displacement produced by a field.

There are two kinds of piezoelectricity, shown in Figure 14.3. In a *ferroelectric* material there is a dipole moment that is intrinsic to the unit cell because of offset interpenetrating lattices. The moment will change its value if the aspect ratio of the unit cell changes. In non-ferroelectric piezoelectricity there is no intrinsic moment, but an applied force can break the symmetry of the local charge distribution and induce one. Ferroelectrics are always piezoelectric, but the converse is not true. Barium titanate ($BaTiO_3$) is an example of the former; quartz the latter. Even bones are piezoelectric, which is how they grow to best resist stress [Fukada & Yasuda, 1957].

Because of the reciprocity relationship, piezoelectrics are equally useful as acoustic detectors and generators. If electrodes are deposited on the top and bottom of a piezo-electric, a mechanically-induced change in the polarization will induce charge in the electrodes, and *vice versa*. Since the material is a charge source rather than a voltage source the response falls off at low frequencies, and piezoelectrics need to be used with high input-impedance instrumentation to not load them.

If the relative charge positions can change in the unit cell, ferroelectrics will have $\vec{E}, \vec{P}$ curves analogous to the $\vec{H}, \vec{M}$ curves of a ferromagnet (hence the name), and a Curie temperature above which their ordering is lost. The hysteresis can be applied in electrically-addressed non-volatile memories [Scott, 1998] and displays [Surguy, 1993].

Ferroelectric piezoelectrics are *poled* by heating the material under an applied field to induce the polarization, possibly along with a mechanical stress to align crystalline do-

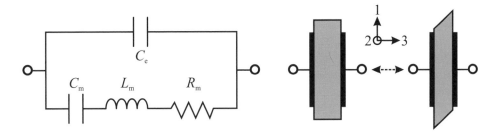

Figure 14.4. The effective circuit for a piezoelectric resonator, and the thickness shear mode.

mains. The anisotropic piezoelectric axes are conventionally defined so that "3" points in the direction of the thickness, "1" points in the extrusion direction, and "2" is transverse. The d_{33} coefficient corresponds to a force applied across the thickness inducing a field in that direction. The common ceramic piezoelectric PZT (Pb(Zr,Ti)O$_3$) has a d_{33} of ~280 pC/N. The polymer PVDF (polyvinylidene fluoride) has a d_{33} of -33 pC/N. While this is smaller than that of PZT, PVDF is attractive because it is thin, flexible, and broadband [Kawai, 1969]. The d_{31} coefficient, which represents a transverse field being produced by in-plane stress, is typically an order of magnitude smaller, but is very useful because the mechanical response can be much greater.

14.3 RELATIVITY

The last assumption to relax is that time and space are independent of the observer. While that might sound more like a matter for philosophy or science fiction than engineering design, relativistic effects have significant implications for applications such as navigation that rely on accurate time-keeping. To make that connection we'll need to first look at how time is measured, then how relativity corrects it, and finally at position measurement.

14.3.1 Clocks

The reciprocal relationship between polarization and displacement in a piezoelectric material means that its electrical behavior depends on its mechanical properties. Figure 14.4 shows the effective circuit for a piezoelectric crystal between electrodes. In parallel with the electrical capacitance C_e, there is a series RLC circuit associated with its mechanical response. The capacitance C_m represents energy storage in physical displacements, the inductance L_m comes from the inertial mass of the material, and the resistance R_m is due to its mechanical dissipation. Because of the piezoelectricity, these all appear as electrical components connected to the electrodes.

Now consider what happens when this circuit is connected in the feedback network of an amplifier. If an amplifier's output is related to its input by inverting gain $y = -Ax$, and if the feedback circuit scales the output by a complex coefficient $x = F(\omega)y$, then $x = -AF(\omega)x$. This is only possible if $AF(\omega) = -1$, therefore the phase of F must be π (or a higher multiple). Problem 14.3 will show that this phase shift occurs around the resonance frequency $\omega = 1/\sqrt{L_m C_m}$. This means that any noise initially in the circuit at that frequency will grow exponentially until it either clips or is intentionally limited. This mechanism is used to generate the clocks for almost all digital systems.

The great advantage of using a piezoelectric crystal rather than an ordinary inductor and capacitor comes in the performance of its effective components. Quartz (SiO_2) is the most commonly-used material, because it is literally as cheap as sand, and because its mechanical parameters can have a weak thermal dependence. The resonant mode used more frequently is a thickness shear, but flexural resonances and higher-order modes can also be excited, and it's possible to use a *Surface Acoustic Wave* (*SAW*) instead of these bulk acoustic waves [Kino, 1987]. For a quartz resonator, C_m is typically on the order of femtofarads and L_m is millihenrys, giving resonant frequencies on the order of megahertz, with overtones up to hundreds of megahertz. Capacitors and inductors in this range would be much more susceptible to parasitic coupling and thermal drift. Even worse, the Q of a conventional LC circuit is ~ 10–100. For a quartz resonator, it's $\sim 10^4$–10^5 because the mechanical damping due to *internal friction* is much smaller than the comparable electrical resistance. Since the Q is equal to $f/\delta f$, the inverse limits the relative uncertainty $\delta t/t$ in the time derived the resonator.

The dominant frequency variation in a quartz resonator comes from thermal drifts changing its stiffness. This can be corrected by measuring the temperature and using it to tune the resonator circuit in a *TCXO* (*Temperature Compensated Crystal Oscillator*), reducing the relative error to $\sim 10^6$. Even better, an *OCXO* (*Oven Compensated Crystal Oscillator*) fixes the temperature of the resonator, and can do that at the temperature where quartz is least sensitive to drift, reducing the short-term relative error to $\sim 10^{-8}$ [Walls & Vig, 1995].

Measuring time to a part in 10^8 may sound impressive, and it is, but it's not hard to need more. An error of 1 ns will be made in $\sim 10^{-9}$ s $\times 10^8 = 0.1$ s. Since electromagnetic radiation travels 1 ft/ns, this means that if a quartz oscillator is used to determine the arrival time of an RF signal, it will be off by the equivalent of 10 feet after running for 1 second, a difference of some concern if you're trying to use it to land an airplane. And as important as the *precision* (variance) is the *accuracy* (bias). The resonant frequency of a quartz oscillator is determined by its mechanical properties. Even if these are sufficiently stable, if the resonator wasn't cut to exactly the right length or if its stiffness changes from aging then the airplane will predictably land in the middle of a field.

Both of these problems are solved by recognizing that macroscopic objects can differ, but the microscopic properties of all atoms are identical. Quantum mechanically there is no way to distinguish between two atoms, so clocks based on atomic resonators will keep the same time. This is an old idea [Kusch, 1949] that has both required and enabled a great deal of new technology [Major, 1998].

Cesium has one stable isotope, ^{133}Cs, with a nuclear spin of $I = 7/2$. There is a single outer electron, which in its ground state is in an s orbital with angular momentum $L = 0$. Therefore, since the electron spin is $S = 1/2$, its total angular momentum $J = L + S$ can be $\pm 1/2$. If the electron is parallel to the nucleus, the angular momentum of them together is $F = I + J = 4$, and if it is antiparallel $F = 3$. These energies will be split because of dipole–dipole coupling between the electron and the nucleus. This is called *fine structure*, because the scale is smaller than transitions between outer electron levels.

Chapter 15 will discuss the addition of quantum angular momentum, and the allowed spin states. The component m aligned with a magnetic field is quantized in integer steps,

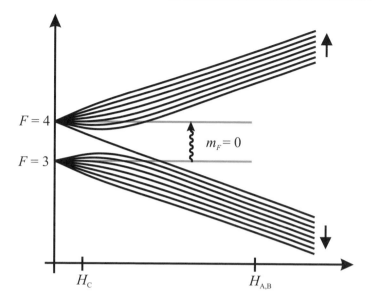

Figure 14.5. Cesium energy levels.

so that for $F = 3, m = -3, -2, \ldots, 2, 3$. The energies of these states are split due to the dipolar coupling to the field, revealing *hyperfine structure* as the field is increased. In the limit of a strong magnetic field the relative orientation of the electron and nuclear spins no longer matters, and the energy depends only on the alignment with the field. This is shown in Figure 14.5.

Now consider the *atomic clock* shown in Figure 14.6. Cesium atoms emerge from an oven into a vacuum chamber, and then pass through a strong magnetic field with a gradient. It's energetically favorable for one group of states to be in a weaker field, and the other in a stronger field, so they split into two beams. An aperture selects just one of these. The cesium atoms then enter into a region with a uniform magnetic field, carefully screened from external fields. Microwaves are applied with a frequency tuned to the $F = 3 \rightarrow 4$ transition between $m_F = 0$ levels, which to first order are independent of the magnetic field. The field used is just strong enough to split the hyperfine states, but not so strong as to start separating the multiplets.

Then the cesium beam passes through another strong gradient field. With no RF applied, the atoms once again are attracted in the direction of the stronger field, pulling them to a detector. But if the RF exactly matches the $F = 3 \rightarrow 4, m_F = 0$ transition frequency, atoms will emerge in the multiplet with the opposite orientation, being attracted to the weaker field and changing the flux reaching the detector. By modulating the RF, and feeding back the detector signal, its frequency can be locked onto the resonance. The frequency of this transition has now become the definition of the second, $\nu = 9\,192\,631\,770$ Hz. The spectral linewidth is limited by the frequency–time uncertainty from passing through the field, which is improved by making the middle region as long as possible. A *Ramsey cavity* reduces the required homogeneity by applying the RF coherently at just the beginning and end of the interval [Ramsey, 1972].

Switching from quartz to a cesium beam reduces the relative uncertainty in the time to

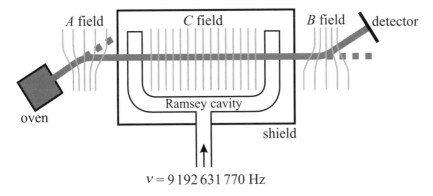

Figure 14.6. Cesium beam atomic clock.

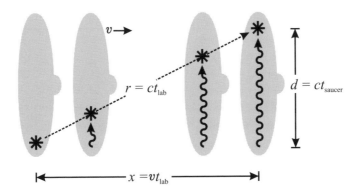

Figure 14.7. Time dilation.

$\sim 10^{-12}$. By using lasers to perform the state selection, and to cool the atoms so that they can be dropped in a gravitational *fountain*, this can be extended to $\sim 10^{-15}$ [Gibble & Chu, 1993]. And this spectacular performance is as stable and repeatable as the quantum states of the atom, which as far as we can tell does not have a cosmological drift.

14.3.2 Time

Atoms might not drift, but time itself can. This shocking conclusion follows from the failure of the Michelson–Morley experiment in 1887 to observe a difference in the speed of light moving parallel and perpendicular to the Earth's orbit that would be caused by motion through a medium carrying electromagnetic waves. This, plus Lorentz's development of transformations between inertial frames that leave Maxwell's equations unchanged, led to the theory of *special relativity* [Einstein, 1916]. The lack of an electromagnetic ether also helped lead to the recognition that electromagnetic radiation is quantized in photons.

Figure 14.7 shows why the experimental observation that the velocity of light is constant regardless of the relative motion of the emitter and detector implies that time must not be. A flying saucer is passing by with a velocity v. At $t = 0$ there is a flash of light, and both a stationary observer in the lab and a moving observer on the flying saucer synchronize their clocks. If the saucer has a diameter d, the flash will reach the far end when $d = ct_{\text{saucer}}$. At that instant the rocket will have traveled $x = vt_{\text{lab}}$, and in the lab frame the light will have traveled a distance $r = ct_{\text{lab}}$. The times in the two frames can be related by reconciling this geometry:

$$r^2 = x^2 + d^2$$
$$c^2 t_{\text{lab}}^2 = v^2 t_{\text{lab}}^2 + c^2 t_{\text{saucer}}^2$$
$$t_{\text{lab}} = \frac{t_{\text{saucer}}}{\sqrt{1 - v^2/c^2}}$$
$$\equiv \gamma t_{\text{rocket}} \quad . \tag{14.54}$$

More time appears to pass in the lab than on the saucer, with time in the moving frame stopping in the limit $v \to c$ (!). This *time dilation* must be the most dramatic application of Pythagoras' Theorem.

Scaling time in turn implies that momentum is also scaled between moving frames,

$$p = mv = m\frac{dx}{dt} \quad \to \quad m\frac{dx}{dt}\gamma = mv\gamma \quad . \tag{14.55}$$

Since energy is the integral of force times displacement, and force is the rate of change of the momentum in a frame, the energy after accelerating from 0 to v is

$$\begin{aligned}
E &= \int_0^v F \, dx \\
&= \int_0^v \frac{dp}{dt} \, dx \\
&= \int_0^v \frac{dx}{dt} \, dp \\
&= \int_0^v v' \, dp \\
&= \int_0^v v' \, d(mv'\gamma) \\
&= m \int_0^v v' \, d\left[\frac{v'}{\left(1 - v'^2/c^2\right)^{1/2}}\right] \\
&= m \int_0^v v' \left[\frac{1}{\left(1 - v'^2/c^2\right)^{1/2}} + \frac{v'^2/c^2}{\left(1 - v'^2/c^2\right)^{3/2}}\right] dv' \\
&= m \int_0^v \frac{v'}{\left(1 - v'^2/c^2\right)^{3/2}} \, dv' \\
&= \frac{mc^2}{\left(1 - v^2/c^2\right)^{1/2}} \\
&= mc^2\gamma \quad . \tag{14.56}
\end{aligned}$$

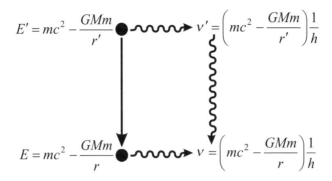

$$E' = mc^2 - \frac{GMm}{r'} \qquad v' = \left(mc^2 - \frac{GMm}{r'}\right)\frac{1}{h}$$

$$E = mc^2 - \frac{GMm}{r} \qquad v = \left(mc^2 - \frac{GMm}{r}\right)\frac{1}{h}$$

Figure 14.8. Gravitational red shift.

In the low-velocity limit this has the famous form

$$E = \frac{mc^2}{\left(1 - v^2/c^2\right)^{1/2}} \approx mc^2 + \frac{1}{2}mv^2 \quad . \tag{14.57}$$

The appearance of a *rest mass* energy mc^2 associated with matter rather than motion had enormous implications for physics (and everything else).

Now consider what happens when a mass m is moved in the gravitational field of a mass M from a radius r' to r, as shown in Figure 14.8. The rest mass energy doesn't change, but the gravitational potential energy does. From quantum mechanics we know that $E = h\nu$, so that in either position it's possible for all of the energy (rest plus potential) to be converted into a photon with frequency E/h. But then there's a problem, because if the photon instead of the mass travels from r' to r, the energies no longer match. By converting the photon back into mass, depending on the direction it travels it's possible to either run the cycle as a perpetual motion machine or cause the mass to vanish. Since that's an appealing but as far as we know entirely impossible outcome, the more plausible conclusion is that the frequency of light must depend on gravity. This is called the *gravitational red shift*.

In the last section we saw that electromagnetic oscillations define time, so the relative time at two radii must also be scaled by the ratio of the frequencies

$$t = \frac{1 - \dfrac{GM}{rc^2}}{1 - \dfrac{GM}{r'c^2}}\, t' \quad . \tag{14.58}$$

Time slows down as gravity increases, one of the most surprising predictions from *general relativity*, the extension of special relativity to include gravity [Einstein, 1905]. The seeds of quantum mechanics were lurking in the invariance of the speed of light in special relativity, and we obtained the red shift by combining quantum mechanics with special relativity, but to date gravity itself has defied a quantum explanation [Hawking, 1993].

The apparent passage of time depends on motion, and on gravity. While not familiar from our non-relativistic everyday experience, these effects are not only measurable by an atomic clock, they are essential corrections needed by practical electromagnetic ranging systems (Problem 14.5).

14.3.3 Position

The history of time is intimately tied to the history of navigation. The latitude of a ship at sea can easily be determined from the stars, but because of the Earth's rotation a reliable time reference is needed to find its longitude. A catastrophic shipwreck in 1707 led the British government to offer a large prize for an improved chronometer, which was eventually claimed by John Harrison in 1765 for a clever mechanical design good to better than 1 s/day, a relative error of 10^{-5} that is competitive with today's quartz clocks [Sobel, 1996].

British ships sailing from London set their chronometers by observing a red ball dropped on a pole above the Royal Observatory in Greenwich at 1 pm every day (chosen to follow noon astronomical observations). This is why balls are dropped to mark new years, and why *UTC* (*Coordinated Universal Time*), originally called *GMT* (*Greenwich Mean Time*), is still measured from the zero degree meridian that passes through the Observatory. While modern position measurement systems have improved on the means for clock distribution, they still rely on timing measurements made relative to a reference event at a known time and location.

Schemes for timing radiation scattered from a passive object include *RADAR* (*RAdio Detection And Ranging*) [Skolnik, 1990], *LIDAR* (*LIght Detection And Ranging*), and *SONAR* (*SOund Navigation And Ranging*). Acoustic ranging is the simplest to implement because of the slow propagation speeds: at 20 °C the speed of sound is 343 m/s in air, 1482 m/s in water, and 5960 m/s in steel. These vary with temperature and ambient conditions, requiring environmental corrections and limiting the accuracy of absolute measurements. The index of refraction for light is a weaker function of atmospheric conditions, and it can propagate much further, but optical ranging requires timing on much faster scales. A *time of flight* measurement determines the distance directly from the arrival delay

$$d = v\,\Delta t \;\Rightarrow\; \Delta t = \frac{d}{v}\quad. \tag{14.59}$$

This can be converted to a phase measurement by modulating the light with an amplitude envelope with a frequency ν, so that an observed modulation phase shift $\Delta\varphi$ is related to the distance by

$$d = \frac{\Delta\varphi}{2\pi}\frac{v}{\nu} \;\Rightarrow\; \Delta\varphi = \frac{2\pi\nu d}{v}\quad. \tag{14.60}$$

Now ν can be selected to bring $\Delta\varphi$ into a convenient range. This is how *laser range-finders* work [Rüeger, 1990].

In all of these systems the transmitter and receiver share a clock so that the timing needs only be reliable over the propagation interval. In a *bistatic* configuration the transmitter is separated from the receiver, putting much more stringent demands on their synchronization. And if an active receiver seeks to directly determine its position from multiple transmitters, they must each have reliable independent time references. The development of atomic clocks that can fly in a satellite makes this possible globally in the *GPS* (*Global Positioning System*) [Bhaskar *et al.*, 1996].

GPS uses the *NAVSTAR* (*NAVigation System with Timing And Ranging*) constellation of 24 satellites at an altitude of 20 200 km in six orbital planes at an inclination to the equator of 55°. Each satellite transmits a message containing its orbital parameters (which are frequently updated from tracking ground stations) and the time from onboard atomic clocks. Knowledge of the distance to one satellite locates the receiver on a sphere, two satellites constrain it to a curve, and three to a pair of points (one of which is usually unphysical). To eliminate the need for the receiver to have an independent atomic clock to determine the range, the signal from a fourth satellite is used to overdetermine the geometry and let the three-dimensional position be found using only relative arrival measurements that can be performed with a quartz oscillator.

The satellites transmit at two frequencies, 1575.42 MHz and 1227.6 MHz, to enable correction of ionospheric propagation delays. To improve both the timing resolution and the interference rejection a spread spectrum modulation is used. There are two codes. The first, *C/A* (*Coarse/Acquisition*), has a sequence of 1023 chips that repeats every millisecond. This is used for the *Standard Positioning Service* (*SPS*) available to anyone, and to transmit the acquisition information for the cryptographically-encoded *Precise Positioning Service* (*PPS*) that has a period of seven days and is restricted to the military.

The resolution of PPS is on the order of 10 m. The possible C/A resolution is about 40 m; this was degraded by the infamous Selective Availability (SA) program to 100 m to limit the strategic use of GPS. These intentional random errors were eliminated in *DGPS* (*Differential GPS*) by using the signal from a ground station with a known location to deduce the correction. With averaging, GPS is used for sub–centimeter measurements of geolotical motion [Herring, 1999].

Problem 14.5 looks at the importance of relativistic corrections to the nanosecond timing of the GPS atomic clocks. Relativity can also be applied directly to measure rotation through the *Sagnac effect*. It will take light a time

$$\tau = \frac{2\pi r}{c} \tag{14.61}$$

to be guided by mirrors or a fiber around a ring of radius r. If the ring is rotating at a rate Ω, the tangential velocity is

$$\Omega \left(\frac{\text{rad}}{\text{s}} \right) \times \frac{1 \text{ cycle}}{2\pi \text{ rad}} \times \frac{2\pi r}{1 \text{ cycle}} = \Omega r \quad . \tag{14.62}$$

The speed of light is the same in the ring frame and in the laboratory frame, but the distance traveled is not. During the time τ the perimeter of the ring will have advanced by a distance $\Omega r \tau$. If the light is traveling in the direction of rotation this distance is added to its path, and if it is traveling in the opposite direction it is subtracted. The difference between the two paths is

$$\Delta l = 2\tau \Omega r = \frac{4\pi \Omega r^2}{c} \quad . \tag{14.63}$$

Beams sent in both directions and then recombined will generate interference fringes from the path difference. This provides a way to measure Ω that is robust because there are no moving parts, and has an enormous dynamic range because of the periodicity of the fringe pattern. The sensitivity can be increased by sending the light around many times, either through multiple fiber optic turns in a *fiber-optic gyroscope*, or by using the ring as the cavity of a laser in a *ring laser gyro* [Chow et al., 1985].

14.4 SELECTED REFERENCES

[Fraden, 1993] Fraden, Jacob. (1993). *AIP Handbook of Modern Sensors: Physics, Designs and Applications*. New York: American Institute of Physics.

∼ 500 pages × 1 sensor/page.

[Tinkham, 1995] Tinkham, Michael. (1995). *Introduction to Superconductivity*. 2nd edn. New York: McGraw-Hill.

Superconductivity, with a strong emphasis on the phenomenology rather than the formal theory.

[Yourgrau *et al.*, 1982] Yourgrau, Wolfgang, van der Merwe, Alwyn, & Raw, Gough. (1982). *Treatise on Irreversible and Statistical Thermophysics: an Introduction to Nonclassical Thermodynamics*. New York: Dover.

Werner Heisenberg "... likes this book very much indeed", and so do I.

[Major, 1998] Major, Fouad G. (1998). *The Quantum Beat: The Physical Principles of Atomic Clocks*. New York: Springer.

All of the physics behind, within, and beyond atomic clocks.

[Taylor & Wheeler, 1992] Taylor, Edwin F., & Wheeler, John Archibald. (1992). *Spacetime Physics: Introduction to Special Relativity*. 2nd edn. New York: W.H. Freeman.

A marvelous guide to special relativity.

[Misner *et al.*, 1973] Misner, C.W., Wheeler, J.A., & Thorne, K.S. (1973). *Gravitation*. New York: W.H. Freeman & Co.

A marvelous guide to general relativity.

14.5 PROBLEMS

(14.1) Do a Taylor expansion of equation (14.6) around $V = 0$.

(14.2) If a SQUID with an area of $A = 1$ cm^2 can detect 1 flux quantum, how far away can it sense the field from a wire carrying 1 A?

(14.3) Typical parameters for a quartz resonator are $C_e = 5$ pF, $C_m = 20$ fF, $L_m = 3$ mH, $R_m = 6$ Ω. Plot, and explain, the dependence of the reactance (imaginary part of the impedance), resistance (real part), and the phase angle of the impedance on the frequency.

(14.4) If a ship traveling on the equator uses one of John Harrison's chronometers to navigate, what is the error in its position after one month? What if it uses a cesium beam atomic clock?

(14.5) GPS satellites orbit at an altitude of 20 180 km.

   (*a*) How fast do they travel?
   (*b*) What is their orbital period?
   (*c*) Estimate the special-relativistic correction over one orbit between a clock on a GPS satellite and one on the Earth. Which clock goes slower?
   (*d*) What is the general-relativistic correction over one orbit? Which clock goes slower?

15 Quantum Computing and Communications

We've seen many ways in which physical laws set profound limits: messages can't travel faster than the speed of light; transistors can't be shrunk below the size of an atom. Conventional scaling of device performance, which has been based on increasing clock speeds and decreasing feature sizes, is already bumping into such fundamental constraints. But these same laws also contain opportunities. The universe offers many more ways to communicate and manipulate information than we presently use, most notably through quantum coherence. Specifying the state of N quantum bits, called *qubits*, requires 2^N coefficients. That's a lot.

Quantum mechanics is essential to explain semiconductor band structure, but the devices this enables perform classical functions. Perfectly good logic families can be based on the flow of a liquid or a gas; this is called *fluidic logic* [Spuhler, 1983] and is used routinely in automobile transmissions. In fact, quantum effects can cause serious problems as VLSI devices are scaled down to finer linewidths, because small gates no longer work the same as their larger counterparts do.

But what happens if the bits themselves are governed by quantum rather than classical laws? Since quantum mechanics is so important, and so strange, a number of early pioneers wondered how a quantum mechanical computer might differ from a classical one [Benioff, 1980; Feynman, 1982; Deutsch, 1985]. Feynman, for example, objected to the exponential cost of simulating a quantum system on a classical computer, and conjectured that a quantum computer could simulate another quantum system efficiently (i.e., with less than exponential resources). We now know this to be true, and it turns out to be just one of the remarkable implications of representing information quantum mechanically. A quantum computer could find prime factors in polynomial time rather than the exponential time required classically, thereby defeating conventional cryptosystems used for electronic commerce and information security. Quantum mechanics also offers an entirely new way to protect information by applying the foundations of quantum measurement rather than number theory, and it can even be used to teleport particles instead of just communicating bits.

To understand the engineering implementation of such science-fiction fantasies it's necessary to understand quantum mechanics (to the extent that's possible). This will entail looking more deeply at the formal structure of the theory than we have in our passing uses of it so far, then extending it to include angular momentum and thermodynamics, and finally building up a language to describe quantum information, protocols, circuits, and algorithms.

15.1   QUANTUM MECHANICS

Of the many attempts to explain the foundations of quantum mechanics, arguably the most successful one is: *because that's the way it is*. Experimental observations have inspired and then justified the development of a theory that is remote from our everyday experience. For this reason, intuition is dangerous in quantum mechanics. At the outset, when in doubt, it's best to focus on the mathematical formalism and leave questions about meaning and interpretation to later when intuition is tempered by experience.

15.1.1   States and Operators

The state of a quantum system is completely specified by its *wave function* $|\psi\rangle$. *Operators* associate transformations of the wave function with observable properties. If an operator $\hat{X}$ acting on a wave function $|n\rangle$ returns the wave function multiplied by a coefficient x_n

$$\hat{X}|n\rangle = x_n|n\rangle \tag{15.1}$$

then $|n\rangle$ is said to be an *eigenvector* of the operator, x_n the corresponding *eigenvalue*, and the index n is the *quantum number* of the state. The spectrum of eigenvalues can be continuous, such as the position of a free particle, or discrete, like the energy levels of an atom.

Quantum operators are linear, which means that

$$\hat{X}\left(\alpha|a\rangle + \beta|b\rangle\right) = \alpha\hat{X}|a\rangle + \beta\hat{X}|b\rangle \quad , \tag{15.2}$$

for arbitrary complex constants α and β. This is an experimental observation, although if quantum mechanics *was* nonlinear then it would be possible to send messages faster than light or solve intractable computational problems [Weinberg, 1989; Abrams & Lloyd, 1998] – remarkable but unlikely capabilities.

A complete set of eigenfunctions forms a basis that can be used to expand an arbitrary wave function

$$|\psi\rangle = \sum_n a_n|n\rangle \quad . \tag{15.3}$$

$|a_n|^2$ is the probability that a measurement of $\hat{X}_n$ will return the value x_n. If this does happen, then the wave function is said to *collapse* into the state $|n\rangle$ regardless of what basis it started in. A quantum measurement necessarily disturbs the system unless it is already in an eigenstate.

The coefficients a_n provide a representation of the wave function, conventionally written as a column vector called the *state vector*. The space of these vectors is called a *Hilbert space*, which will be finite- or infinite-dimensional for discrete or continuous eigenvalues respectively. Choosing a different set of eigenfunctions gives a different representation for the state vector. Operators on state vectors are matrices.

In the state vector representation, quantum mechanical manipulations are just linear algebra [Strang, 1988]. Associated with each column vector $|\psi\rangle$ is a row vector $\langle\psi|$

$$\langle\psi| = |\psi\rangle^{\dagger} \quad , \tag{15.4}$$

where the dagger operator is defined to be the complex conjugate of the transpose

$$\begin{pmatrix} a & b \\ c & d \\ e & f \end{pmatrix}^{\dagger} = \begin{pmatrix} a^* & c^* & e^* \\ b^* & d^* & f^* \end{pmatrix} \tag{15.5}$$

and is called the *Hermitian adjoint.* Taking the Hermitian adjoint of a product reverses the order

$$\left(\hat{X}\hat{Y}\right)^{\dagger} = \hat{Y}^{\dagger}\hat{X}^{\dagger} \tag{15.6}$$

and so in particular

$$\left(\hat{X}|\psi\rangle\right)^{\dagger} = \langle\psi|\hat{X}^{\dagger} \quad . \tag{15.7}$$

The adjoint of a scalar is just the complex conjugate

$$\left(x|\psi\rangle\right)^{\dagger} = \langle\psi|x^* \quad . \tag{15.8}$$

The *inner product* of two state vectors is

$$\langle\psi|\varphi\rangle = (\psi_1^* \ \psi_2^* \ \cdots) \times \begin{pmatrix} \varphi_1 \\ \varphi_2 \\ \vdots \end{pmatrix}$$

$$= \psi_1^*\varphi_1 + \psi_2^*\varphi_2 + \cdots \quad . \tag{15.9}$$

Dirac named $\langle\psi|$ a *bra*, $|\varphi\rangle$ a *ket*, and $\langle\psi|\varphi\rangle$ a ... *bracket* (get it?).

Eigenfunctions are orthonormal,

$$\langle j|k\rangle = \delta_{ij} \quad . \tag{15.10}$$

Wave functions are therefore normalized because the sum of the probabilities of measurement outcomes must equal 1:

$$\langle\psi|\psi\rangle = \sum_j \langle j|a_j^* a_j|j\rangle$$

$$= \sum_j |a_j|^2$$

$$= 1 \quad . \tag{15.11}$$

The *expectation value* of an operator is found from

$$\langle\hat{X}\rangle = \langle\psi|\hat{X}|\psi\rangle \quad , \tag{15.12}$$

with an adjoint

$$\left(\langle\psi|\hat{X}|\psi\rangle\right)^{\dagger} = \langle\psi|\hat{X}^{\dagger}|\psi\rangle \quad . \tag{15.13}$$

Taking the expectation value of an eigenfunction

$$\langle n|\hat{X}|n\rangle = \langle n|x_n|n\rangle$$

$$= x_n\langle n|n\rangle$$

$$= x_n \quad , \tag{15.14}$$

and its adjoint

$$\left(\langle n|\hat{X}|n\rangle \right)^{\dagger} = \langle n|\hat{X}^{\dagger}|n\rangle$$
$$= \langle n|x_n^*|n\rangle$$
$$= x_n^* \langle n|n\rangle$$
$$= x_n^* \quad , \tag{15.15}$$

shows that if the observable is real valued ($x_n = x_n^*$) then $\hat{X}^{\dagger} = \hat{X}$. An operator with this property is said to be *Hermitian*. Since the real world has real-valued observables, measurable quantities are associated with Hermitian operators.

The *outer product* of two state vectors is an operator

$$|\psi\rangle\langle\varphi| = \begin{pmatrix} \psi_1 \\ \psi_2 \\ \vdots \end{pmatrix} \times (\varphi_1^* \ \varphi_2^* \ \cdots)$$

$$= \begin{pmatrix} \psi_1\varphi_1^* & \psi_1\varphi_2^* & \cdots \\ \psi_2\varphi_1^* & \psi_2\varphi_2^* & \cdots \\ \vdots & \vdots & \end{pmatrix} \quad . \tag{15.16}$$

The outer product of two eigenvectors

$$\hat{P}_n = |n\rangle\langle n| \tag{15.17}$$

is called a *projector* because of its action on a wave function

$$\hat{P}_n|\psi\rangle = |n\rangle\langle n| \sum_m a_m|m\rangle$$
$$= \sum_m a_m|n\rangle\langle n|m\rangle$$
$$= a_n|n\rangle\langle n|n\rangle$$
$$= a_n|n\rangle \quad . \tag{15.18}$$

The expectation value of a projector is just the likelihood that a measurement will result in that state

$$\langle\psi|\hat{P}_n|\psi\rangle = \sum_m \langle m|a_m^*|n\rangle\langle n|a_m|m\rangle$$
$$= \sum_m |a_m|^2|\langle n|m\rangle|^2$$
$$= |a_n|^2 \quad . \tag{15.19}$$

Projectors can be used to write an operator in a *spectral representation* in terms of its eigenvalues

$$\hat{X} = \sum_n x_n|n\rangle\langle n| \quad . \tag{15.20}$$

This provides a natural way to define a function of an operator as

$$f(\hat{X}) = \sum_n f(x_n)|n\rangle\langle n| \quad . \tag{15.21}$$

The *commutator* of two operators is defined to be

$$[\hat{A}, \hat{B}] = \hat{A}\hat{B} - \hat{B}\hat{A} \quad . \tag{15.22}$$

If the commutator vanishes then the eigenvectors of $\hat{A}$ and $\hat{B}$ will be shared,

$$\hat{A}\hat{B} - \hat{B}\hat{A} = 0 \;\Rightarrow\; \hat{A}\hat{B}|ab\rangle = \hat{B}\hat{A}|ab\rangle$$
$$\hat{A}b|ab\rangle = \hat{B}a|ab\rangle$$
$$ab|ab\rangle = ba|ab\rangle \quad , \tag{15.23}$$

but in general this need not be the case.

Consider two operators that do not commute, $[\hat{A}, \hat{B}] = i\hat{C}$, where $\hat{C}$ is written with a prefactor of i for convenience in the following calculation. If we define new operators by subtracting off the expectation value $\delta\hat{A} \equiv \hat{A} - \langle\hat{A}\rangle$, $\delta\hat{B} \equiv \hat{B} - \langle\hat{B}\rangle$, then these operators will satisfy the same commutation relation $[\delta\hat{A}, \delta\hat{B}] = i\hat{C}$ because the scalar expectation values of the operators do commute. The expected value of the commutator is then

$$\langle\delta\hat{A}\,\delta\hat{B}\rangle - \langle\delta\hat{B}\,\delta\hat{A}\rangle = i\langle\hat{C}\rangle$$
$$\left[\langle\delta\hat{A}\,\delta\hat{B}\rangle - \langle\delta\hat{B}\,\delta\hat{A}\rangle\right]^2 = -\langle\hat{C}\rangle^2$$
$$\langle\delta\hat{A}\,\delta\hat{B}\rangle^2 + \langle\delta\hat{B}\,\delta\hat{A}\rangle^2$$
$$-2\langle\delta\hat{A}\,\delta\hat{B}\rangle\langle\delta\hat{B}\,\delta\hat{A}\rangle = -\langle\hat{C}\rangle^2$$
$$\langle\delta\hat{A}\,\delta\hat{B}\rangle\langle\delta\hat{B}\,\delta\hat{A}\rangle = \frac{1}{2}\left[\langle\delta\hat{A}\,\delta\hat{B}\rangle^2 + \langle\delta\hat{B}\,\delta\hat{A}\rangle^2 + \langle\hat{C}\rangle^2\right]$$
$$= \frac{1}{2}\left[\langle\delta\hat{A}\,\delta\hat{B} + \delta\hat{B}\,\delta\hat{A}\rangle^2 - 2\langle\delta\hat{A}\,\delta\hat{B}\rangle\langle\delta\hat{B}\,\delta\hat{A}\rangle + \langle\hat{C}\rangle^2\right]$$
$$= \frac{1}{4}\left[\langle\delta\hat{A}\,\delta\hat{B} + \delta\hat{B}\,\delta\hat{A}\rangle^2 + \langle\hat{C}\rangle^2\right] \quad . \tag{15.24}$$

Using the Cauchy–Schwarz inequality yet again, the expectation of a product will be bounded by the product of the expectations

$$\langle\delta\hat{A}\,\delta\hat{B}\rangle\langle\delta\hat{B}\,\delta\hat{A}\rangle \leq \langle\delta\hat{A}^2\rangle\langle\delta\hat{B}^2\rangle \quad . \tag{15.25}$$

Therefore

$$\langle\delta\hat{A}^2\rangle\langle\delta\hat{B}^2\rangle \geq \frac{1}{4}\left[\langle\delta\hat{A}\,\delta\hat{B} + \delta\hat{B}\,\delta\hat{A}\rangle^2 + \langle\hat{C}\rangle^2\right] \quad . \tag{15.26}$$

If the anticommutator on the right hand side vanishes it sets a lower bound for the inequality of

$$\langle\delta\hat{A}^2\rangle\langle\delta\hat{B}^2\rangle \geq \frac{1}{4}\langle\hat{C}\rangle^2 \tag{15.27}$$

or

$$\langle\delta\hat{A}^2\rangle^{1/2}\langle\delta\hat{B}^2\rangle^{1/2} \equiv \Delta A\Delta B \geq \frac{1}{2}|\langle\hat{C}\rangle| \quad . \tag{15.28}$$

This is the *Heisenberg uncertainty principle*. The left hand side is the product of the expected spread around the expectation value of the operators, and the right hand side is the expectation value of the commutator. If two operators do not commute then determining one of them more precisely imposes less certainty on the distribution of outcomes for the other. Because position and momentum don't commute (equation 15.48)

$$\Delta x \Delta p \geq \frac{\hbar}{2} \quad . \tag{15.29}$$

If we know exactly where a particle is then we have no idea how fast it is moving, or *vice versa*.

The most important operator of all is the *Hamiltonian*, the sum of the operators for the potential $\hat{T}$ and kinetic energies $\hat{V}$

$$\hat{\mathcal{H}} = \hat{T} + \hat{V} \quad . \tag{15.30}$$

In terms of it, *Schrödinger's equation* gives the time evolution of a wave function,

$$i\hbar \frac{d}{dt} |\psi(t)\rangle = \hat{\mathcal{H}} |\psi(t)\rangle \quad . \tag{15.31}$$

Because $\hat{\mathcal{H}}$ is Hermitian, the adjoint of the Hamiltonian is

$$-i\hbar \frac{d}{dt} \langle \psi(t)| = \langle \psi(t)|\hat{\mathcal{H}}^{\dagger}$$
$$= \langle \psi(t)|\hat{\mathcal{H}} \quad . \tag{15.32}$$

The time derivative of the expectation value of an operator that does not depend on time can therefore be found from the chain rule,

$$i\hbar \frac{d}{dt} \langle \hat{A} \rangle = i\hbar \frac{d}{dt} \langle \psi(t)|\hat{A}|\psi(t)\rangle$$
$$= i\hbar \langle \psi(t)|\hat{A} \frac{d|\psi(t)\rangle}{dt} + i\hbar \frac{d\langle\psi(t)|}{dt} \hat{A}|\psi(t)\rangle$$
$$= \langle \psi(t)|\hat{A}\hat{\mathcal{H}}|\psi(t)\rangle - \langle \psi(t)|\hat{\mathcal{H}}\hat{A}|\psi(t)\rangle$$
$$= \langle \psi(t)|[\hat{A},\hat{\mathcal{H}}]|\psi(t)\rangle$$
$$= \langle [\hat{A},\hat{\mathcal{H}}] \rangle \quad . \tag{15.33}$$

This is *Ehrenfest's Theorem*; Problem 15.1 extends it to time-dependent operators. One consequence is that if an operator commutes with the Hamiltonian its observable is a constant of the motion. Another consequence follows from combining equations (15.28) and (15.33) to find

$$\Delta \hat{A} \Delta \hat{\mathcal{H}} \geq \frac{\hbar}{2} \left| \frac{d\langle \hat{A} \rangle}{dt} \right| \quad . \tag{15.34}$$

In terms of this derivative, a time Δt can be defined as

$$\left| \frac{d\langle \hat{A} \rangle}{dt} \right| \equiv \frac{\Delta \hat{A}}{\Delta t} = \frac{\langle \delta \hat{A}^2 \rangle^{1/2}}{\Delta t} \quad . \tag{15.35}$$

Δt is the time it takes the expected value of an operator, at its instantaneous rate of change, to differ by its standard deviation. Recognizing that $\Delta \hat{\mathcal{H}} \equiv \Delta E$ is the expected energy spread,

$$\Delta \hat{A} \Delta E \geq \frac{\hbar}{2} \frac{\Delta \hat{A}}{\Delta t} \quad , \tag{15.36}$$

and cancelling $\Delta \hat{A}$ gives

$$\Delta E \Delta t \geq \frac{\hbar}{2} \quad . \tag{15.37}$$

This is the *energy–time uncertainty relation*. It fundamentally differs from the other uncertainty relationships because time appears in quantum mechanics as a parameter rather than an operator. Equation (15.37) can be interpreted as putting a quantum limit on how the energy distribution in a system bounds the rate at which an observable property of the system can change, although there are *many* other ways to (mis)define and (mis)interpret it [Aharonov & Bohm, 1961; Peres, 1993].

The *evolution operator* $\hat{U}(t)$ is the one that advances a wave function in time:

$$|\psi(t)\rangle = \hat{U}(t)|\psi(0)\rangle \quad . \tag{15.38}$$

Differentiating both sides and comparing to equation (15.31) shows that $\hat{U}$ satisfies

$$i\hbar \frac{d}{dt}|\psi(t)\rangle = i\hbar \frac{d}{dt}\hat{U}(t)|\psi(0)\rangle$$

$$\hat{\mathcal{H}}|\psi(t)\rangle = i\hbar \frac{d}{dt}\hat{U}(t)|\psi(0)\rangle$$

$$\hat{\mathcal{H}}\hat{U}(t)|\psi(0)\rangle = i\hbar \frac{d}{dt}\hat{U}(t)|\psi(0)\rangle \quad . \tag{15.39}$$

The evolution operator therefore is a solution to

$$i\hbar \frac{d}{dt}\hat{U}(t) = \hat{\mathcal{H}}\hat{U}(t) \quad . \tag{15.40}$$

If the Hamiltonian does not explicitly depend on time this can be integrated to find that

$$\hat{U}(t) = e^{-i\hat{\mathcal{H}}t/\hbar} \quad , \tag{15.41}$$

where the exponential of an operator can be defined by equation (15.21) or a power series expansion

$$e^{-i\hat{\mathcal{H}}t/\hbar} = \hat{1} - \frac{it}{\hbar}\hat{\mathcal{H}} + \frac{1}{2!}\left(\frac{it}{\hbar}\right)^2 \hat{\mathcal{H}}\hat{\mathcal{H}} + \cdots \tag{15.42}$$

(where $\hat{1}$ is the identity operator).

We know that wave functions always start out normalized:

$$\langle \psi(0)|\psi(0)\rangle = 1 \quad . \tag{15.43}$$

At a future time the inner product is

$$\langle \psi(t)|\psi(t)\rangle = \langle \psi(0)|\hat{U}(t)^\dagger \hat{U}(t)|\psi(0)\rangle \quad , \tag{15.44}$$

which must still equal 1. Therefore

$$\hat{U}^\dagger \hat{U} = \hat{I} \quad , \tag{15.45}$$

where $\hat{I}$ is the identity operator (matrix). An operator with this property is said to be *unitary*, and corresponds to a rotation in Hilbert space.

15.1.2 Angular Momentum

The non-obvious properties of quantum mechanical angular momentum are essential to understanding quantum information. They are found by generalizing the classical definition. The quantum position operator is simple in the position basis, just giving the coordinates

$$\hat{\vec{x}} = \vec{x} \quad . \tag{15.46}$$

The momentum operator is less trivial in the position basis; it is equal to the spatial gradient of the wave function

$$\hat{\vec{p}} = -i\hbar\nabla \tag{15.47}$$

(while this can be motivated, it's simplest to take it as an experimental fact). Therefore position and momentum do not commute:

$$[\hat{\vec{x}}, \hat{\vec{p}}]|\psi\rangle = -\vec{x}i\hbar\nabla|\psi\rangle + i\hbar\nabla\vec{x}|\psi\rangle$$
$$= -\vec{x}i\hbar\nabla|\psi\rangle + i\hbar|\psi\rangle + i\hbar\vec{x}\nabla|\psi\rangle$$
$$= i\hbar|\psi\rangle$$
$$[\hat{\vec{x}}, \hat{\vec{p}}] = i\hbar \quad . \tag{15.48}$$

The angular momentum operator is found from

$$\hat{\vec{L}} = \hat{\vec{x}} \times \hat{\vec{p}} \quad . \tag{15.49}$$

Writing out the components shows that

$$[\hat{L}_l, \hat{x}_m] = i\hbar\epsilon_{lmn}\hat{x}_n \quad , \tag{15.50}$$

$$[\hat{L}_l, \hat{p}_m] = i\hbar\epsilon_{lmn}\hat{p}_n \quad , \tag{15.51}$$

and

$$[\hat{L}_l, \hat{L}_m] = i\hbar\epsilon_{lmn}\hat{L}_n \quad . \tag{15.52}$$

The last commutator is satisfied by any quantity that behaves like an angular momentum, including the spin degrees of freedom.

Because the components of angular momentum do not commute it is not possible for a wave function to simultaneously be in an eigenstate of all of them. By convention, the direction that is an angular momentum eigenstate is taken to be z. But evaluating the components does show that the angular momentum operator commutes with the total angular momentum $\hat{L}^2 = \hat{\vec{L}} \cdot \hat{\vec{L}}$

$$[\hat{L}^2, \hat{\vec{L}}] = 0 \quad . \tag{15.53}$$

Therefore a wave function can simultaneously be an eigenfunction of the total angular momentum and its component in one direction. If l is the quantum number for the total angular momentum and m the quantum number for the z component, an angular momentum eigenstate can be written as $|l, m\rangle$. For the coming calculations it will be convenient to define these indices by

$$\hat{L}^2|l, m\rangle = \hbar^2 l(l+1)|l, m\rangle$$
$$\hat{L}_z|l, m\rangle = \hbar m|l, m\rangle \quad . \tag{15.54}$$

Two useful new operators $\hat{L}_+$ and $\hat{L}_-$ can be defined in terms of the orthogonal directions

$$\hat{L}_\pm = \hat{L}_x \pm i\hat{L}_y \quad . \tag{15.55}$$

From this definition its straightforward to show that

$$\hat{L}_-^\dagger = \hat{L}_+ \quad , \tag{15.56}$$

$$\hat{L}_+\hat{L}_- = \hat{L}^2 - \hat{L}_z^2 + \hbar\hat{L}_z \quad , \tag{15.57}$$

$$[\hat{L}_z, \hat{L}_\pm] = \pm\hbar\hat{L}_\pm \quad , \tag{15.58}$$

and

$$[\hat{L}_+, \hat{L}_-] = 2\hbar\hat{L}_z \quad . \tag{15.59}$$

Because of the adjoint relationship, the product of these operators must be positive

$$\begin{aligned}
\langle l, m|\hat{L}_+\hat{L}_-|l, m\rangle &= \langle l, m|\hat{L}_-^\dagger \hat{L}_-|l, m\rangle \\
&= |\hat{L}_-|l, m\rangle|^2 \\
&\geq 0 \quad .
\end{aligned} \tag{15.60}$$

But we also know that

$$\begin{aligned}
\langle l, m|\hat{L}_+\hat{L}_-|l, m\rangle &= \langle l, m|\hat{L}^2 - \hat{L}_z^2 + \hbar\hat{L}_z|l, m\rangle \\
&= \hbar^2[l(l + 1) - m^2 + m] \langle l, m|l, m\rangle \\
&= \hbar^2[l(l + 1) - m^2 + m] \quad .
\end{aligned} \tag{15.61}$$

Therefore

$$l(l + 1) - m^2 + m \geq 0 \quad . \tag{15.62}$$

And taking the operators in the opposite order gives

$$l(l + 1) - m^2 - m \geq 0 \quad . \tag{15.63}$$

That means that

$$l^2 + l \geq m^2 \pm m \quad , \tag{15.64}$$

which can only be satisfied if

$$-l \leq m \leq l \quad . \tag{15.65}$$

Now consider the following sequence of operators:

$$\begin{aligned}
\hat{L}_z\hat{L}_-|l, m\rangle &= (\hat{L}_-\hat{L}_z - \hbar\hat{L}_-)|l, m\rangle \\
&= \hbar(m - 1)\hat{L}_-|l, m\rangle \quad .
\end{aligned} \tag{15.66}$$

$\hat{L}_-$ gives a new eigenfunction of $\hat{L}_z$ with m reduced by 1; for this reason it is called a *lowering operator*. The normalization can be found from equation (15.61) to be

$$\hat{L}_-|l, m\rangle = \hbar\sqrt{l(l + 1) - m(m - 1)}|l, m - 1\rangle \quad . \tag{15.67}$$

Likewise, $\hat{L}_+$ is a *raising operator*

$$\hat{L}_+|l, m\rangle = \hbar\sqrt{l(l + 1) - m(m + 1)}|l, m + 1\rangle \quad . \tag{15.68}$$

Since $-l \leq m \leq l$, and we've just found that m can be changed by integer increments n,

$$-l + nm = l$$
$$\frac{nm}{2} = l \quad . \tag{15.69}$$

l must be either an integer $(0, 1, \ldots)$ or a half-integer $(1/2, 3/2, \ldots)$.

A spin-$1/2$ particle like an electron or a proton has $l = 1/2$, and two possible m states which can be written equivalently as

$$|l, m\rangle = |1/2, 1/2\rangle = |\uparrow\rangle = \begin{pmatrix} 1 \\ 0 \end{pmatrix}$$

$$|l, m\rangle = |1/2, -1/2\rangle = |\downarrow\rangle = \begin{pmatrix} 0 \\ 1 \end{pmatrix} \quad . \tag{15.70}$$

In the last basis the angular momentum operator can be written in terms of the *Pauli spin matrices*

$$\hat{\vec{L}} = \frac{\hbar}{2}\hat{\vec{\sigma}}$$

$$\hat{\sigma}_x = \begin{pmatrix} 0 & 1 \\ 1 & 0 \end{pmatrix} \quad \hat{\sigma}_y = \begin{pmatrix} 0 & -i \\ i & 0 \end{pmatrix} \quad \hat{\sigma}_z = \begin{pmatrix} 1 & 0 \\ 0 & -1 \end{pmatrix} \quad , \tag{15.71}$$

which can be verified by checking their commutation relations.

The spin magnetic moment $\vec{\mu}$ of a particle is related to its angular momentum $\vec{L}$ by the gyromagnetic ratio $\vec{\mu} = \gamma\vec{L}$ (Section 9.4). The moment can thus be determined by the operator

$$\hat{\vec{\mu}} = \gamma\hat{\vec{L}}$$
$$= \gamma\frac{\hbar}{2}\hat{\vec{\sigma}} \quad . \tag{15.72}$$

If the spin density is n, the macroscopic spin magnetization is therefore

$$\vec{M} = n\gamma\frac{\hbar}{2}\langle\psi|\hat{\vec{\sigma}}|\psi\rangle \quad . \tag{15.73}$$

Now consider two systems labelled a and b, with angular momentum operators $\hat{\vec{L}}(a)$ and $\hat{\vec{L}}(b)$. Because they satisfy the angular momentum relations individually, their sum

$$\hat{\vec{L}} = \hat{\vec{L}}(a) + \hat{\vec{L}}(b) \tag{15.74}$$

will also. But because

$$\hat{L}^2 = \left[\hat{\vec{L}}(a) + \hat{\vec{L}}(b)\right]^2$$

$$= \hat{L}^2(a) + \hat{L}^2(b) + 2\hat{\vec{L}}(a) \cdot \hat{\vec{L}}(b)$$

$$= \hat{L}^2(a) + \hat{L}^2(b) + 2\hat{L}_z(a)\hat{L}_z(b) + \hat{L}_+(a)\hat{L}_-(b) + \hat{L}_-(a)\hat{L}_+(b) \tag{15.75}$$

the eigenfunctions for systems a and b will in general not be angular momentum eigenfunctions of the combined system. For spin $1/2$ there are four eigenfunctions of the individual systems

$$| \uparrow\uparrow \rangle , | \uparrow\downarrow \rangle , | \downarrow\uparrow \rangle , | \downarrow\downarrow \rangle \quad , \tag{15.76}$$

and the magnitude of the total angular momentum is from equation (15.75)

$$\hat{L}^2 = \frac{3}{4} + 2\hat{L}_z(a)\hat{L}_z(b) + \hat{L}_+(a)\hat{L}_-(b) + \hat{L}_-(a)\hat{L}_+(b) \quad . \tag{15.77}$$

Substitution shows that the eigenfunctions of the combined systems are

$$|l = 0, m = 0\rangle = \frac{1}{\sqrt{2}}(|\uparrow\downarrow\rangle - |\downarrow\uparrow\rangle)$$

$$|l = 1, m = -1\rangle = |\downarrow\downarrow\rangle$$

$$|l = 1, m = 0\rangle = \frac{1}{\sqrt{2}}(|\uparrow\downarrow\rangle + |\downarrow\uparrow\rangle)$$

$$|l = 1, m = 1\rangle = |\uparrow\uparrow\rangle \quad . \tag{15.78}$$

The $l = 0$ state is called a *singlet*, and the other three are called *triplet* states. These are the symmetric and antisymmetric spin states that we used in Chapter 12 to explain magnetism. For the more general case of adding arbitrary angular momenta it is necessary to use *Clebsch–Gordon coefficients* to find the expansion of the total angular momentum eigenfunctions in terms of the individual basis functions.

The angular momentum operator, not surprisingly, has an intimate connection with rotations. Consider a *rotation operator* $\hat{R}(\vec{\theta})$ that returns the wave function rotated by an angle θ around the axis of $\vec{\theta}$. Rotating the wave function forward is the same as rotating its argument backwards, and in the limit of a small angle the coordinate rotation matrix can be approximated by a cross product

$$\hat{R}(\delta\vec{\theta})|\psi(\vec{x})\rangle \approx |\psi(\vec{x} - \delta\vec{\theta} \times \vec{x})\rangle$$

$$\approx |\psi(\vec{x})\rangle - (\delta\vec{\theta} \times \vec{x}) \cdot \nabla|\psi(\vec{x})\rangle$$

$$= |\psi(\vec{x})\rangle - \delta\vec{\theta} \cdot \vec{x} \times \nabla|\psi(\vec{x})\rangle$$

$$= |\psi(\vec{x})\rangle - \delta\vec{\theta} \cdot \vec{x} \times \frac{1}{-i\hbar}\hat{\vec{p}}|\psi(\vec{x})\rangle$$

$$= |\psi(\vec{x})\rangle - \frac{i}{\hbar}\delta\vec{\theta} \cdot \hat{\vec{L}}|\psi(\vec{x})\rangle$$

$$\hat{R}(\delta\vec{\theta}) = \hat{I} - \frac{i}{\hbar}\delta\vec{\theta} \cdot \hat{\vec{L}} \quad . \tag{15.79}$$

The angular momentum operator has appeared as the generator of an infinitesimal rotation. A finite rotation can be built up from the product of many inifinitesimal rotations, so that if we write $\vec{\theta} = N\,\delta\vec{\theta}$ and use the limit

$$\lim_{N \to \infty} \left[1 + \frac{x}{N}\right]^N = e^x \tag{15.80}$$

we find the rotation operator to be

$$\hat{R}(\vec{\theta}) = e^{-i\vec{\theta} \cdot \vec{L}/\hbar} \quad . \tag{15.81}$$

Plugging in the Pauli spin matrices gives the rotation operators for each axis

(Problem 15.4):

$$\hat{R}_x(\theta) = \begin{pmatrix} \cos\frac{\theta}{2} & -i\sin\frac{\theta}{2} \\ -i\sin\frac{\theta}{2} & \cos\frac{\theta}{2} \end{pmatrix}$$

$$\hat{R}_y(\theta) = \begin{pmatrix} \cos\frac{\theta}{2} & -\sin\frac{\theta}{2} \\ \sin\frac{\theta}{2} & \cos\frac{\theta}{2} \end{pmatrix}$$

$$\hat{R}_z(\theta) = \begin{pmatrix} e^{-i\theta/2} & 0 \\ 0 & e^{i\theta/2} \end{pmatrix} \quad . \tag{15.82}$$

These are going to be essential ingredients in implementing quantum logic.

15.1.3 Density Matrices

Uncertainty is central to quantum mechanics through the unpredictability of the outcome of a measurement. It is also central to thermodynamics, in the probability distribution over allowed states. These have very different characters, though: the former is a fundamental limit while the latter reflects incomplete knowledge about a system with many degrees of freedom. These can be combined by using the *density matrix* $\hat{\rho}$.

A *pure state* is one for which we know the exact state $|n\rangle$; its density matrix is just defined to be the projector

$$\hat{\rho} = |n\rangle\langle n| \quad . \tag{15.83}$$

A *mixed state* is one that has a classical probability distribution over possible quantum states. If the classical probability for a mixed state to be in the quantum state $|\psi_i\rangle$ is p_i then the density matrix is a weighted sum over the projection operators

$$\hat{\rho} = \sum_i p_i |\psi_i\rangle\langle\psi_i| \quad . \tag{15.84}$$

The *trace* of an operator is defined by

$$\text{Tr}(\hat{X}) = \sum_n \langle n|\hat{X}|n\rangle \quad . \tag{15.85}$$

It has the property of *cyclic invariance*

$$\text{Tr}(\hat{X}\hat{Y}) = \text{Tr}(\hat{Y}\hat{X}) \quad , \tag{15.86}$$

and is unchanged under a change of basis. In a matrix representation of an operator the trace is equal to the sum of the diagonal elements. Evaluating the trace of a density matrix in the basis in which it is diagonal shows that it is normalized:

$$\text{Tr}(\hat{\rho}) = \sum_n \rho_{nn}$$

$$= \sum_n \langle n|\hat{\rho}|n\rangle$$

$$= \sum_n \langle n| \sum_{n'} p_{n'} |n'\rangle\langle n'|n\rangle$$

$$= \sum_n p_n$$

$$= 1 \quad . \tag{15.87}$$

But $\mathrm{Tr}(\hat{\rho}^2) \leq 1$, with equality holding for pure states (Problem 15.2). This provides a way to distinguish between the density matrices for pure and mixed states.

In terms of the trace over $\hat{\rho}$, the expectation value of an observable is

$$\begin{aligned}
\langle \hat{X} \rangle &= \mathrm{Tr}(\hat{\rho}\hat{X}) \\
&= \sum_n \langle n| \sum_i p_i |\psi_i\rangle\langle\psi_i| \hat{X} |n\rangle \\
&= \sum_n \langle n| \sum_i p_i \sum_{n'} a_{in'} |n'\rangle\langle n'| a_{in'}^* \hat{X} |n\rangle \\
&= \sum_n \sum_i p_i a_{in} \langle n| a_{in'}^* \hat{X} |n\rangle \\
&= \sum_i p_i \langle\psi_i| \hat{X} |\psi_i\rangle \quad .
\end{aligned} \tag{15.88}$$

This is a classically weighted sum over the quantum expectation.

The time derivative of the density matrix

$$\begin{aligned}
\frac{d\hat{\rho}}{dt} &= \sum_i p_i \left[\frac{d|\psi_i\rangle}{dt} \langle\psi_i| + |\psi_i\rangle \frac{d\langle\psi_i|}{dt} \right] \\
&= \sum_i p_i \left[\frac{1}{i\hbar} \hat{\mathcal{H}} |\psi_i\rangle\langle\psi_i| + |\psi_i\rangle \frac{1}{-i\hbar} \langle\psi_i| \hat{\mathcal{H}} \right] \\
&= \frac{1}{i\hbar} [\hat{\mathcal{H}}, \hat{\rho}]
\end{aligned} \tag{15.89}$$

is given by its commutator with the Hamiltonian. This is the *Liouville–von Neumann* evolution equation. The time dependence of the density matrix can also be found from equation (15.41) to be the unitary transformation

$$\begin{aligned}
\hat{\rho}(t) &= \sum_i p_i |\psi_i(t)\rangle\langle\psi_i(t)| \\
&= \sum_i p_i e^{-i\hat{\mathcal{H}}t/\hbar} |\psi_i(0)\rangle\langle\psi_i(0)| e^{i\hat{\mathcal{H}}t/\hbar} \\
&= e^{-i\hat{\mathcal{H}}t/\hbar} \hat{\rho}(0) e^{i\hat{\mathcal{H}}t/\hbar} \\
&= \hat{U}(t)\hat{\rho}(t)\hat{U}^\dagger(t) \quad .
\end{aligned} \tag{15.90}$$

Unitary evolution is a serious constraint on quantum dynamics. It cannot change the trace, eigenvalue spectrum, or entropy of a density matrix. This means that information cannot be created or erased, just rearranged. How can this be reconciled with our everyday experience of these things? The answer is that while the universe as a whole is believed to be unitary, subsystems need not be.

If there are two quantum systems with state vectors $|u\rangle$ and $|v\rangle$, then a state vector in the combined system is given by the *tensor product* $|u\rangle \otimes |v\rangle$, which can be abbreviated as $|u\rangle|v\rangle$ or just $|uv\rangle$. Operators distribute over the tensor product

$$\left(\hat{U} \otimes \hat{V} \right) \left(|u\rangle \otimes |v\rangle \right) = \hat{U}|u\rangle \otimes \hat{V}|v\rangle \quad ; \tag{15.91}$$

in a matrix representation the tensor product of two operators is

$$
\begin{pmatrix} a & b \\ c & d \end{pmatrix} \otimes \begin{pmatrix} \alpha & \beta \\ \gamma & \delta \end{pmatrix} = \begin{pmatrix} a\alpha & a\beta & b\alpha & b\beta \\ a\gamma & a\delta & b\gamma & b\delta \\ c\alpha & c\beta & d\alpha & d\beta \\ c\gamma & c\delta & d\gamma & d\delta \end{pmatrix} \quad . \tag{15.92}
$$

Taking a *partial trace* over a subsystem returns the density matrix for the other part:

$$
\begin{aligned}
\mathrm{Tr}_v(\hat{\rho}_{uv}) &= \mathrm{Tr}_v \left(\sum_{ij} p_{ij} |u_i v_j\rangle\langle u_i v_j| \right) \\
&= \sum_n \langle v_n| \sum_{ij} p_{ij} |u_i v_j\rangle\langle u_i v_j||v_n\rangle \\
&= \sum_{ij} p_{ij} |u_i\rangle\langle u_i| \\
&= \sum_i p_i |u_i\rangle\langle u_i| \\
&= \hat{\rho}_u \quad .
\end{aligned} \tag{15.93}
$$

The expectation value of an observable in a subsystem is thus found by tracing out the rest of the system

$$
\begin{aligned}
\mathrm{Tr}_u(\hat{U}\hat{\rho}_u) &= \mathrm{Tr}_u \left(\hat{U}\mathrm{Tr}_v(\hat{\rho}_{uv}) \right) \\
&= \mathrm{Tr}_{uv} \left((\hat{U} \otimes \hat{I}_v)\hat{\rho}_{uv} \right) \quad .
\end{aligned} \tag{15.94}
$$

While unitary evolution cannot create or destroy information, it can move it between subsystems. If a partial trace is taken over just a subsystem of interest, its dynamics no longer must be unitary. One system could be a computer, and the other the rest of the universe. When the computer appears to erase a bit it really just moves the information to unobserved degrees of freedom, so that the combined system remains unitary.

Through this mechanism the operators, like projectors, that describe measurements need not be unitary because information can be exchanged between the system being measured and the measurement apparatus. Let $\hat{M}_n$ be the measurement operator corresponding to obtaining the value n, for example a projector $|\uparrow\rangle\langle\uparrow|$ onto the up-state of a spin. The probability of this outcome is found from the magnitude

$$
\begin{aligned}
p(n) &= \left| \hat{M}_n |\psi\rangle \right| \\
&= \langle\psi|\hat{M}_n^\dagger \hat{M}_n|\psi\rangle \quad .
\end{aligned} \tag{15.95}
$$

If this does occur, the new state vector is found by normalizing the result of this operator

$$
\begin{aligned}
|\psi\rangle &\xrightarrow{n} \frac{\hat{M}_n|\psi\rangle}{\left| \hat{M}_n|\psi\rangle \right|} \\
&= \frac{\hat{M}_n|\psi\rangle}{\left[\langle\psi|\hat{M}_n^\dagger \hat{M}_n|\psi\rangle \right]^{1/2}} \quad .
\end{aligned} \tag{15.96}
$$

Likewise, the density matrix becomes

$$\hat{\rho} \xrightarrow{n} \frac{\hat{M}_n \hat{\rho} \hat{M}_n^\dagger}{\mathrm{Tr}\left(\hat{M}_n \hat{\rho} \hat{M}_n^\dagger\right)} \quad . \tag{15.97}$$

In thermal equilibrium the state probabilities are just given by Boltzmann factors, so a thermalized density matrix written in terms of energy eigenstates $|n\rangle$ with eigenvalues E_n is

$$\begin{aligned}
\hat{\rho}_{\text{thermal}} &= \sum_n p_n |n\rangle \langle n| \\
&= \sum_n \frac{e^{-\beta E_n}}{\sum_m e^{-\beta E_m}} |n\rangle \langle n| \\
&= \frac{e^{-\beta \hat{\mathcal{H}}}}{\mathrm{Tr}(e^{-\beta \hat{\mathcal{H}}})} \\
&= \frac{e^{-\beta \hat{\mathcal{H}}}}{\mathcal{Z}} \quad ,
\end{aligned} \tag{15.98}$$

where $\beta = 1/kT$ and $\mathcal{Z}$ is the partition function. In this basis the density matrix is diagonal. The possibility of processing information quantum mechanically rests on the availability of all of the off-diagonal terms. These represent quantum *coherence*, which eventually *decoheres* by interaction with the environment [Zurek, 1998]. This is the essential challenge in working with quantum information: while it's necessary to manipulate it to do anything useful, it must carefully be isolated to protect it from decoherence. You will break a quantum computer if you look at it too hard, which given the fragility of quantum coherence is not very hard at all.

15.2 INFORMATION

The quantum theory of information in many ways looks like its classical counterpart, with the classical entropy sum

$$H(p) = -\sum_n p_n \log p_n \tag{15.99}$$

replaced by a trace over the density matrix

$$S(\hat{\rho}) = -\mathrm{Tr}(\hat{\rho} \log \hat{\rho}) \quad . \tag{15.100}$$

Using this *von Neumann entropy*, corresponding noiseless and noisy coding theorems hold [Jozsa & Schumacher, 1994; Lloyd, 1997].

Don't be lulled into a false sense of security, however: quantum information behaves profoundly unlike its classical counterpart. A bit of quantum of information is called a *qubit*. It might reside in two orthogonal polarization states of a linearly polarized photon, or the spin-up and spin-down orientations of a proton in a magnetic field. Such qubit eigenstates can be written as $|0\rangle$ and $|1\rangle$, where 0 and 1 index the two orthogonal basis states. Unlike a classical bit which must be a 0 or 1, or an analog degree of freedom that could be between 0 and 1, a qubit can be in a *superposition*

$$|\psi\rangle = \alpha|0\rangle + \beta|1\rangle \tag{15.101}$$

in which it simultaneously has $|0\rangle$ and $|1\rangle$ components.

The configuration of a two-bit classical register is specified by giving the value of the bits (which can be 00, 01, 10, or 11); describing the state of a two-qubit register requires $2^2 = 4$ coefficients $\alpha|00\rangle + \beta|01\rangle + \gamma|10\rangle + \delta|11\rangle$. An N-bit classical register can store N bits, but the configuration of N qubits is specified by 2^N coefficients ($2^N - 1$ actually, because of the normalization constraint). This ability to be in an arbitrary superposition lets a qubit represent exponentially more information than a classical bit. It introduces a quantum notion of parallelism because a quantum computer can operate on all possible inputs at the same time.

A second difference between quantum information and classical information is that it cannot be copied. Consider an "amplifier" operator $\hat{A}$ that (if it could exist) would take an input state and produces two identical output states

$$\hat{A}|\psi\rangle = |\psi\rangle|\psi\rangle \quad . \tag{15.102}$$

If we ask it to amplify a superposition we should get two copies

$$\begin{aligned} \hat{A}\left(\alpha|0\rangle + \beta|1\rangle\right) &= \left(\alpha|0\rangle + \beta|1\rangle\right)\left(\alpha|0\rangle + \beta|1\rangle\right) \\ &= \alpha^2|0\rangle|0\rangle + \beta^2|1\rangle|1\rangle + \alpha\beta\left(|0\rangle|1\rangle + |1\rangle|0\rangle\right) \quad . \end{aligned} \tag{15.103}$$

The problem is that quantum mechanical operators must be linear. This means that our amplifier operator must also satisfy

$$\begin{aligned} \hat{A}\left(\alpha|0\rangle + \beta|1\rangle\right) &= \alpha\hat{A}|0\rangle + \beta\hat{A}|1\rangle \\ &= \alpha|0\rangle|0\rangle + \beta|1\rangle|1\rangle \quad , \end{aligned} \tag{15.104}$$

a contradiction with equation (15.103). The only way for these equations to agree is if the state to be amplified is not a superposition in the basis used by the amplifier. But this means that its value can be determined with certainty and hence a quantum amplifier isn't even needed to generate copies. Therefore, it is not possible to copy an arbitrary quantum state. This *no-clone* theorem [Wooters & Zurek, 1982] is the foundation of quantum cryptography. It is a consequence of the unitarity of quantum mechanics, which is reversible and hence allows qubits to be rearranged but not created or destroyed (although we will see that there is a loophole that permits error correction). Fortunately, the classical study of reversible computation showed that this is indeed possible with just a small overhead in complexity [Bennett, 1988].

The most remarkable property of all is *entanglement*. This lies at the heart of the apparent mystery of quantum mechanics, establishing a spooky connection among quantum systems after they interact. If a system comprises two particles, a and b, it is said to be in a *product state* if the total wave function can be written as a product of the individual particle wave functions

$$|\psi\rangle = |a\rangle|b\rangle \quad . \tag{15.105}$$

This is not always the case. Consider a radioactive decay process the emits two spin-1/2 particles. If the system starts out with no net angular momentum it must end up with no net angular momentum, and so it must be in the singlet state

$$|\psi\rangle = \frac{1}{\sqrt{2}}\left(|\uparrow\rangle_a|\downarrow\rangle_b - |\downarrow\rangle_a|\uparrow\rangle_b\right) \quad . \tag{15.106}$$

Because this cannot be written as a product state it is said to be *entangled*. Now make a measurement of the spin orientation of particle a. Before the measurement there's an equal chance for it to be up or down; after the measurement it must be either up or down. But something unexpected has happened to b after the measurement: its state is also determined. If a turns out to be up then b is down, and *vice versa*. This holds instantaneously and independent of the distance between the particles. According to the laws of quantum mechanics b could have been sent to Mars and its state would still have been determined by a measurement of a on Earth. This disturbing effect has been confirmed experimentally with entangled photons emitted in a multiple-photon process [Aspect *et al.*, 1981; Tittel *et al.*, 1998]

This is called the *Einstein–Podolsky–Rosen* paradox [Einstein *et al.*, 1935], and because it appears to violate the locality required by relativity it is one of the reasons why Einstein didn't like quantum mechanics. Such entangled particles are called an *EPR pair* in honor of the first people to be bothered by them. There are almost as many interpretations as there are interpreters; among the popular ones are:

- Nothing is wrong. No one disagrees with the predictions of quantum mechanics; the only problem is our intuition, which has no place here.
- Since the particles have a shared origin, they contain *hidden variables* that we don't know how to access but which define the outcome of the measurements. This would be a satisfying explanation, but a series of results starting with *Bell's Theorem* appear to rule it out [Bell, 1964; Greenberger *et al.*, 1990; Peres, 1990; Mermin, 1993]. Problem 15.3 works through a distressingly simple example.
- After particle a's interaction a message can travel backwards in time to the shared origin and then forwards in time to b in the present.
- The wave function for the universe splits following a measurement into non-interacting branches, selecting out the one with the correct answer for particle b. This is Everett's *Many-Worlds* Theory [Everett, 1957].

Amidst these and the even more bizarre explanations that have been put forward [Mermin, 1985], it is important to keep in mind that

- The predictions of quantum mechanics have been experimentally verified to fabulous precision. This is a paradox about interpretation, not about what is actually calculated and observed.
- Entanglement cannot be used to send messages faster than the speed of light, much to the consternation of some research grantors and grantees. b knows the state of a following the measurement, but the state of a can't be set in advance.

Entanglement is a channel that carries purely quantum information. We will use it for teleportation in communications, and as a kind of interconnect in Hilbert space for computation.

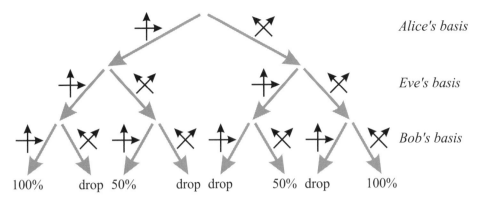

Figure 15.1. Quantum key distribution with an eavesdropper, showing the probability of Eve remaining undetected for each of the possible combinations of polarization bases.

15.3 COMMUNICATIONS

Given the tools to describe quantum information, we now turn to explore the possibilities afforded by quantum mechanical communications, from new ways to do old tasks (like cryptography) to ideas that until recently were matters of science fiction rather than fact (like teleportation).

15.3.1 Cryptography

Secure electronic transactions rely on cryptography, classical cryptography relies on number theory (Section 13.3.4), and this branch of number theory ultimately rests on very little. While there are good reasons to believe that finding prime factors is necessarily an exponentially difficult task, and hence for all practical purposes can be considered intractable, there is no proof of this. That is why there is growing appreciation of the insight that the theory of quantum measurement can be applied to protect information by physical laws rather than number theory [Wiesner, 1983].

The most important application is to the weakest link in the cryptographic chain, distributing private keys to establish a shared secret. The initial scheme of [Bennett & Brassard, 1984] is shown in Figure 15.1. Alice wants to send a key to Bob, and Eve wants to intercept it (these names are required in any discussion of cryptographic protocols).

Alice is going to communicate by sending one photon per bit to Bob, using a source that can emit single photons [Brunel *et al.*, 1999]. The bit will be represented by one of two orthogonal polarization directions, and for each bit she chooses one of two sets of orientations of the axes. Bob will likewise independently choose one of the two possible orientations for his axes to measure the polarization state. After a string of bits is sent, Alice and Bob can publicly report which orientations they used. Bits for which these don't match are discarded because the result of Bob's measurements would be random. Alice and Bob then publicly compare the values of some of the bits that were generated and measured using the same axis orientation. If there is no eavesdropping then these will match.

Now consider Eve's influence. Since each bit is represented by a single photon, because of the no-clone theorem Eve cannot surreptitiously determine the value of the bit without performing a measurement on the photon. Since she has no advance information about the random orientation of Alice's axes, she can do no better than randomly choose her own orientation, and following a measurement send a new matching photon on to Bob. If Eve is lucky enough to guess Alice's basis then she can reliably determine the polarization state and generate an identical photon to send to Bob. But if her axes don't match Alice's then Eve will obtain a random result. She has no way of telling which is which.

If Bob and Eve both match Alice's basis then Bob will receive the correct bit value and not detect Eve's presence. But if Bob's basis matches Alice's when Eve's doesn't, half of Bob's measurement on Eve's bit copies will randomly match what Alice sent, and half won't. This means that Eve has a total probability of 3/4 of being undetected on a single bit. That's pretty good odds for one bit, but if Alice and Bob compare two bits her probability of being undetected on both is $(3/4)^2$, and for N bits $(3/4)^N$. This exponential growth means that Alice and Bob need compare only a short string of values to be confident of detecting Eve's presence.

Unlike conventional cryptographic schemes that seek to obscure information, this one detects tampering with the message after it happens. While it suceeds only statistically rather than with certainty, its power comes from the exponential bound put on the probability of the eavesdropper remaining undetected. Like cryptosystems based on the (presumed) difficulty of factoring, a linear increase in effort by the sender leads to an exponential increase in the difficulty for the eavesdropper. Beyond resting on the firm foundation of the impossibility of copying quantum information, one of the appeals of quantum cryptography is the ease of understanding the scheme, although ruling out attacks by a quantum adversary has been a much more challenging task [Lo & Chau, 1999].

From the first experimental demonstration of a quantum cryptosystem [Bennett & Brassard, 1989], implementations have matured to include links over tens of kilometers using commercial telecommunications fibers [Muller *et al.*, 1996], and through the atmosphere in daylight [Hughes *et al.*, 2000].

15.3.2 Circuits

Quantum channels, like classical ones, require coding to take full advantage of their capabilities. In Chapter 10 we saw that a nonlinear gate plus an inverter is sufficient to generate any logical function. This can't be true quantum mechanically because the constraint of unitarity requires that the primitive gates be invertible so that it's possible to deduce their inputs from their outputs. But an analogous result does hold: any unitary transformation can be decomposed into a combination of unitary operations on a single qubit and a nonlinear two-qubit operator [Barenco *et al.*, 1995].

Because a unitary transformation of a qubit preserves its norm, it can be written as a rotation about an arbitrary axis $\vec{\theta}$ by an angle θ, along with a possible overall phase factor

$$\hat{U} = e^{i\varphi}\hat{R}(\vec{\theta}) \quad . \tag{15.107}$$

The rotation can in turn be decomposed into rotations about an orthogonal set of axes by using the rotation operators in equation (15.82). This is the quantum analog of a NOT gate, but the continuous parameterization lets it do many more things. Problem 15.5

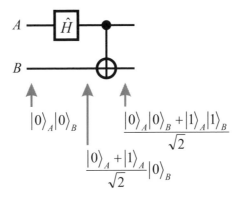

Figure 15.2. Hadamard and CNOT gates in a quantum circuit.

constructs a $\sqrt{\text{NOT}}$ gate that serves as an inverter when applied twice; a particularly useful single qubit gate will be the *Hadamard* transform

$$\hat{H} = \frac{1}{\sqrt{2}} \begin{bmatrix} 1 & 1 \\ 1 & -1 \end{bmatrix} \quad . \tag{15.108}$$

An example of a nonlinear two-input quantum gate is the CNOT (*Controlled-NOT*) transformation

$$\text{CNOT} = \begin{bmatrix} 1 & 0 & 0 & 0 \\ 0 & 1 & 0 & 0 \\ 0 & 0 & 0 & 1 \\ 0 & 0 & 1 & 0 \end{bmatrix} \begin{matrix} |00\rangle \\ |01\rangle \\ |10\rangle \\ |11\rangle \end{matrix} \quad . \tag{15.109}$$

If the first bit is down nothing happens to the second, and if the first bit is up the second one is flipped. This is a quantum generalization of XOR, made reversible by retaining the control input.

Figure 15.2 shows a simple circuit made up of a Hadamard transform followed by a CNOT. By convention, qubits A and B enter at the left and time advances along the wires towards the right, and the symbol for the CNOT indicates that A is the control qubit and B is the target. Working through the circuit, the Hadamard first puts A into a superposition. After the CNOT, if A is in the $|0\rangle$ state nothing happens to B, but if it is a $|1\rangle$ then B will be inverted. Because A is in a superposition, B is conditionally flipped based on the state of A. The value of B is not determined until that of A is (or *vice versa*). This circuit creates an entangled pair.

15.3.3 Teleportation

Quantum channels can send quantum as well as classical information, with remarkable consequences. Consider the circuit shown in Figure 15.3. It starts with Hadamard and CNOT gates to create an EPR pair from Alice's and Bob's qubits. Then, Alice stays on the Earth while Bob flies to the Moon (for example). Alice now wishes to send a new qubit

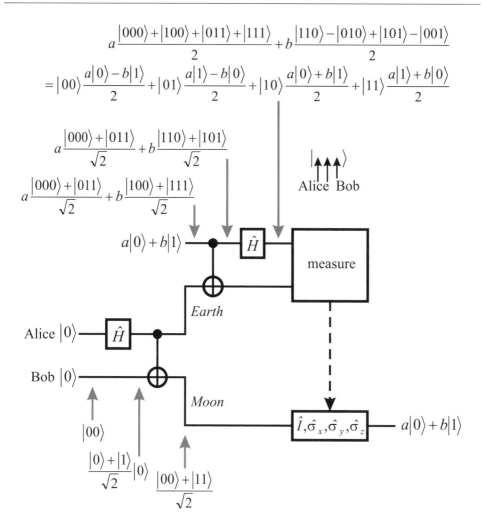

Figure 15.3. Quantum teleportation circuit.

to Bob. She can't measure it and communicate its state classically because that would force it into the measurement basis. But she can use a second CNOT gate to entangle the qubit with her half of the EPR pair, which is in turn entangled with Bob's qubit. She does this in the opposite order of the EPR circuit, following the CNOT with a Hadamard transform. Finally, Alice does perform a measurement to determine the state of her two qubits.

The overall state after the second Hadamard transform can be written suggestively as shown. Through the entanglement, the coefficients of her qubit have moved over to Bob's qubit. When Alice measures her qubits she learns nothing about the value of a and b. But the outcome does force Bob's qubit into the state tagged by the result of her measurement. Alice can use a classical channel to communicate to Bob which of the four possible outcomes she obtained, and then Bob can use that information to change the

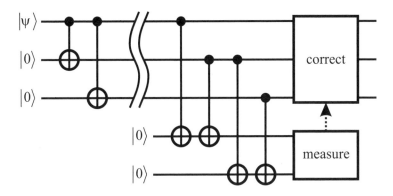

Figure 15.4. Quantum bit-flip error correction.

sign or swap the terms in his qubit as needed according to

$$|00\rangle \rightarrow \hat{\sigma}_z = \begin{pmatrix} 1 & 0 \\ 0 & -1 \end{pmatrix}$$

$$|01\rangle \rightarrow i\hat{\sigma}_y = \begin{pmatrix} 0 & 1 \\ -1 & 0 \end{pmatrix}$$

$$|10\rangle \rightarrow \hat{I} = \begin{pmatrix} 1 & 0 \\ 0 & 1 \end{pmatrix}$$

$$|11\rangle \rightarrow \hat{\sigma}_x = \begin{pmatrix} 0 & 1 \\ 1 & 0 \end{pmatrix} \quad . \tag{15.110}$$

Once he's finished his qubit has the coefficients of the hers. Because quantum information can't be copied, this really *is* her qubit; the one she started out with was consumed by her measurement. At the instant it disappears its quantum information is transferred to Bob. For obvious reasons this has come to be called *teleportation* [Bennett *et al.*, 1993]. While there's a long way to go before this is a practical way to reach your spaceship, it has been demonstrated experimentally [Furusawa *et al.*, 1998].

15.3.4 Error Correction

Decoherence was once widely believed to be a terminal obstacle to practical applications of quantum information [Unruh, 1995]. After all, any attempt to measure the state of a quantum system in order to check for errors would necessarily disturb it. It was thus a great surprise to discover that not only could quantum errors be corrected, but it could be done using methods from Chapter 13.

Figure 15.4 shows a simple scheme that protects a qubit $|\psi\rangle = a|0\rangle + b|1\rangle$ from bit flips. Like classical error correction, extra bits (here called *ancilla*) are introduced to add redundancy. Following the first two CNOTs, the qubit is encoded as

$$|\psi\rangle = a|000\rangle + b|111\rangle \quad . \tag{15.111}$$

After the possible introduction of errors, two more ancilla are used. The sequential CNOTs flip these if the first bit does not match the second, and if the second bit does not match the third. Because of the entanglement, measurement of the state of these ancilla then forces the encoded qubits into one of these states.

Consider what happens if the first bit is flipped by an error

$$|\psi\rangle \to a|100\rangle + b|011\rangle \quad . \tag{15.112}$$

Following the ancilla measurement the first bit is found to disagree with the second and third. This can unambiguously be determined to be due to an error in the first bit, which can then be flipped back without learning anything about a or b. This is just a quantum version of a three-bit majority code.

This alone is not sufficient because qubits, unlike classical bits, have a phase as well as an amplitude. If there is a phase-flip error that changes the relative signs of the first qubit,

$$|\psi\rangle \to a|000\rangle - b|111\rangle \quad , \tag{15.113}$$

this cannot be recognized by the code and will erroneously appear as an error in the coefficient.

Phase-flip errors can be corrected by using Hadamards to encode the qubit in superpositions

$$
\begin{aligned}
|\psi\rangle = {} & a \left(|0\rangle + |1\rangle\right) \left(|0\rangle + |1\rangle\right) \left(|0\rangle + |1\rangle\right) \\
& + b \left(|0\rangle - |1\rangle\right) \left(|0\rangle - |1\rangle\right) \left(|0\rangle - |1\rangle\right) \quad .
\end{aligned}
\tag{15.114}
$$

Now if there is a relative sign error in the first qubit

$$
\begin{aligned}
|\psi\rangle \to {} & a \left(|0\rangle - |1\rangle\right) \left(|0\rangle + |1\rangle\right) \left(|0\rangle + |1\rangle\right) \\
& + b \left(|0\rangle + |1\rangle\right) \left(|0\rangle - |1\rangle\right) \left(|0\rangle - |1\rangle\right)
\end{aligned}
\tag{15.115}
$$

it can be recognized and corrected once again by checking to see if the pairs of encoded states match. But now a bit flip will cause a coefficient error

$$
\begin{aligned}
|\psi\rangle \to {} & a \left(|1\rangle + |0\rangle\right) \left(|0\rangle + |1\rangle\right) \left(|0\rangle + |1\rangle\right) \\
& + b \left(|1\rangle - |0\rangle\right) \left(|0\rangle - |1\rangle\right) \left(|0\rangle - |1\rangle\right) \\
= {} & a \left(|0\rangle + |1\rangle\right) \left(|0\rangle + |1\rangle\right) \left(|0\rangle + |1\rangle\right) \\
& - b \left(|0\rangle - |1\rangle\right) \left(|0\rangle - |1\rangle\right) \left(|0\rangle - |1\rangle\right) \quad .
\end{aligned}
\tag{15.116}
$$

Both kinds of errors can be prevented by using both schemes, encoding the state in nine qubits as

$$
\begin{aligned}
|\psi\rangle = {} & a \left(|000\rangle + |111\rangle\right) \left(|000\rangle + |111\rangle\right) \left(|000\rangle + |111\rangle\right) \\
& + b \left(|000\rangle - |111\rangle\right) \left(|000\rangle - |111\rangle\right) \left(|000\rangle - |111\rangle\right) \quad .
\end{aligned}
\tag{15.117}
$$

Now a bit flip

$$
\begin{aligned}
|\psi\rangle \to {} & a \left(|100\rangle + |011\rangle\right) \left(|000\rangle + |111\rangle\right) \left(|000\rangle + |111\rangle\right) \\
& + b \left(|100\rangle - |011\rangle\right) \left(|000\rangle - |111\rangle\right) \left(|000\rangle - |111\rangle\right)
\end{aligned}
\tag{15.118}
$$

can be corrected within one of the encoded states, and a phase flip

$$
\begin{aligned}
|\psi\rangle \to {} & a \left(|000\rangle - |111\rangle\right) \left(|000\rangle + |111\rangle\right) \left(|000\rangle + |111\rangle\right) \\
& + b \left(|000\rangle + |111\rangle\right) \left(|000\rangle - |111\rangle\right) \left(|000\rangle - |111\rangle\right)
\end{aligned}
\tag{15.119}
$$

can be corrected between them. An arbitrary quantum error can be written in terms of bit and phase flips, hence this scheme will repair any kind of damage to the qubit [Shor, 1995].

The possibility of quantum error correction was lurking in equation (13.33), which showed that in a classical block code the syndrome identifies an error without providing any information about the message. In quantum language this is called a *non-demolition measurement*. Through this insight much of the classical theory of error correction has been lifted into a quantum setting [Steane, 1996; Calderbank & Shor, 1996].

The importance of error correction extends far beyond communications. Because a computer or a storage device can be considered to be a channel, error correction is essential to their reliable operation. The existence of a channel capacity in these settings has the profound implication that as long as the physical error rate is below a threshold, arbitrarily long computations and storage times can be realized with imperfect components [von Neumann, 1956; Winograd & Cowan, 1963]. The same thing is true quantum mechanically in *fault-tolerant* quantum computing [Shor, 1996; Knill *et al.*, 1998b].

Assume that a logical operation has a probability p of introducing an error. If instead the same operation is performed on qubits that have been encoded in an error-correction scheme that can fix n errors, the probability of failure after the subsequent decoding becomes Cp^{n+1}. The exponent is $n + 1$ because correction fails if there are that many errors, and C is a factor that accounts for the overhead of the extra steps needed for the encoding and decoding. If the error probability for a single step is below

$$Cp^{n+1} < p \;\Rightarrow\; p < C^{-1/n} \tag{15.120}$$

then the added work of the error correction is more than compensated for, and can be applied iteratively to improve the performance. A clever way to do this is with a *concatenated* code. This encodes qubits, which themselves are encoded qubits, which in turn are encoded qubits, and so forth. If one block encoding fails with a probability of Cp^{n+1}, the second level will fail if $n + 1$ blocks have an error, with a total probability of $C(Cp^{n+1}t)^{n+1}$, and a three-level encoding will fail with a probability of $C(C(Cp^{n+1})^{n+1})^{n+1}$. As long as the initial error is small enough, this provides a super-exponential improvement (with a corresponding increase in the number of extra qubits).

15.4 COMPUTATION

The quantum circuits we've considered so far have been used for coding, but because they can evaluate arbitrary functions it's natural to wonder whether they also have logical capabilities beyond their classical counterparts. The answer is a dramatic yes.

To understand what a computer can do, it's necessary to know something about the kinds of questions that can be asked of it. There are some problems that are known to be impossible to solve, like deciding if a computer program will halt without actually having to run it [Turing, 1936]. All of the remaining problems are theoretically soluble, but these solutions may not be practically usable given engineering constraints such as the length of a lifespan. While the execution time of a program will depend on the details of the machine used to run it, an essential distinction is between problems requiring a number of steps that is exponential versus polynomial in the problem size. For non-trivial questions the size of an exponential will almost always exceed the amount of one's patience, hence these are usually considered to be intractable. Because the *Church–Turing thesis* assures us that reasonably-defined computers can execute each other's programs with a polynomial-time prefactor to run an emulator, the distinction between polynomial and exponential is invariant over a broad class of architectures.

Problems such as arithmetic or sorting that can be solved in a time that is polynomial in the problem size are said to belong to the class **P**. There is a larger class of problems that have solutions that can be checked in a polynomial time, but that might require longer than that to find an answer. This is the class **NP**. It seems apparent that $\mathbf{P} \neq \mathbf{NP}$ because the latter class could include many more problems, but proving this remains the greatest open problem in computer science [Garey & Johnson, 1979].

Within the class **NP** there is a subset of problems that are known to be as hard as any other one. If it's possible to efficiently solve any one of these **NP**-*complete* problems, then all of the other ones in **NP** can be solved efficiently. *Cook's Theorem* is the proof of this remarkable fact [Cook, 1971]. It applies to the *satisfiability* (*SAT*) problem of deciding whether there is an argument that satisfies a Boolean expression; other **NP**-complete problems include the traveling salesman problem of finding the shortest route connecting a group of cities, and coloring a graph with distinct colors on the vertices.

Efficient quantum algorithms have not been found for solving **NP**-complete problems, and there are some problems for which it's known that quantum mechanics provides no speed-up [Farhi *et al.*, 1998]. While the possibility of finding an **NP**-complete algorithm remains an open question, there do exist good quantum algorithms for many important problems, including searching, factoring, and quantum simulation.

15.4.1 Searching

A very general way to formulate computational problems is in terms of an *oracle* $f(x)$ that equals 1 for arguments x that solve a problem of interest. The x might be entries in a database and $f(x)$ a test for answers to a query, or x could be a pair of integers and $f(x)$ a check for prime factors.

If there are N possible values of x, and they are unsorted and nothing is known about the inner workings of the oracle, then the only possible algorithm is to go through all arguments exhaustively to find an answer in $\mathcal{O}(N)$ steps. Think of finding a name in a phone book given a phone number, or randomly dialing a padlock to find the combination. Lov Grover astounded the world by showing that a quantum computer could do the problem in $\sqrt{N}$ steps [Grover, 1997; Grover, 1998].

Start by preparing a superposition of all possible logical states, which can be accomplished in a single step by applying a Hadamard transform to each of the qubits used to represent the states (Problem 15.6)

$$|\psi\rangle = \frac{1}{\sqrt{N}} \sum_{x=1}^{N} |x\rangle \quad . \tag{15.121}$$

Now assume that there are M values of x for which $f(x) = 1$. Define $|1\rangle$ to be the superposition of these states

$$|1\rangle = \frac{1}{\sqrt{M}} \sum_{f(x)=1} |x\rangle \tag{15.122}$$

and $|0\rangle$ to be what's left

$$|0\rangle = \frac{1}{\sqrt{N-M}} \sum_{f(x)=0} |x\rangle \quad , \tag{15.123}$$

so that

$$|\psi\rangle = \sqrt{\frac{N-M}{N}} \, |0\rangle + \sqrt{\frac{M}{N}} \, |1\rangle \quad . \tag{15.124}$$

If an angle θ is defined by

$$\sin \theta = \sqrt{\frac{M}{N}} \tag{15.125}$$

then the superposition state $|\psi\rangle$ can be rewritten in this basis as

$$|\psi\rangle = \cos \theta |0\rangle + \sin \theta |1\rangle = \begin{pmatrix} \sin \theta \\ \cos \theta \end{pmatrix} \quad . \tag{15.126}$$

Next define an operator

$$\begin{aligned} \hat{U}_1 &= \hat{I} - 2|1\rangle\langle 1| \\ &= \begin{bmatrix} 1 & 0 \\ 0 & 1 \end{bmatrix} - 2 \begin{bmatrix} 1 & 0 \\ 0 & 0 \end{bmatrix} \\ &= \begin{bmatrix} -1 & 0 \\ 0 & 1 \end{bmatrix} \quad . \end{aligned} \tag{15.127}$$

This flips the sign of the states that satisfy the oracle. While this might seem to require knowing the answer to the problem in advance, quantum mechanics allows the oracle to be applied to all arguments simultaneously. The problem is that flipping the signs alone is of no use because there are still N states to check to find the desired ones. To help uncover them, define a second operator

$$\begin{aligned} \hat{U}_\psi &= 2|\psi\rangle\langle\psi| - \hat{I} \\ &= 2 \begin{bmatrix} \sin \theta \sin \theta & \sin \theta \cos \theta \\ \cos \theta \sin \theta & \cos \theta \cos \theta \end{bmatrix} - \begin{bmatrix} 1 & 0 \\ 0 & 1 \end{bmatrix} \\ &= \begin{bmatrix} 2\sin^2 \theta - 1 & 2\sin \theta \cos \theta \\ 2\cos \theta \sin \theta & 2\cos^2 \theta - 1 \end{bmatrix} \end{aligned} \tag{15.128}$$

that inverts the coefficients of an arbitrary state around their mean value. Now look what happens when these two opeartors are applied sequentially:

$$\begin{aligned} \hat{U}_\psi \hat{U}_1 &= \begin{bmatrix} 1 - 2\sin^2 \theta & 2\sin \theta \cos \theta \\ -2\sin \theta \cos \theta & 2\cos^2 \theta - 1 \end{bmatrix} \\ &= \begin{bmatrix} \cos 2\theta & \sin 2\theta \\ -\sin 2\theta & \cos 2\theta \end{bmatrix} \quad . \end{aligned} \tag{15.129}$$

This is just a rotation by $2\theta = 2\sqrt{M/N}$ in the $|0\rangle, |1\rangle$ space. $|\psi\rangle$ started out with an angle of θ in that space due to the small fraction of states that satisfy the oracle. If $\hat{U}_\psi \hat{U}_1$ is applied iteratively I times, all of the probability will be in the $|1\rangle$ state after a rotation of $\pi/2$, requiring a number of iterations

$$\theta + I2\theta = \frac{\pi}{2}$$

$$I = \frac{\pi}{4}\frac{1}{\theta} - \frac{1}{2}$$

$$= \frac{\pi}{4}\sqrt{\frac{N}{M}} - \frac{1}{2} \quad . \tag{15.130}$$

After $\mathcal{O}(\sqrt{N})$ iterations the system will be in the $|x\rangle$ state with certainty, after which a single measurement can obtain an answer that satisfies the oracle!

If an answer could be found in $\log N$ time then this algorithm would efficiently solve **NP**-complete problems, but regrettably $\sqrt{N}$ is known to be a lower bound on the possible speed-up [Boyer *et al.*, 1998]. Nevertheless this can still be a very significant improvement, particularly given the broad applicability of oracle problems; searching a database of a million entries in a thousand steps is an enormous quantitative if not qualitative advantage.

15.4.2 Transforms and Factoring

The most significant result in quantum computing, and perhaps in all of computational complexity theory, was Peter Shor's proof that a quantum computer could find prime factors in polynomial time [Shor, 1997]. Factoring is one of the hardest problems that is not an **NP**-complete problem. Not only was it widely believed to require exponential time (Section 13.3.4), existing systems for secure electronic transactions rely on that belief. Unlike most results in the study of computational complexity that rest on problem classifications that remain open questions, here was a constructive demonstration of how to turn a problem from intractable to tractable.

Shor's result, and related algorithms, use a *Quantum Fourier Transform (QFT)*. This is defined by applying a classical *Discrete Fourier Transform (DFT)* to the coefficients of a quantum state

$$\sum_{x=0}^{N-1} a_n |x\rangle \mapsto \sum_{k=0}^{N-1}\sum_{n=0}^{N-1} a_n e^{2\pi i k x/N}|k\rangle \quad . \tag{15.131}$$

A classical DFT requires $\mathcal{O}(N^2)$ steps because of the double sum over N indices. A *Fast Fourier Transform (FFT)* reduces this to $\mathcal{O}(N \log N)$ by recognizing that $e^{2\pi i k x/N}$ does not need to be evaluated for values of kx/N that are a power of 2. To turn this into an algorithm (quantum or classical), note that because x and k are integer indices they can be written in a binary expansion with coefficients that can be either 0 or 1

$$x = x_0 + 2^1 x_1 + \cdots + 2^{n-1} x_{n-1}$$
$$k = k_0 + 2^1 k_1 + \cdots + 2^{n-1} k_{n-1} \tag{15.132}$$

where $N = 2^n$. Then discarding all of the terms in xk with an exponent of 2^n or higher

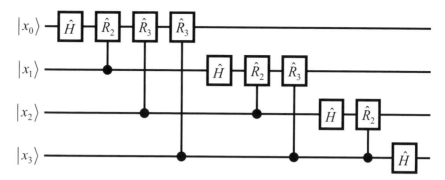

Figure 15.5. Circuit for a quantum Fourier transform.

leaves the non-trivial terms

$$\frac{kx}{2^n} = k_{n-1}\left(2^{-1}x_0\right) + k_{n-2}\left(2^{-1}x_1 + 2^{-2}x_0\right) + \cdots \quad . \tag{15.133}$$

This expansion can be plugged back into the definition of the QFT to find its action on a basis vector

$$|x\rangle = |x_0 x_1 \cdots x_{n-1}\rangle$$

$$\mapsto \frac{1}{\sqrt{N}} \sum_{k=0}^{N-1} e^{2\pi i k x / N} |k\rangle$$

$$= \frac{1}{2^{n/2}} \sum_{k_{n-1}=0}^{1} \sum_{k_{n-2}=0}^{1} e^{2\pi i \left[k_{n-1}\left(2^{-1}x_0\right) + k_{n-2}\left(2^{-1}x_1 + 2^{-2}x_0\right) + \cdots\right]} |k_{n-1} \cdots k_1 k_0\rangle$$

$$= \frac{1}{2^{n/2}} \left(|0\rangle + e^{2\pi i \left(2^{-1}x_0\right)}|1\rangle\right) \left(|0\rangle + e^{2\pi i \left(2^{-1}x_1 + 2^{-2}x_0\right)}|1\rangle\right) + \cdots \quad . \tag{15.134}$$

The product representation in equation (15.134) has a natural implementation in a quantum circuit, shown in Figure 15.5. Input is the state $|x_0 x_1 \cdots x_{n-1}\rangle$. After the first Hadamard this becomes

$$\left(|0\rangle + e^{2\pi i \left(2^{-1}x_0\right)}|1\rangle\right) |x_1 \cdots x_{n-1}\rangle \tag{15.135}$$

because the sign in front of the $|1\rangle$ is a $+$ if $x_0 = 0$ and a $-$ if $x_0 = 1$ (equation 15.108). Then comes a conditional phase shift defined by

$$\hat{R}_k = \begin{bmatrix} 1 & 0 & 0 & 0 \\ 0 & 1 & 0 & 0 \\ 0 & 0 & 1 & 0 \\ 0 & 0 & 0 & e^{2\pi i/2^k} \end{bmatrix} . \tag{15.136}$$

This does nothing to the $|0\rangle$ state, but adds an extra phase shift to the $|1\rangle$ state based on the value of x_1, leaving

$$\left(|0\rangle + e^{2\pi i \left(2^{-1}x_0 + 2^{-2}x_1\right)}|1\rangle\right) |x_1 \cdots x_{n-1}\rangle \quad . \tag{15.137}$$

After applying all of the conditional phase shifts, and then the Hadamard to the next bit, the state is

$$\left(|0\rangle + e^{2\pi i\left(2^{-1}x_0+2^{-2}x_1+\cdots\right)}|1\rangle\right)\left(|0\rangle + e^{2\pi i\left(2^{-1}x_1\right)}|1\rangle\right)|x_2\cdots x_{n-1}\rangle \quad . \tag{15.138}$$

Continuing in this manner reproduces equation (15.134).

One Hadamard and $n-1$ phase gates are applied to $|x_0\rangle$ for a total of n operations, then there are $n-1$ gates for $|x_1\rangle$, and so forth, adding up to a total of $n + (n-1) + \cdots + 1 = n(n+1)/2$ steps. This is for a single basis vector, but because of the linearity of quantum mechanics it will also work on an arbitrary superposition, therefore the QFT reduces the total number of steps to $\mathcal{O}\left(n^2\right) = \mathcal{O}\left((\log N)^2\right)$, quite a savings over the FFT. While this is not directly applicable to classical signal-processing problems because of the difficulty of loading in and reading out the data, it can be used as a primitive in a number of quantum algorithms, most notably factoring.

The QFT appears in Shor's algorithm as a way to use quantum interference for period-finding:

- Given an integer N to factor, pick a power of two $K = 2^L$ such that $N^2 < K < 2N^2$.
- Choose random integer $x < N$.
- Initialize a pair of K-qubit registers

$$\frac{1}{\sqrt{K}} \sum_{k=0}^{K-1} |0\rangle|0\rangle \quad . \tag{15.139}$$

- Create the uniform superposition in the first register

$$\frac{1}{\sqrt{K}} \sum_{k=0}^{K-1} |k\rangle|0\rangle \quad . \tag{15.140}$$

- Perform a *modular exponentiation* of the first register on the second

$$\frac{1}{\sqrt{K}} \sum_{k=0}^{K-1} |k\rangle|x^k \bmod N\rangle \quad . \tag{15.141}$$

- Make a measurement on the second register. If the result is

$$x^k \bmod N = y \tag{15.142}$$

then the first register will collapse into those values of x compatible with y. Because the modular exponentiation is periodic in k with an *order* r [Koblitz, 1994], this leaves the first register in the state

$$\frac{1}{\sqrt{M}} \sum_{m=0}^{M-1} |l + mr\rangle \quad , \tag{15.143}$$

where l is an offest that depends on y, and M is the number of values of k satisfying equation (15.142).

- To find r, perform a QFT

$$\frac{1}{\sqrt{M}} \sum_{m=0}^{M-1} \frac{1}{\sqrt{K}} \sum_{n=0}^{K-1} e^{2\pi i(l+mr)n/K} |n\rangle \quad . \tag{15.144}$$

- Measure the first register. The probability of obtaining a value n is proportional to its coefficient

$$\left| \sum_{m=0}^{M-1} e^{2\pi i(l+mr)n/K} \right|^2 = \left| \sum_{m=0}^{M-1} e^{2\pi imrn/K} \right|^2 \quad . \tag{15.145}$$

Since the terms in the sum will add incoherently unless rn/K is an integer, the result is $n = KC/r$, where C is a random integer.

- From KC/r find the order r, from which the factors of N can be found [Ekert & Jozsa, 1996].

A complete implementation of this algorithm requires $\mathcal{O}\left((\log N)^3\right)$ gates, dominated by the calculation of the modular exponentiation [Beckman *et al.*, 1996].

15.4.3 Simulation

One of the original questions that inspired the study of quantum information was whether a quantum computer could efficiently simulate the evolution of another quantum system [Feynman, 1982]. If, for example, an N-spin system is being studied, modeling its state requires a state vector with 2^N elements, and operators on it are $2^N \times 2^N$ matrices. Since $2^{40} = 10^{12}$, simulating just 40 spins would exceed the capacity of the largest digital computers.

Not all quantum systems can be modeled efficiently on a quantum computer, but a tremendous simplification is possible if the Hamiltonian can be written as a sum [Lloyd, 1996]

$$e^{i\hat{\mathcal{H}}t/\hbar} = e^{i\sum_n \hat{\mathcal{H}}_n t/\hbar} \quad . \tag{15.146}$$

This will be the usual case if there are short-range interactions that extend near neighbors. We then want to find a numerical method to appoximate equation (15.146) on a quantum computer, but unlike classical numerical methods [Gershenfeld, 1999a] this one must recognize that the Hamiltonian terms in general won't commute.

For two terms, the evolution over a time Δt is to second order

$$e^{i(\hat{A}+\hat{B})\Delta t}$$
$$= \hat{1} + i(\hat{A} + \hat{B})\Delta t - \frac{1}{2}(\hat{A} + \hat{B})(\hat{A} + \hat{B})\Delta t^2 + \mathcal{O}(\Delta t^3) \tag{15.147}$$
$$= \hat{1} + i\hat{A}\Delta t - \frac{1}{2}\hat{A}^2\Delta t^2 + i\hat{B}\Delta t - \frac{1}{2}\hat{B}^2\Delta t^2 + \frac{1}{2}\hat{A}\hat{B}\Delta t^2 + \frac{1}{2}\hat{B}\hat{A}\Delta t^2 + \mathcal{O}(\Delta t^3) \quad .$$

An obvious guess would be to just apply each evolution separately, but this is incorrect at second order

$$e^{i\hat{A}\Delta t}e^{i\hat{B}\Delta t}$$
$$= \left(\hat{1} + i\hat{A}\Delta t - \frac{1}{2}\hat{A}^2\Delta t^2 + \mathcal{O}(\Delta t^3) \right)\left(\hat{1} + i\hat{B}\Delta t - \frac{1}{2}\hat{B}^2\Delta t^2 + \mathcal{O}(\Delta t^3) \right)$$
$$= \hat{1} + i\hat{A}\Delta t - \frac{1}{2}\hat{A}^2\Delta t^2 + i\hat{B}\Delta t - \frac{1}{2}\hat{B}^2\Delta t^2 - \hat{A}\hat{B}\Delta t^2 + \mathcal{O}(\Delta t^3) \quad . \tag{15.148}$$

That can be fixed by splitting the first operator as

$$e^{i\hat{A}\Delta t/2}e^{i\hat{B}\Delta t}e^{i\hat{A}\Delta t/2}$$

$$= \left(\hat{1} + \frac{i}{2}\hat{A}\Delta t - \frac{1}{8}\hat{A}^2\Delta t^2 + \mathcal{O}(\Delta t^3)\right)\left(\hat{1} + i\hat{B}\Delta t - \frac{1}{2}\hat{B}^2\Delta t^2 + \mathcal{O}(\Delta t^3)\right)$$

$$\left(\hat{1} + \frac{i}{2}\hat{A}\Delta t - \frac{1}{8}\hat{A}^2\Delta t^2 + \mathcal{O}(\Delta t^3)\right) \tag{15.149}$$

$$= \hat{1} + i\hat{A}\Delta t - \frac{1}{2}\hat{A}^2\Delta t^2 + i\hat{B}\Delta t - \frac{1}{2}\hat{B}^2\Delta t^2 - \frac{1}{2}\hat{A}\hat{B}\Delta t^2 - \frac{1}{2}\hat{B}\hat{A}\Delta t^2 + \mathcal{O}(\Delta t^3) \ .$$

Asymptotically, this leads to the *Trotter formula*

$$\lim_{N\to\infty}\left(e^{i\hat{A}t/N}e^{i\hat{B}t/N}\right)^N = e^{i(\hat{A}+\hat{B})t} \quad . \tag{15.150}$$

The second-order cancellation can also be obtained by grouping the terms in the *Campbell–Baker–Hausdorf formula*

$$e^{i(\hat{A}+\hat{B})\Delta t} = e^{i\hat{A}\Delta t}e^{i\hat{B}\Delta t}e^{-[\hat{A},\hat{B}]\Delta t^2/2} + \mathcal{O}(\Delta t^3) \quad . \tag{15.151}$$

These relationships are the basis of quantum simulation. Quantum circuits are devised that emulate the local Hamiltonian, and then these are applied iteratively to model the global evolution. A linear number of applications corresponds to an exponential amount of work in the Hilbert space of the quantum computer, which can be a physical system unrelated to the one being modeled [Somaroo *et al.*, 1999].

Although quantum simulation has received less attention than other quantum algorithms because prospective users are people who already study quantum mechanics, it offers an exponential improvement with broad applicability to many areas of technological interest such as molecular structure and quantum effects in VLSI scaling.

15.4.4 Experimental Implementation

This is a fitting final section. Quantum computing is one of the most exciting opportunities in all of information processing, both for its fundamental implications and practical applications. The explanation of how these devices function crosses the traditional division between the theory of computation and physical theory, and experimental implementations draw on ideas from almost every page in this book.

Of all the candidate technologies to continue scaling beyond the VLSI era, quantum logic has one unique feature. Moore's Law is exponential; any classical approach demands an exponential increases in space or time. Even the Avogadro's number of elements in a DNA computer is quickly limited by the size of an exponential problem [Adleman, 1994]. Quantum computing accesses Hilbert space, the one exponential resource that has been untapped for computation. What's so remarkable about algorithms such as Shor's and Grover's is that they provide answers with certainty, rather than a probability of success that decreases inversely with the problem size.

Early theoretical studies of quantum computing used "designer" Hamiltonians with desired properties. The development of experimental approaches began with the recognition that physically-plausible local interactions were adequate for realizing arbitrary quantum algorithms [Lloyd, 1993]. Initial attempts to apply this insight ran afoul of the essential paradox in quantum logic: the system must be protected from the external environment to remain quantum, but must remain accessible to the external environment for I/O and programming. Most physical systems fit into one of two categories, those that are externally isolated and internally uncoupled and have quantum coherence times that can be on the order of the age of the universe, and strongly-coupled interacting ones with subnanosecond coherence times.

Heroic experimental efforts are beginning to develop controllable quantum nonlinearities. Among the candidate systems are ions in laser-addressed optical traps [Cirac & Zoller, 1995; Wieman *et al.*, 1999], atoms in optical cavities small enough to feel quantum electrodynamic effects [Ye *et al.*, 1999], dilute spins in semiconductors with exchange interactions controlled by carriers induced by a gate voltage [Kane, 1998], and superconducting current loops [Mooij *et al.*, 1999].

Prospects for their practical implementation improved significantly with the development of quantum error correction and fault-tolerant circuit design that can use imperfect components. While quantitative estimates depend on device and algorithm details, the required relative loss of coherence per gate for steady-state operation is in the experimentally accessible range of 10^{-3}–10^{-6} [Knill *et al.*, 1998b].

A second insight that significantly simplifies the device design is to recognize that a macroscopic ensemble of microscopic quantum computers can be used. By representing each logical qubit in many physical qubits, intentional and unintentional external interactions become *weak measurements* that can paradoxically obtain continuous information about the state of the logical qubits without projecting them. Post-processing can be added to quantum algorithms so that a classical channel can be used to obtain quantum results from an ensemble [Gershenfeld & Chuang, 1997].

Nuclear spins in molecular liquids are screened and protected by rapid tumbling averaging out inter-molecular forces, giving coherence times on the order of seconds. Their exchange interactions through bonds provide a nonlinearity that can be controlled with RF fields. And mature chemical synthesis techniques can be used to produce macroscopic samples of molecules of interest. These desirable attributes led to the development of nuclear magnetic resonance pulse sequences for computation [Cory *et al.*, 1997; Chuang *et al.*, 1998a].

Two spins in a strong magnetic field will have a Hamiltonian [Ernst *et al.*, 1994]

$$\hat{\mathcal{H}} = \hbar\omega_A \frac{\hat{\sigma}_A}{2} + \hbar\omega_B \frac{\hat{\sigma}_B}{2} + \hbar\omega_{AB} \frac{\hat{\sigma}_A}{2}\frac{\hat{\sigma}_B}{2} \quad . \tag{15.152}$$

ω_A and ω_B are the precession frequencies of the spins in the magnetic field, possibly differing because of a *chemical shift* in the local field strength. The ω_{AB} term represents nonlinear coupled evolution due to the exchange interaction. Free precession under this Hamiltonian corresponds to linear and nonlinear rotations, which along with rotation by transverse RF pulses provides the operators needed for univeral quantum circuits. These were used for the first demonstrations of non-trivial quantum algorithms [Chuang *et al.*, 1998b; Jones & Mosca, 1998] and error correction [Cory *et al.*, 1998].

Because the *Zeeman splitting* between spin eigenstates due to dipole coupling to a magnetic field is 10^{-5} smaller than kT at room temperature in fields of a few tesla, thermal spins are in a weakly-polarized high-temperature limit. Even though their evolution is quantum, the pure states needed for algorithm initial conditions cannot be obtained from an equilibrium distribution (equation 15.98) by using unitary RF pulses. NMR techniques simulate non-unitary operators by adding extra degrees of freedom (spins, space, or time [Knill *et al.*, 1998a]), but these do not scale beyond small demonstrations because of an exponential loss in signal strength as qubits are added due to the partition function denominator in equation (15.98) [Schack & Caves, 1999]. Overcoming this will require order unity spin polarization, which has been accomplished in simpler atomic systems by *optical pumping* [Song *et al.*, 1999].

Whether or not any one of these schemes can eventually be scaled to defeat number-theoretic cryptosystems, their development is already having a significant impact on the study of both computation and physical systems. For many years it's been clear that computers can and will be constructed on atomic scales, the domain of *nanotechnology* [Feynman, 1992; Drexler, 1992; Merkle, 1998]. The demonstration by NMR of quantum algorithms in nuclear spin evolution shows that not only is this possible, but that nature is already a very powerful computer if it is interrogated in the right way, and that the kinds of computations it can perform go far beyond what can be conceived of by our classical intuition alone.

The interchange between the study of nature and computation increasingly operates in both directions. Physical theory has provided improved devices to compute with, but computation is also providing an improved language to described devices with. By viewing natural systems in a computational framework it can be possible to understand how to "program" them to obtain desired behavior. An example is the traditional difficulty in NMR of exchanging the quantum coefficients between a spin of interest and a more sensitive accessible one; it's been shown that this can be accomplished by a pulse sequence that implements the circuit for the SWAP operation (Problem 15.7) [Linden *et al.*, 1999]. Another example is writing an arbitrary wave function into an atom by putting a pump laser in a feedback loop with a computer running a machine learning algorithm [Weinacht *et al.*, 1999]. Such cross-fertilization could equally well be called *The Information of Physics Technology*, enhancing our understanding of, and ability to shape, the world around us by merging the descriptions of the information in a system with that of its physical properties.

15.5 SELECTED REFERENCES

[Baym, 1973] Baym, Gordon. (1973). *Lectures on Quantum Mechanics*. Reading: W.A. Benjamin.

A good intuitive introduction to quantum mechanics.

[Peres, 1993] Peres, Asher. (1993). *Quantum Theory: Concepts and Methods*. Boston: Kluwer Academic.

Modern quantum theory.

[Balian, 1991] Balian, Roger. (1991). *From Microphysics to Macrophysics: Methods and Applications of Statistical Physics*. New York: Springer–Verlag. Translated by D. ter Haar and J.F. Gregg, 2 volumes.

Statistical mechanics with a strong quantum flavor.

[quant-ph] http://xxx.lanl.gov

    The quantum physics e-print archive, where most results first appear.

[Nielsen & Chuang, 2000] Nielsen, M.A., & Chuang, I.L. (2000). *Quantum Computation and Quantum Information.* New York: Cambridge University Press.

    A monumental compendium of most everything there is to know about quantum mechanics, and computation.

15.6 PROBLEMS

(15.1) How is Ehrenfest's theorem changed if the observable has an explicit time dependence?

(15.2) Show that that $\text{Tr}(\hat{\rho}^2) \leq 1$.

(15.3) (*a*) Using the Pauli matrices in the z eigenbasis of equations (15.71), find the eigenvectors for $\hat{\sigma}_x$ and $\hat{\sigma}_y$.

    (*b*) Using the result of the previous problem, apply to the singlet state $|\psi\rangle = \left(|\uparrow\rangle_1|\downarrow\rangle_2 - |\downarrow\rangle_1|\uparrow\rangle_2\right)/\sqrt{2}$ a projector $|m_x = 1/2\rangle_1\langle m_x = 1/2|_1$ onto the $m_{1x} = 1/2$ state of the first spin. What state is the second spin left in? Repeat this in the y direction.

    (*c*) How are the products of eigenvalues $m_{1x}m_{2y}$ and $m_{1y}m_{2x}$ related?

    (*d*) Recognizing that $\hat{\vec{\sigma}}_1$ commutes with $\hat{\vec{\sigma}}_2$, what does the last result imply for the relationship between $\langle\psi|\hat{\sigma}_{1x}\hat{\sigma}_{2y}|\psi\rangle$ and $\langle\psi|\hat{\sigma}_{1y}\hat{\sigma}_{2x}|\psi\rangle$?

    (*e*) In the z eigenbasis, work out the tensor products to evaluate $\langle\psi|\hat{\sigma}_{1x}\hat{\sigma}_{2y} + \hat{\sigma}_{1y}\hat{\sigma}_{2x}|\psi\rangle$.

    (*f*) Compare the results of the last two parts. What happened?

(15.4) Using the spectral representation, work out the rotation operators associated with the Pauli spin matrices.

(15.5) Find the matrix representation of the $\sqrt{\text{NOT}}$ gate that when applied twice gives a NOT gate (don't worry about getting the signs of the final state right).

(15.6) Show that applying Hadamard transformations individually to N qubits each in the $|0\rangle$ state puts them into an equal superposition of the 2^N possible logical states.

(15.7) What does this circuit do?

Appendix 1 Problem Solutions

A1.1 INTRODUCTION

There are no problems in Chapter 1.

A1.2 INTERACTIONS, UNITS, AND MAGNITUDES

(2.1) (*a*) *How many atoms are there in a yoctomole?*

$$6 \times 10^{23} \times 10^{-24} = 0.6 \approx 1 \quad . \tag{A1.1}$$

(*b*) *How many seconds are there in a nanocentury? Is the value near that of any important constants?*

$$\frac{100 \text{ years}}{1 \text{ century}} \times \frac{1 \text{ century}}{10^9 \text{ nanocentury}} \times \frac{365 \text{ days}}{1 \text{ year}} \times \frac{24 \text{ hours}}{1 \text{ day}}$$

$$\times \frac{60 \text{ minutes}}{1 \text{ hour}} \times \frac{60 \text{ seconds}}{1 \text{ minute}} = 3.1536 \approx \pi \quad . \tag{A1.2}$$

(2.2) *A large data storage systems holds on the order of a terabyte. How tall would a 1 terabyte stack of floppy disks be? How does that compare to the height of a tall building?*

$$\frac{10^{12} \text{ byte}}{1 \text{ terabyte}} \times \frac{1 \text{ floppy}}{10^6 \text{ byte}} \times \frac{0.003 \text{ m}}{1 \text{ floppy}} = 3000 \text{ m} \quad . \tag{A1.3}$$

A tall building has ~ 100 stories, and each is ~ 5 m tall, and so it is ~ 500 m (the Sears Tower is 443 m). Therefore, the stack of floppies equals ~ 6 tall buildings.

(2.3) *If all the atoms in our universe were used to write an enormous binary number, using one atom per bit, what would that number be (converted to base 10)?*

Take $\sim 10^{70}$ atoms in the universe. If each atom is used as a binary bit, the number is:

$$2^{10^{70}} = 2^{10 \times 10^{69}} = \left(2^{10}\right)^{10^{69}} \approx \left(10^3\right)^{10^{69}} = 10^{3 \times 10^{69}} \quad . \tag{A1.4}$$

(2.4) *Compare the gravitational acceleration due to the mass of the Earth at its surface to that produced by a 1 kg mass at a distance of 1 m. Express their ratio in decibels.*

The magnitude of the gravitational force between two masses is $F = Gm_1m_2/r^2$, or the acceleration due to one mass is $a = Gm/r^2$. Therefore,

$$a_{\text{Earth}} = \frac{6.67\times10^{-11}\ \text{m}^3\cdot\text{kg}^{-1}\cdot\text{s}^{-2} \times 5.98\times10^{24}\ \text{kg}}{(6.38\times10^6\ \text{m})^2}$$

$$= 9.8\ \text{m}\cdot\text{s}^{-2}$$

$$a_{1\ \text{kg}} = \frac{6.67\times10^{-11}\ \text{m}^3\cdot\text{kg}^{-1}\cdot\text{s}^{-2} \times 1\ \text{kg}}{(1\ \text{m})^2}$$

$$= 6.67\times10^{-11}\ \text{m}\cdot\text{s}^{-2}$$

$$20\log_{10}\left(\frac{9.8}{6.67\times10^{-11}}\right) = 223\ \text{dB} \quad . \tag{A1.5}$$

This is of course a tiny effect, at the limit of detection by the most sensitive instrument for measuring gravity, an atomic interferometer [Peters *et al.*, 1999].

(2.5) The following examples show that knowing almost nothing about chemistry and nuclear physics it is still possible to make surprisingly good estimates based solely on the relevant energy scales.

(a) *Approximately estimate the chemical energy in a ton of TNT. You can assume that nitrogen is the primary component; think about what kind of energy is released in a chemical reaction, where it is stored, and how much there is.*

Assume that the TNT consists entirely of nitrogen, and that in the explosion each atom releases a typical bond's worth of energy (1 eV):

$$\frac{10^6\ \text{g}}{1\ \text{ton}} \times \frac{6\times10^{23}\ \text{N atoms}}{14\ \text{g}} \times \frac{1\ \text{eV}}{\text{N atom}} \times \frac{1.6\times10^{-19}\ \text{J}}{1\ \text{eV}}$$

$$= 6\times10^9\ \text{J} \quad . \tag{A1.6}$$

The real answer is 4.184×10^9 J.

(b) *Estimate how much uranium would be needed to make a nuclear explosion equal to the energy in a chemical explosion in 10 000 tons of TNT (once again, think about where the energy is stored).*

Now assume that each nucleon releases a typical nuclear excitation (1 MeV):

$$10^4\ \text{tons} \times \frac{6\times10^9\ \text{J}}{1\ \text{ton}} = 6\times10^{13}\ \text{J}$$

$$6\times10^{13}\ \text{J} \times \frac{1\ \text{eV}}{1.6\times10^{-19}\ \text{J}} \times \frac{1\ ^{235}\text{U atom}}{235\times10^6\ \text{eV}} \times \frac{235\ \text{g}}{6\times10^{23}\ ^{235}\text{U atoms}}$$

$$= 625\ \text{g} \quad . \tag{A1.7}$$

Quite a difference. In reality, this is an underestimate, because not all the nucleons participate in the explosion; actual fission warheads use 1–10 kg of fissile material. 10 kilotons is an important size because that is used as the trigger for fusion bombs.

(c) *Compare this to the* rest mass energy $E = mc^2$ *of that amount of material (Chapter 14), which gives the maximum amount of energy that could be liberated from it.*

$$0.625 \text{ kg} \times (3{\times}10^8 \text{ m}\cdot\text{s}^{-1})^2 = 5.6 \times 10^{16} \text{ J} \quad . \tag{A1.8}$$

The difference between this and 6×10^{13} J is the energy locked up in subnuclear structure.

(2.6) (a) *What is the approximate de Broglie wavelength of a thrown baseball?*

$$\lambda = \frac{h}{p} = \frac{6.626{\times}10^{-34} \text{ J}\cdot\text{s}}{0.1 \text{ kg} \times 100 \text{ km}\cdot\text{h}^{-1} \times 3600^{-1} \text{ h}\cdot\text{s}^{-1} \times 10^3 \text{ m}\cdot\text{km}^{-1}}$$

$$= 2.38{\times}10^{-34} \text{ m} = 2.38{\times}10^{-24} \text{ Å} \tag{A1.9}$$

(b) *Of a molecule of nitrogen gas at room temperature and pressure? (This requires either the result of Section 3.4.2, or dimensional analysis.)*

Since there are three spatial degrees of freedom, the average kinetic energy can be found from the equipartition theorem:

$$\frac{3}{2}kT = \frac{1}{2}mv^2 \Rightarrow p = mv = \sqrt{3mkT} \quad . \tag{A1.10}$$

This result could also have been estimated on dimensional grounds alone by setting equal the two characteristic energies in the problem, kT and $mv^2/2$. Therefore,

$$\lambda = \frac{h}{p} = \frac{h}{\sqrt{3mkT}} \tag{A1.11}$$

$$= \frac{6.626{\times}10^{-34} \text{ J}\cdot\text{s}}{[3 \times 0.014 \text{ kg} \times (6.022{\times}10^{23})^{-1} \times 1.38{\times}10^{-23} \text{ J}\cdot\text{K}^{-1} \times 300 \text{ K}]^{1/2}}$$

$$= 3.9{\times}10^{-11} \text{ m} = 0.39 \text{ Å} \quad .$$

(c) *What is the typical distance between the molecules in this gas?*

The typical distance is

$$d = \left(\frac{V}{N_A}\right)^{1/3} = \left(\frac{22.4 \text{ liter} \times 10^{-3} \text{ m}^3 \cdot \text{liter}^{-1}}{6.022{\times}10^{23}}\right)^{1/3}$$

$$= 3.3{\times}10^{-9} \text{ m} = 33 \text{ Å} \quad .$$

(d) *If the volume of the gas is kept constant as it is cooled, at what temperature does the wavelength become comparable to the distance between the molecules?*

If the volume is kept constant, the spacing remains at the value found in the previous problem. Therefore,

$$\lambda = \frac{h}{\sqrt{3mkT}} \tag{A1.12}$$

$$\Rightarrow T = \frac{h^2}{3mk\lambda^2}$$

$$= \frac{(6.626 \times 10^{-34} \text{ J} \cdot \text{s})^2}{3 \times 2.3 \times 10^{-26} \text{ kg} \times 1.38 \times 10^{-23} \text{ J} \cdot \text{K}^{-1} \times (3.3 \times 10^{-9} \text{ m})^2}$$

$$= 0.04 \text{ K} \quad .$$

Quantum effects are clearly irrelevant for playing baseball, but can be important for a cold gas (although nitrogen liquefies at 77 K, well before this temperature is reached).

(2.7) (a) *The potential energy of a mass m a distance r from a mass M is* $-GMm/r$. *What is the* escape velocity *required to climb out of that potential?*

$$\frac{GMm}{r} = \frac{1}{2}mv^2$$

$$\frac{2GM}{r} = v^2 \quad . \tag{A1.13}$$

(b) *Since nothing can travel faster than the speed of light (Chapter 14), what is the radius within which nothing can escape from the mass?*

$$\frac{2GM}{c^2} = r \quad . \tag{A1.14}$$

This is the *Schwarzschild radius* of a *black hole*.

(c) *If the rest energy of a mass M is converted into a photon, what is its wavelength?*

$$E = Mc^2 = \frac{hc}{\lambda}$$

$$\lambda = \frac{h}{Mc} \quad . \tag{A1.15}$$

This is the *Compton wavelength*.

(d) *For what mass does its equivalent wavelength equal the size within which light cannot escape?*

$$M = \sqrt{\frac{hc}{2G}} \simeq \sqrt{\frac{hc}{G}} = 5.46 \times 10^{-8} \text{ kg} \quad . \tag{A1.16}$$

This is the *Planck mass*, found from combining the constants that are fundamental to quantum mechanics, electromagnetics, and gravitation.

(e) *What is the corresponding size?*

$$\lambda = \frac{h}{c}\sqrt{\frac{G}{hc}} = \sqrt{\frac{Gh}{c^3}} = 4.05 \times 10^{-35} \text{ m} \quad . \tag{A1.17}$$

The *Planck distance* is the shortest length scale in physics; below it quantum mechanics and gravity merge and the structure of space breaks down.

(f) *What is the energy?*

$$E = hc\sqrt{\frac{c^3}{Gh}} = \sqrt{\frac{c^5 h}{G}} = 4.91 \times 10^9 \text{ J} = 3.07 \times 10^{28} \text{ eV} \quad . \tag{A1.18}$$

Beyond this *Planck energy* all the forces are unified.

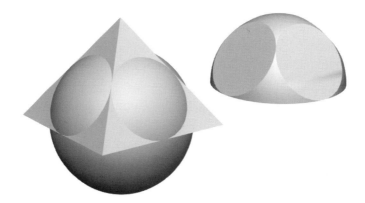

Figure A1.1. Sphere and pyramid, and the common volume.

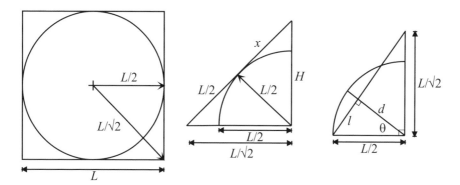

Figure A1.2. Top, side, and 45° views of the sphere inscribed in the pyramid.

(g) *What is the period?*

$$E = h\nu = \frac{h}{\tau}$$

$$\tau = h\sqrt{\frac{G}{c^5 h}} = \sqrt{\frac{Gh}{c^5}} = 1.35 \times 10^{-43} \text{ s} \quad . \tag{A1.19}$$

The *Planck time* is when quantum corrections cease to be important to general relativity following the *Big Bang*. These Planck scale numbers set the ultimate limits on the performance of a computer [Lloyd, 2000], or at least on our understanding of physics that could be applied to computation, whichever comes first.

(2.8) *Consider a pyramid of height H and a square base of side length L. A sphere is placed so that its center is at the center of the square at the base of the pyramid, and so that it is tangent to all of the edges of the pyramid (intersecting each edge at just one point).*

See Figures A1.1 and A1.2.

(*a*) *How high is the pyramid in terms of L?*

$$\frac{L^2}{2} + H^2 = \left(\frac{L}{2} + x\right)^2 \quad \text{and} \quad x^2 + \frac{L^2}{4} = H^2 \tag{A1.20}$$

$$\Rightarrow H = \frac{L}{\sqrt{2}} \quad .$$

(*b*) *What is the volume of the space common to the sphere and the pyramid?*

$$\cos\theta = \frac{2d}{L}, \quad \cos(90 - \theta) = \sin\theta = \frac{\sqrt{2}d}{L} \Rightarrow \tan\theta = \frac{1}{\sqrt{2}} = \frac{l}{d}$$

$$d^2 + l^2 = \frac{L^2}{4} \Rightarrow d = \frac{L}{\sqrt{6}}$$

$$V_{\text{cap}} = \int_{L/\sqrt{6}}^{L/2} \pi\left(\frac{L^2}{4} - x^2\right) \, dx = \pi L^3 \left[\frac{1}{12} - \frac{7}{36\sqrt{6}}\right]$$

$$V_{\text{common}} = \frac{1}{2} \cdot \frac{4}{3}\pi \left(\frac{L}{2}\right)^3 - 4V_{\text{cap}} = \pi L^3 \left[\frac{7}{9\sqrt{6}} - \frac{1}{4}\right] \quad .$$

A1.3 NOISE IN PHYSICAL SYSTEMS

(3.1) (*a*) *Derive equation (3.16) from the binomial distribution and Stirling's approximation,*

$$p_n(x) = \binom{n}{x} p^x (1 - p)^{n-x}$$

$$\log p_n(x) = \log n! - \log(n - x)! - \log x! + x \log p$$
$$+ (n - x)\log(1 - p)$$
$$\approx n \log n - (n - x) \underbrace{\log(n - x)}_{\approx \log n - x/n} - \log x!$$

$$+x \log p + (n - x)\underbrace{\log(1 - p)}_{\approx -p}$$

$$\approx x \log(np) - (np) - \log x!$$

$$\Rightarrow p_n(x) \approx \frac{(np)^x e^{-np}}{x!} \tag{A1.21}$$

(using $p \ll 1$ and $n \gg x$, and $\log(1 + x) \approx x$ for small x).

(*b*) *use it to derive equation (3.18), and*

$$\langle x(x - 1)\cdots(x - m + 1)\rangle$$
$$= \sum_{x=0}^{\infty} \frac{e^{-N}N^x}{x!} \underbrace{x(x - 1)\cdots(x - m + 1)}_{0 \text{ for } x < m}$$

$$= \sum_{x=m}^{\infty} \frac{e^{-N}N^x}{x!} x(x-1)\cdots(x-m+1)$$

$$= \sum_{x=m}^{\infty} \frac{e^{-N}N^x}{(x-m)(x-m-1)\cdots 1}$$

$$= N^m \sum_{y=0}^{\infty} \frac{e^{-N}N^y}{y!} \qquad (y \equiv x - m)$$

$$= N^m \qquad\qquad\qquad (A1.22)$$

(c) *use that to derive equation (3.19).*

$$N = \langle x \rangle$$
$$N^2 = \langle x^2 - x \rangle = \langle x^2 \rangle - \langle x \rangle \quad .$$

Therefore,

$$\sigma^2 = \langle x^2 \rangle - \langle x \rangle^2$$
$$= N^2 + N - N^2$$
$$= N \quad . \qquad\qquad (A1.23)$$

(3.2) *Assume that photons are generated by a light source independently and randomly with an average rate N per second. How many must be counted by a photodetector per second to be able to determine the rate to within 1%? To within 1 part per million? How many watts do these cases correspond to for visible light?*

$$\frac{1}{\sqrt{N}} = 0.01 \Rightarrow N = 10^4$$

$$\frac{1}{\sqrt{N}} = 10^{-6} \Rightarrow N = 10^{12} \quad . \qquad (A1.24)$$

Using $E = h\nu$,

$$6.6\times10^{-34}\ \text{J}\cdot\text{s} \times 3\times10^8\ \frac{\text{m}}{\text{s}} \times \frac{1}{6\times10^{-7}\ \text{m}}$$

$$= 3.3\times10^{-19}\ \frac{\text{J}}{\text{photon}}$$

$$3.3\times10^{-19}\ \frac{\text{J}}{\text{photon}} \times 10^4\ \frac{\text{photon}}{\text{s}} = 3.3\times10^{-15}\ \text{W}$$

$$3.3\times10^{-19}\ \frac{\text{J}}{\text{photon}} \times 10^{12}\ \frac{\text{photon}}{\text{s}} = 3.3\times10^{-7}\ \text{W} \quad . \qquad (A1.25)$$

(3.3) *Consider an audio amplifier with a 20 kHz bandwidth.*

(a) *If it is driven by a voltage source at room temperature with a source impedance of 10 kΩ how large must the input voltage be for the SNR with respect to the source Johnson noise to be 20 dB?*

$$\langle V_{\text{noise}}^2 \rangle = 4kTR\Delta f$$

$$= 4 \times 1.38\times10^{-23}\ \frac{\text{J}}{\text{K}} \times 300\ \text{K} \times 10^4\ \Omega \times 2\times10^4\ \text{s}^{-1}$$

$$= 3.3 \times 10^{-12} \text{ V}^2$$

$$V_{\text{RMS noise}} = 1.8 \times 10^{-6} \text{ V}$$

$$10 \log_{10} \left(\frac{\langle V_{\text{signal}}^2 \rangle}{3.3 \times 10^{-12} \text{ V}^2} \right) = 20 \implies V_{\text{RMS signal}} = 1.8 \times 10^{-5} \text{ V} \quad . \quad \text{(A1.26)}$$

(b) *What size capacitor has voltage fluctuations that match the magnitude of this Johnson noise?*

$$\frac{1}{2} C \langle V^2 \rangle = \frac{1}{2} kT$$

$$C = \frac{kT}{\langle V^2 \rangle} = \frac{1.38 \times 10^{-23} \text{ J} \cdot \text{K}^{-1} \times 300 \text{ K}}{3.3 \times 10^{-12} \text{ V}^2} = 1.2 \text{ nF} \quad . \quad \text{(A1.27)}$$

(c) *If it is driven by a current source, how large must it be for the RMS shot noise to be equal to 1% of that current?*

$$\langle I_{\text{noise}}^2 \rangle = 2q \langle I \rangle \Delta f$$

$$= 2 \times 1.6 \times 10^{-19} \text{ C} \times \langle I \rangle \frac{\text{C}}{\text{s}} \times 2 \times 10^4 \text{ s}^{-1}$$

$$= 6.4 \times 10^{-15} \times \langle I \rangle$$

$$I_{\text{RMS noise}} = 8 \times 10^{-8} \times \langle I \rangle^{1/2}$$

$$0.01 = \frac{I_{\text{RMS noise}}}{\langle I \rangle} = \frac{8 \times 10^{-8}}{\langle I \rangle^{1/2}}$$

$$\implies \langle I \rangle = 6.4 \times 10^{-11} \text{ A} \quad . \quad \text{(A1.28)}$$

The Johnson noise limit is typically encountered before shot noise becomes significant.

(3.4) *This problem is much harder than the others. Consider a stochastic process $x(t)$ that randomly switches between $x = 0$ and $x = 1$. Let $\alpha \, dt$ be the probability that it makes a transition from 0 to 1 during the interval dt if it starts in $x = 0$, and let $\beta \, dt$ be the probability that it makes a transition from 1 to 0 during dt if it starts in $x = 1$.*

(a) *Write a matrix differential equation for the change in time of the probability $p_0(t)$ to be in the 0 state and the probability $p_1(t)$ to be in the 1 state.*

α is the rate at which $p_0(t)$ decreases due to transitions from 0 to 1, and is also the rate at which $p_1(t)$ increases from these transitions out of $p_0(t)$. Similarly, β is the rate for $1 \to 0$ transitions. This can be written as a matrix differential equation

$$\frac{d}{dt} \underbrace{\begin{pmatrix} p_0(t) \\ p_1(t) \end{pmatrix}}_{\vec{p}} = \underbrace{\begin{pmatrix} -\alpha & \beta \\ \alpha & -\beta \end{pmatrix}}_{\mathbf{A}} \begin{pmatrix} p_0(t) \\ p_1(t) \end{pmatrix}$$

$$\frac{d\vec{p}}{dt} = \mathbf{A} \cdot \vec{p} \quad . \quad \text{(A1.29)}$$

(b) *Solve this by diagonalizing the 2 × 2 matrix.*

If solving linear systems of equations is not familiar, see a linear algebra text such as [Strang, 1988]. The first step is to find the eigenvalues of **A** from the characteristic equation

$$
\begin{aligned}
0 &= |\mathbf{A} - \lambda \mathbf{I}| \\
&= (-\alpha - \lambda)(-\beta - \lambda) - \alpha\beta \\
&= \lambda^2 + \lambda(\alpha + \beta) \\
&\Rightarrow \lambda = 0, -(\alpha + \beta) \quad .
\end{aligned}
\tag{A1.30}
$$

By inspection, the corresponding eigenvectors are

$$
\begin{pmatrix} 1 \\ \alpha/\beta \end{pmatrix} \text{ for } \lambda_0 = 0
\tag{A1.31}
$$

and

$$
\begin{pmatrix} 1 \\ -1 \end{pmatrix} \text{ for } \lambda_1 = -(\alpha + \beta) \quad .
\tag{A1.32}
$$

We next form a matrix with these as the columns

$$
\mathbf{M} = \begin{pmatrix} 1 & 1 \\ \alpha/\beta & -1 \end{pmatrix} \quad ,
\tag{A1.33}
$$

which has an inverse

$$
\mathbf{M}^{-1} = \begin{pmatrix} \dfrac{1}{1 + \alpha/\beta} & \dfrac{1}{1 + \alpha/\beta} \\[3mm] \dfrac{\alpha/\beta}{1 + \alpha/\beta} & \dfrac{-1}{1 + \alpha/\beta} \end{pmatrix} \quad .
\tag{A1.34}
$$

In terms of new variables $\vec{q} = \mathbf{M}^{-1}\vec{p}$, the differential equation becomes

$$
\begin{aligned}
\frac{d\vec{q}}{dt} &= \mathbf{M}^{-1} \cdot \mathbf{A} \cdot \mathbf{M} \cdot \vec{q} \\
&= \begin{pmatrix} 0 & 0 \\ 0 & -(\alpha + \beta) \end{pmatrix} \vec{q}(t) \quad .
\end{aligned}
\tag{A1.35}
$$

This has a solution

$$
\vec{q}(t) = \begin{pmatrix} q_0(0) \\ q_1(0)e^{-(\alpha+\beta)t} \end{pmatrix} \quad .
\tag{A1.36}
$$

If

$$
\vec{p}(0) = \begin{pmatrix} 0 \\ 1 \end{pmatrix}
\tag{A1.37}
$$

(the reason for this choice will be seen in part (*c*), then

$$
\vec{q}(0) = \mathbf{M}^{-1} \cdot \vec{p}(0) = \begin{pmatrix} \dfrac{1}{1 + \alpha/\beta} \\[3mm] \dfrac{-1}{1 + \alpha/\beta} \end{pmatrix}
\tag{A1.38}
$$

and

$$\vec{q}(t) = \begin{pmatrix} \dfrac{1}{1 + \alpha/\beta} \\[2mm] \dfrac{-1}{1 + \alpha/\beta} e^{-(\alpha+\beta)t} \end{pmatrix}.$$ (A1.39)

Therefore,

$$\vec{p}(t) = \mathbf{M} \cdot \vec{q}(t) = \frac{1}{1 + \alpha/\beta} \begin{pmatrix} 1 - e^{-(\alpha+\beta)t} \\[1mm] \alpha/\beta + e^{-(\alpha+\beta)t} \end{pmatrix}.$$ (A1.40)

The limit $t \to \infty$ gives the average probability to be in either state

$$\vec{p}(\infty) = \begin{pmatrix} \dfrac{\beta}{\alpha + \beta} \\[2mm] \dfrac{\alpha}{\alpha + \beta} \end{pmatrix}.$$ (A1.41)

(c) *Use this solution to find the autocorrelation function* $\langle x(t)x(t + \tau)\rangle$.

$$\begin{aligned}
\langle x_t x_{t+\tau}\rangle &= \sum_{x_t}\sum_{x_{t+\tau}} x_t x_{t+\tau} p(x_t, x_{t+\tau}) \\
&= 0 \times 0 \times p(x_t = 0, x_{t+\tau} = 0) \\
&\quad + 0 \times 1 \times p(x_t = 0, x_{t+\tau} = 1) \\
&\quad + 1 \times 0 \times p(x_t = 1, x_{t+\tau} = 0) \\
&\quad + 1 \times 1 \times p(x_t = 1, x_{t+\tau} = 1) \\
&= p(x_t = 1, x_{t+\tau} = 1) \\
&= p(x_{t+\tau} = 1|x_t = 1)\, p(x_t = 1) \\
&= \frac{1}{1 + \alpha/\beta}\left(\frac{\alpha}{\beta} + e^{-(\alpha+\beta)|\tau|}\right)\frac{\alpha}{\alpha + \beta} \\
&= \frac{\alpha\beta}{(\alpha + \beta)^2}\left(\frac{\alpha}{\beta} + e^{-(\alpha+\beta)|\tau|}\right).
\end{aligned}$$ (A1.42)

Note that $|\tau|$ is used instead of τ because the system is stationary, and so the calculation in part (b) is invariant under time reversal.

(d) *Use the autocorrelation function to show that the power spectrum is a Lorentzian.*

From the Wiener–Khinchin theorem the power spectrum is the Fourier transform of the autocorrelation function:

$$\begin{aligned}
S(f) &= \int_{-\infty}^{\infty} e^{i2\pi f\tau}\frac{\alpha\beta}{(\alpha + \beta)^2}\left(\frac{\alpha}{\beta} + e^{-(\alpha+\beta)|\tau|}\right) d\tau \\
&= \left(\frac{\alpha}{\alpha + \beta}\right)^2 \delta(f) + \frac{\alpha\beta}{(\alpha + \beta)^2} \\
&\quad \times \left[\int_{-\infty}^{0} e^{\tau[i2\pi f+(\alpha+\beta)]}\, d\tau + \int_{0}^{\infty} e^{\tau[i2\pi f-(\alpha+\beta)]}\, d\tau\right]
\end{aligned}$$

$$= \left(\frac{\alpha}{\alpha + \beta}\right)^2 \delta(f) + \frac{\alpha\beta}{(\alpha + \beta)^2}$$

$$\times \left[\frac{1}{i2\pi f + (\alpha + \beta)} - \frac{1}{i2\pi f - (\alpha + \beta)}\right]$$

$$= \left(\frac{\alpha}{\alpha + \beta}\right)^2 \delta(f) + \frac{\alpha\beta}{(\alpha + \beta)^2} \frac{2/(\alpha + \beta)}{1 + \left(\frac{2\pi f}{\alpha + \beta}\right)^2} \quad . \tag{A1.43}$$

The first term is the DC component in the spectrum because this is not a zero-mean process. The second term is a Lorentzian, the Fourier transform of an exponential.

(e) *At what frequency is the magnitude of the Lorentzian reduced by half relative to its low-frequency value?*

Ignoring the DC component,

$$S(f) = \frac{1}{2}S(0) \Rightarrow f = \frac{\alpha + \beta}{2\pi} \quad . \tag{A1.44}$$

This is called the *knee frequency*.

(f) *For a thermally activated process, show that a flat distribution of barrier energies leads to a distribution of switching times $p(\tau) \propto 1/\tau$, and in turn to $S(f) \propto 1/f$.*

If $y = f(x)$, then $p(x) = p(y)|dy/dx|$ [Gershenfeld, 1999a]. So,

$$\tau = \tau_0 e^{E/kT}$$

$$\Rightarrow p(\tau) = \frac{p(E)}{|d\tau/dE|}$$

$$= \frac{p(E)}{\tau_0 e^{E/kT}/kT}$$

$$= \frac{kT\, p(E)}{\tau} \quad . \tag{A1.45}$$

If $p(E) = E_0$, then $p(\tau) = kTE_0/\tau$. Plugging this in to equation (3.36) gives

$$S(f) = \int_0^\infty \frac{2\tau}{1 + (2\pi f\tau)^2}\, p(\tau)\, d\tau$$

$$= \int_0^\infty \frac{2\tau}{1 + (2\pi f\tau)^2} \frac{kTE_0}{\tau}\, d\tau$$

$$= \left. \frac{kTE_0 \tan^{-1}(2\pi f\tau)}{\pi f}\right|_0^\infty$$

$$= \frac{kTE_0}{2f} \quad . \tag{A1.46}$$

A1.4   INFORMATION IN PHYSICAL SYSTEMS

(4.1) *Verify that the entropy function satisfies the required properties of continuity, non-negativity, monotonicity, and independence.*

- *Continuity*
  Since the sum of continuous functions is also continuous, if suffices to show that $p \log p$ is continuous in p. This will be so if for any $\epsilon > 0$ there exists a $\delta > 0$ for all p such that

$$|(p + \Delta) \log(p + \Delta) - p \log p| < \epsilon \qquad (A1.47)$$

  whenever

$$0 < |(p + \Delta) - p| = |\Delta| < \delta \quad . \qquad (A1.48)$$

  Since

$$\lim_{\Delta \to 0} |(p + \Delta) \log(p + \Delta) - p \log p| = 0 \qquad (A1.49)$$

  for all p, this can always be satisfied.
- *Non-negativity*

$$0 \le p \le 1 \Rightarrow \log p \le 0$$
$$\Rightarrow -p \log p \ge 0 \quad . \qquad (A1.50)$$

- *Boundedness*
  The extremal values of an equation $f(x_1, \ldots, x_N)$ under the constraints

$$g_1(x_1, \ldots, x_N) = a_1$$
$$g_2(x_1, \ldots, x_N) = a_2$$
$$\ldots$$
$$g_M(x_1, \ldots, x_N) = a_M \qquad (A1.51)$$

  are found by solving the N equations for the Lagrange multipliers λ_j

$$\frac{\partial f}{\partial x_i} - \sum_{j=1}^{M} \lambda_j \frac{\partial g_j}{\partial x_i} = 0 \quad .$$

  For this problem we want to find the p_i that maximize the entropy $-\sum_{i=1}^{N} p_i \log p_i$ subject to the constraint $\sum_{i=1}^{N} p_i = 1$. First, find p_i in terms of λ:

$$-\frac{\partial}{\partial p_j} \sum_{i=1}^{N} p_i \log p_i = \lambda \frac{\partial}{\partial p_j} \sum_{i=1}^{N} p_i$$
$$-\log p_j - 1 = \lambda$$
$$p_i = e^{-(1+\lambda)} \quad . \qquad (A1.52)$$

  Then substitute back into the normalization constraint to evaluate λ:

$$\sum_{i=1}^{N} e^{-(1+\lambda)} = 1$$
$$\Rightarrow \lambda = \log N - 1$$
$$\Rightarrow p_i = \frac{1}{N}$$
$$\Rightarrow -\sum_{i=1}^{N} p_i \log p_i = \log N \quad . \qquad (A1.53)$$

- *Additivity*

$$H(p,q) = -\sum_{x,y} p(x)q(y)\log(p(x)q(y))$$

$$= -\sum_{x,y} p(x)q(y)\log p(x) - \sum_{x,y} p(x)q(y)\log q(y)$$

$$= -\sum_{x} p(x)\log p(x) - \sum_{y} q(y)\log q(y)$$

$$= H(p) + H(q) \quad . \tag{A1.54}$$

(4.2) *Prove the relationships in equation (4.10).*

$$\sum_{x,y} p(x,y)\log\frac{p(x,y)}{p(x)p(y)} = \sum_{x,y} p(x,y)\log p(x,y) - \sum_{x,y} p(x,y)\log p(x)$$

$$- \sum_{x,y} p(x,y)\log p(y)$$

$$= \sum_{x,y} p(x,y)\log p(x,y) - \sum_{x} p(x)\log p(x)$$

$$- \sum_{y} p(y)\log p(y)$$

$$= H(x) + H(y) - H(x,y) \tag{A1.55}$$

$$\sum_{x,y} p(x,y)\log\frac{p(x,y)}{p(x)p(y)} = \sum_{x,y} p(x,y)\log\frac{p(x|y)p(y)}{p(x)p(y)}$$

$$= \sum_{x,y} p(x,y)\log p(x|y) - \sum_{x,y} p(x,y)\log p(x)$$

$$= H(x) - H(x|y) \quad . \tag{A1.56}$$

(4.3) *Calculate the differential entropy of a Gaussian process.*

$$p(x) = \frac{1}{\sqrt{2\pi\sigma^2}}e^{-(x-x_0)^2/2\sigma^2}$$

$$H = -\int_{-\infty}^{\infty} p(x)\log p(x)\,dx$$

$$= -\int_{-\infty}^{\infty} p(x)\left[-\ln\sqrt{2\pi\sigma^2} - \frac{(x-x_0)^2}{2\sigma^2}\right]$$

$$= \frac{1}{2}\ln(2\pi\sigma^2) + \frac{\langle(x-x_0)^2\rangle}{2\sigma^2}$$

$$= \frac{1}{2}\ln(2\pi\sigma^2) + \frac{1}{2}$$

$$= \frac{1}{2}\ln(2\pi e\sigma^2)\ \text{(nats)}$$

$$= \frac{1}{2}\log_2(2\pi e\sigma^2)\ \text{(bits)} \tag{A1.57}$$

Remember that the units are defined solely by the base of the logarithm; it is not necessary (or correct) to change the functional form of the entropy when changing bases. Also, the $2\pi e$ factor appears as an additive constant in the entropy and hence can be ignored in comparing differential entropies among Gaussian processes.

(4.4) *A standard telephone line has a bandwidth of 3300 Hz and an SNR of 20 dB.*

(a) *What is the capacity?*

$$C = \Delta f \log_2 \left(1 + \frac{S}{N} \right)$$

$$= (3300 \text{ s}^{-1}) \, \log_2 \left(1 + 10^{20/10} \right)$$

$$= 22\,000 \, \frac{\text{bits}}{\text{s}} \quad . \tag{A1.58}$$

The fastest modems achieve steady-state rates $\sim\!50\,000$ bits/s; accomplishing this requires adaptive compression and error correction that can handle much more than uncorrelated Gaussian errors. The noise in a channel sets a limit on how close states can be in amplitude and frequency and still reliably be distinguished; the deep theory of optimal *sphere packing* is used to guide the selection of the particular states that a modem sends.

(b) *What SNR would be necessary for the capacity to be 1 Gbit/s?*

$$10^9 = 3300 \log_2 \left(1 + \frac{S}{N} \right)$$

$$\Rightarrow \frac{S}{N} = 2^{10^9/3300} - 1$$

$$\approx 2^{3 \times 10^5}$$

$$\approx 10^{10^5}$$

$$\Rightarrow \text{SNR} = 10^6 \text{ dB} \quad . \tag{A1.59}$$

This is unlikely to be achieved any time soon, but in Chapter 6 we will see that the bandwidth of ordinary coaxial cable can extend above 1 GHz, hence the interest in using the cable system instead of the phone system for home Internet access. The linear dependence of the capacity on bandwidth is much more powerful than the logarithmic dependence on SNR.

(4.5) *Let $(x_1, x_2, \ldots, x_n)$ be drawn from a Gaussian distribution $N(x_0, \sigma^2)$ with variance σ^2 and unknown mean value x_0. Show that $f(x_1, \ldots x_n) = n^{-1} \sum_{i=1}^{n} x_i$ is an estimator for x_0 that is unbiased and achieves the Cramér–Rao lower bound.*

First check for bias:

$$\langle f \rangle = \left\langle \frac{1}{n} \sum_{i=1}^{n} x_i \right\rangle$$

$$= \frac{1}{n} \sum_{i=1}^{n} \langle x_i \rangle$$

$$= x_0 \quad . \tag{A1.60}$$

Now find the variance:

$$\langle (f - x_0)^2 \rangle = \left\langle \left[\frac{1}{n} \sum_{i=1}^{n} (x_i - x_0) \right]^2 \right\rangle$$

$$= \left\langle \left[\frac{1}{n} \sum_{i=1}^{n} (x_i - x_0) \right] \left[\frac{1}{n} \sum_{j=1}^{n} (x_j - x_0) \right] \right\rangle$$

$$= \frac{1}{n^2} \sum_{i=1}^{n} \langle (x_i - x_0)^2 \rangle$$

$$= \frac{\sigma^2}{n} \qquad \qquad (A1.61)$$

(because the variables are uncorrelated). Finally, calculate the Fisher information for the set of variables

$$J_n(x_0) = n J(x_0)$$

$$= n \left\langle \left[\partial_{x_0} \log N(x_0, \sigma^2) \right]^2 \right\rangle$$

$$= n \left\langle \left[\frac{x - x_0}{\sigma^2} \right]^2 \right\rangle$$

$$= \frac{n}{\sigma^2} \qquad . \qquad (A1.62)$$

Therefore,

$$\langle (f - x_0)^2 \rangle = \frac{1}{J_n(x_0)} \qquad , \qquad (A1.63)$$

thus achieving the Cramér–Rao lower bound.

A1.5 ELECTROMAGNETIC FIELDS AND WAVES

(5.1) *Prove the BAC–CAB rule*

$$\vec{A} \times (\vec{B} \times \vec{C}) = \vec{B}(\vec{A} \cdot \vec{C}) - \vec{C}(\vec{A} \cdot \vec{B}) \qquad (A1.64)$$

by writing it out in the summations convention, and use it to show that

$$\nabla \times (\nabla \times \vec{E}) = \nabla(\nabla \cdot \vec{E}) - \nabla^2 \vec{E} \qquad . \qquad (A1.65)$$

$$[\vec{A} \times (\vec{B} \times \vec{C})]_i = \epsilon_{ijk} A_j \epsilon_{klm} B_l C_m$$

$$= \epsilon_{ijk} \epsilon_{klm} A_j B_l C_m$$

$$= (\delta_{il}\delta_{jm} - \delta_{im}\delta_{jl}) A_j B_l C_m$$

$$= \delta_{il} A_j B_l C_j - \delta_{im} A_j B_j C_m$$

$$= B_i(\vec{A} \cdot \vec{C}) - C_i(\vec{A} \cdot \vec{B})$$

$$\vec{A} \times (\vec{B} \times \vec{C}) = \vec{B}(\vec{A} \cdot \vec{C}) - \vec{C}(\vec{A} \cdot \vec{B}) \qquad . \qquad (A1.66)$$

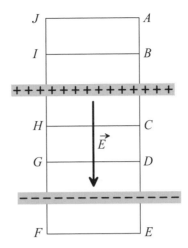

Figure A1.3. Cross-sectional integration volumes for Problem 5.2.

Plugging in $\vec{A} = \vec{B} = \nabla$,

$$\nabla \times \nabla \times \vec{E} = \nabla(\nabla \cdot \vec{E}) - \nabla^2 \vec{E} \quad . \tag{A1.67}$$

(5.2) (*a*) *Use Gauss' Law to find the capacitance between two parallel plates of area A at a potential difference V and with a spacing d. Neglect the fringing fields by assuming that this is a section of an infinite capacitor.*

Consider the integration volumes in Figure A1.3, seen in cross-section. By symmetry, the electric field cannot have a horizontal component (the horizontal contribution from any point will be canceled by that from a point on the opposite side).

$$\int_{CDGHC} \vec{D} \cdot d\vec{A} = 0 \tag{A1.68}$$

⇒ the field is constant inside the capacitor

$$\int_{BCHIB} \vec{D} \cdot d\vec{A} = \int_{ACHJA} \vec{D} \cdot d\vec{A}$$

⇒ the field is constant outside the capacitor

$$\int_{BEFIB} \vec{D} \cdot d\vec{A} = 0$$

⇒ the field is the same on both sides of the capacitor

$$\vec{D} = 0 \text{ at infinity}$$

⇒ the field vanishes outside the capacitor

$$\int_{BCHIB} \vec{D} \cdot d\vec{A} = \int_{BCHIB} \rho \, dV$$

$$\epsilon E A = Q$$

$$E = \frac{Q}{\epsilon A}$$

(where A is the area of the plate in the integration volume, and Q is the charge enclosed). Integrate the field to find the potential:

$$
\begin{aligned}
V &= - \int \vec{E} \cdot d\vec{l} \\
&= Ed \\
&= \frac{Qd}{\epsilon A} \quad ,
\end{aligned}
\tag{A1.69}
$$

which gives the capacitance:

$$
\begin{aligned}
C &= \frac{Q}{V} \\
&= \frac{\epsilon A}{d} \quad .
\end{aligned}
\tag{A1.70}
$$

(b) *Show that when a current flows through the capacitor, the integral over the internal displacement current is equal to the external electrical current.*

$$
\int_{CH} \frac{\partial \vec{D}}{\partial t} \cdot d\vec{A} = \int_{CH} \epsilon \frac{\partial \vec{E}}{\partial t} \cdot d\vec{A}
\tag{A1.71}
$$

$$
\begin{aligned}
&= \epsilon \frac{1}{d} \frac{dV}{dt} A \\
&= C\dot{V} = \dot{Q} = I \quad .
\end{aligned}
\tag{A1.72}
$$

(c) *Integrate the energy density to find the stored energy at a fixed potential. The answer should be expressed in terms of the capacitance.*

$$
\begin{aligned}
U &= \frac{1}{2} \int_V \vec{D} \cdot \vec{E} \, dV \\
&= \frac{1}{2} \int_V \epsilon E^2 \, dV \\
&= \frac{1}{2} \epsilon \frac{Q^2}{\epsilon^2 A^2} \, d\,A \\
&= \frac{1}{2} CV^2 \quad .
\end{aligned}
\tag{A1.73}
$$

(d) *Batteries are rated by amp-hours, the current they can supply at the design voltage for an hour. Consider a 10 V laptop battery that provides $10\ A \cdot h$. Assuming a plate spacing of 10^{-6} m $\equiv 1\ \mu m$ and a vacuum dielectric, what area would a capacitor need to be able to store this amount of energy? If such plates were 10 cm on a side and stacked vertically, how tall would the stack have to be to provide this total area?*

$$
10\ V \times 10\ A = 100\ W
$$

$$
100\ \frac{J}{s} \times 3600\ s = 3.6 \times 10^5\ J
$$

$$
\frac{1}{2} C (10\ V)^2 = 3.6 \times 10^5\ J \Rightarrow C = 7200\ F
$$

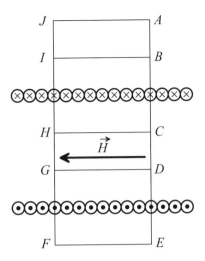

Figure A1.4. Integration paths for Problem 5.3.

$$7220 \text{ F} = \frac{(8.854{\times}10^{-12} \text{ F} \cdot \text{m}^{-1}) \, A}{10^{-6} \text{ m}} \Rightarrow A = 8{\times}10^8 \text{ m}^2$$

$$\frac{8{\times}10^8 \text{ m}^2}{(0.1 \text{ m})^2} \times 10^{-6} \text{ m} = 8{\times}10^4 \text{ m} \quad . \tag{A1.74}$$

This is slightly pessimistic. The largest commonly used capacitors are ~ 1 F; to achieve this capacitance in a reasonable size they use insulation materials with large dielectric constants, thin spacings, and they increase the surface area of the electrodes by using materials with rough textures. Nevertheless, it is unlikely that supercapacitors will be able to provide $\sim 10^5$ F in the same volume as a conventional chemical laptop battery. And while capacitors are much easier than batteries to charge and discharge, because they are leaky they are useful only for short-term storage.

(5.3) (a) *Use Stokes' Law to find the magnetic field of an infinite solenoid carrying a current I with n turns/meter.*

Consider the integration paths in Figure A1.4. By similar arguments to those used in the previous problem, the magnetic field must be horizontal, constant inside the solenoid, and vanish outside.

$$\oint \vec{H} \cdot d\vec{l} = \int_S \vec{J} \cdot d\vec{A}$$
$$Hl = nlI$$
$$H = nI \quad . \tag{A1.75}$$

(b) *Integrate the energy density to find the energy stored in a solenoid of radius r and length l, once again, neglecting fringing fields.*

$$U = \frac{1}{2} \int_V \vec{B} \cdot \vec{H} \, dV$$

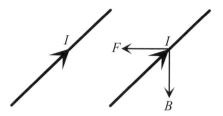

Figure A1.5. Definition of the ampere.

$$= \frac{1}{2} \int_V \mu H^2 \, dV$$

$$= \frac{1}{2} \mu n^2 I^2 \pi r^2 l \quad . \tag{A1.76}$$

In the next chapter we will see that the inductance of a solenoid is $L = \mu n^2 l \pi r^2$, therefore $U = LI^2/2$.

(c) *Consider a 10 T MRI magnet with a bore diameter of 1 m and a length of 2 m. What is the outward force on the magnet? Remember – force is the gradient of potential for a conservative force.*

$$F = \frac{\partial U}{\partial r} = \mu_0 N^2 I^2 \pi r l$$

$$= \frac{\pi r L B^2}{\mu_0}$$

$$= \frac{\pi \times 0.5 \text{ m} \times 2 \text{ m} \times (10 \text{ T})^2}{4\pi \times 10^{-7} \text{ H} \cdot \text{m}^{-1}}$$

$$= 2.5 \times 10^8 \text{ N}$$

$$\sim 28,000 \text{ tons} \quad . \tag{A1.77}$$

The furry rubber bands have a lot of fur.

(5.4) *Calculate the force per meter between two parallel wires one meter apart, each carrying a current of one ampere (this is the geometry used to define the ampere).*

Applying equation (5.86) for Figure A1.5,

$$d\vec{F} = I \, d\vec{l} \times \vec{B}$$

$$= \frac{\mu_0 I^2}{2\pi r} \, dl$$

$$F = \frac{4\pi \times 10^{-7} \text{ H} \cdot \text{m}^{-1} \times (1 \text{ A})^2 \times 1 \text{ m}}{2\pi \times 1 \text{ m}}$$

$$= 2 \times 10^{-7} \text{ N} \quad . \tag{A1.78}$$

(5.5) (a) *Assume that sunlight has an power density of 1 kW/m² (this is a peak number; the typical average value in the continental USA is ~ 200 W/m²). Estimate the electric field strength associated with this radiation.*

From equation (5.102), the time average power being transported by radiation is equal to the Poynting vector averaged over a cycle:

$$\langle |P| \rangle = \langle |\vec{E} \times \vec{H}| \rangle = \langle \sqrt{\epsilon_0/\mu_0} E^2 \rangle \quad . \tag{A1.79}$$

If $\vec{E} = \vec{E}_0 e^{i(\vec{k}\cdot x - \omega t)}$, then since

$$|e^{i\vec{k}\cdot\vec{x}}| = 1 \tag{A1.80}$$

and

$$\langle [\mathrm{Re}(e^{-i\omega t})]^2 \rangle = \langle [\cos(\omega t)^2 \rangle = \frac{1}{2} \quad , \tag{A1.81}$$

the time average Poynting vector is

$$\langle |\vec{P}| \rangle = \frac{1}{2}\sqrt{\frac{\epsilon_0}{\mu_0}} E_0^2 \quad . \tag{A1.82}$$

Therefore

$$10^3 \text{ W} = \int_S \vec{P} \times d\vec{A}$$

$$= \frac{1}{2}\sqrt{\frac{\epsilon_0}{\mu_0}} (1 \text{ m})^2 E_0^2$$

$$\Rightarrow E_0^2 = \frac{2 \times 10^3 \text{ W}}{\sqrt{\epsilon_0/\mu_0} \ (1 \text{ m})^2}$$

$$\Rightarrow E_0 = 868 \ \frac{\text{V}}{\text{m}} \quad . \tag{A1.83}$$

The force from this oscillating field is how sunlight heats materials.

(b) *If 1 W of power is focused in a laser beam to a square millimeter, what is the field strength? What about if it is focused to the diffraction limit of $\sim 1 \ \mu\mathrm{m}^2$?*

$$E_0 = \left[\frac{2 \cdot 1 \text{ W}}{\sqrt{\epsilon_0/\mu_0} \ (10^{-3} \text{ m})^2} \right]^{1/2} = 2.7 \times 10^4 \ \frac{\text{V}}{\text{m}}$$

$$E_0 = \left[\frac{2 \cdot 1 \text{ W}}{\sqrt{\epsilon_0/\mu_0} \ (10^{-6} \text{ m})^2} \right]^{1/2} = 2.7 \times 10^7 \ \frac{\text{V}}{\text{m}} \quad . \tag{A1.84}$$

The availability of such enormous electric fields in the laboratory has opened up many new areas of research. For example, the field within an atom is a small perturbation compared to this strength and so a strong laser can be used to drive atoms into novel states [Weinacht *et al.*, 1999].

A1.6 CIRCUITS, TRANSMISSION LINES, AND WAVE GUIDES

(6.1) *Cables designed to carry a low-frequency signal with minimum pickup of interference often consist of a twisted pair of conductors surrounded by a grounded shield. Why the twist? Why the shield?*

Twisting minimizes inductive pickup by reducing the flux linking the wires

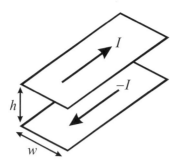

Figure A1.6. Transmission line for Problem 6.4.

and by alternately reversing the sign of the contribution, and shielding minimizes capacitative pickup through electrostatic screening.

(6.2) *Salt water has a conductivity $\sim 4\ S/m$. What is the skin depth at $10^4\ Hz$?*

$$\delta = \frac{1}{\sqrt{\pi \nu \mu \sigma}}$$

$$= \frac{1}{\sqrt{\pi \times 10^4\ \mathrm{s^{-1}} \times 4\pi \times 10^{-7}\ \mathrm{H \cdot m^{-1}} \times 4\ \mathrm{S \cdot m^{-1}}}}$$

$$= 2.5\ \mathrm{m} \quad . \tag{A1.85}$$

Even at this low frequency the skin depth is just a few meters. This is why submarines must communicate at extremely low frequencies (and therefore at a very low bit rate), requiring huge antennas.

(6.3) *Integrate Poynting's vector $\vec{P} = \vec{E} \times \vec{H}$ to find the power flowing across a cross-sectional slice of a coaxial cable, and relate the answer to the current and voltage in the cable.*

$$\vec{P} = \vec{E} \times \vec{H}$$

$$= \frac{Q}{2\pi\epsilon r}\hat{r} \times \frac{I}{2\pi r}\hat{\theta}$$

$$= \frac{IQ}{(2\pi)^2 \epsilon r^2}\hat{z}$$

$$P = \int_0^{2\pi} \int_{r_{\mathrm{i}}}^{r_{\mathrm{o}}} \frac{IQ}{(2\pi)^2 \epsilon r^2}\ dr\ r\ d\theta$$

$$= \frac{IQ}{2\pi\epsilon} \int_{r_{\mathrm{i}}}^{r_{\mathrm{o}}} \frac{1}{r}\ dr$$

$$= \frac{IQ}{2\pi\epsilon} \ln \frac{r_{\mathrm{o}}}{r_{\mathrm{i}}}$$

$$= IV \quad . \tag{A1.86}$$

(6.4) *Find the characteristic impedance and signal velocity for a transmission line consisting of two parallel strips with a width w and a separation h (Figure 6.4). You can ignore fringing fields by assuming that they are sections of conductors infinitely wide.*

For a section of an infinite parallel plate capacitor of width w and spacing d the capacitance per meter is

$$\mathcal{C} = \frac{\epsilon h}{d} \left(\frac{\text{F}}{\text{m}}\right) \quad . \tag{A1.87}$$

The magnetic field is found from a loop around one of the plates:

$$I = Hw \Rightarrow H = \frac{I}{h} \Rightarrow B = \mu \frac{I}{h} \quad . \tag{A1.88}$$

The flux linking a cross-sectional area of height d and length l is

$$\Phi = \int \vec{B} \cdot d\vec{A} = \mu \frac{Ild}{h} \quad , \tag{A1.89}$$

and therefore the inductance per meter is

$$\mathcal{L} = \frac{\Phi}{Il} = \mu \frac{d}{h} \left(\frac{\text{H}}{\text{m}}\right) \quad . \tag{A1.90}$$

This gives the characteristic impedance

$$Z = \sqrt{\frac{\mathcal{L}}{\mathcal{C}}} = \sqrt{\frac{\mu d^2}{\epsilon h^2}} \quad (\Omega) \tag{A1.91}$$

and the velocity

$$v = \frac{1}{\sqrt{\mathcal{L}\mathcal{C}}} = \sqrt{\frac{h}{\mu d}\frac{d}{\epsilon h}} = \frac{1}{\sqrt{\mu\epsilon}} \left(\frac{\text{m}}{\text{s}}\right) \quad . \tag{A1.92}$$

Such *striplines* are commonly used for guiding signals on high-frequency circuit boards.

(6.5) *The most common coaxial cable, RG58/U, has a dielectric with a relative permittivity of 2.26, an inner radius of 0.406 mm, and an outer radius of 1.48 mm.*

(a) *What is the characteristic impedance?*

$$Z = \sqrt{\frac{\mathcal{L}}{\mathcal{C}}}$$

$$= \left[\frac{\mu_0}{2\pi} \ln\left(\frac{r_\text{o}}{r_\text{i}}\right) \frac{1}{2\pi\epsilon_r\epsilon_0} \ln\left(\frac{r_\text{o}}{r_\text{i}}\right)\right]^{1/2}$$

$$= \left[\frac{4\pi \times 10^{-7}}{2\pi} \ln\left(\frac{1.48}{0.406}\right) \frac{1}{2\pi(2.26)(8.854 \times 10^{-12})}\right.$$

$$\left. \times \ln\left(\frac{1.48}{0.406}\right)\right]^{1/2}$$

$$= 51.59 \ \Omega \approx 50 \ \Omega \quad . \tag{A1.93}$$

(b) *What is the transmission velocity?*

$$v = \frac{1}{\sqrt{\mathcal{L}\mathcal{C}}} = \left[\frac{2\pi}{\mu_0 \ln(r_\text{o}/r_\text{i})} \frac{\ln(b/a)}{2\pi\epsilon}\right]^{1/2} = \frac{1}{\sqrt{\mu_0\epsilon_r\epsilon_0}}$$

$$= [(4\pi \times 10^{-7})(2.26)(8.854 \times 10^{-12})]^{-1/2}$$

$$\approx 2 \times 10^8 \ \text{m/s}$$

$$\approx 0.7c \quad . \tag{A1.94}$$

This corresponds to ~1 ns/ft. That is enormously faster than the drift velocity of electrons in a wire, which is measured in meters per second rather than the speed of light. Signals in cables travel as waves rather than by electron transport, much like waves in the ocean don't carry water with them.

(c) *If a computer has a clock speed of 1 ns, how long can a length of RG58/U be and still deliver a pulse within one clock cycle?*

$$2 \times 10^8 \; \mathrm{m \cdot s^{-1}} \times 10^{-9} \; \mathrm{s} = 0.2 \; \mathrm{m} \quad . \tag{A1.95}$$

This is why computers become smaller as their clock speeds increase: conventional architectures require that signals arrive within one clock cycle everywhere.

(d) *It is often desirable to use thinner coaxial cable to minimize size or weight but still match the impedance of RG58/U to minimize reflections. If such a cable has an outer diameter of 30 mils (a mil is a thousandth of an inch), what is the inner diameter?*

$$\frac{1 \; \mathrm{in}}{10^3 \; \mathrm{mil}} \times \frac{0.0254 \; \mathrm{m}}{1 \; \mathrm{in}} = 25.4 \; \frac{\mu\mathrm{m}}{\mathrm{mil}} \tag{A1.96}$$

$$
\begin{aligned}
50 \; \Omega = \bigg[& \frac{4\pi \times 10^{-7}}{2\pi} \ln \left(\frac{0.762 \; \mathrm{mm}}{r_\mathrm{i}} \right) \\
& \times \frac{1}{2\pi \times 2.226 \times 8.854 \times 10^{-12}} \ln \left(\frac{0.762 \; \mathrm{mm}}{r_\mathrm{i}} \right) \bigg]^{1/2} \\
\Rightarrow r_\mathrm{i} = {} & 0.22 \; \mathrm{mm} = 8.7 \; \mathrm{mil} \quad .
\end{aligned}
\tag{A1.97}
$$

(e) *For RG58/U, at what frequency does the wavelength become comparable to the diameter?*

$$2 \times 10^8 \; \frac{\mathrm{m}}{\mathrm{s}} \times \frac{1}{2.96 \times 10^{-3} \; \mathrm{m}} \approx 6.7 \times 10^{10} \; \frac{1}{\mathrm{s}}$$
$$= 67 \; \mathrm{GHz} \quad . \tag{A1.98}$$

(6.6) *Consider a 10 Mbit/s ethernet signal traveling in a RG58/U cable.*

(a) *What is the physical length of a bit?*

$$2 \times 10^8 \; \frac{\mathrm{m}}{\mathrm{s}} \times 10^{-7} \; \frac{\mathrm{s}}{\mathrm{bit}} = 20 \; \frac{\mathrm{m}}{\mathrm{bit}} \quad . \tag{A1.99}$$

Ethernet is a *CSMA* protocol (*Carrier Sense Multiple Access*). Stations wanting to transmit listen until there is no traffic, then start transmitting, and if they receive a signal while they are transmitting they abort and try again later. If a network is longer than the length of a packet of data, the chance of two stations simultaneously trying to transmit increases significantly.

(b) *Now consider what would happen if a "T" connector was used to connect one ethernet coaxial cable to two other ones. Estimate the reflection coefficient for a signal arriving at the T.*

The incoming 50 Ω line sees two parallel outgoing 50 Ω lines, therefore the total impedance it is driving is

$$\frac{1}{Z_\mathrm{L}} = \frac{1}{Z} + \frac{1}{Z} \Rightarrow Z_\mathrm{T} = \frac{Z}{2} = 25 \; \Omega \quad . \tag{A1.100}$$

This gives a reflection coefficient of

$$R = \frac{Z_{\mathrm{L}} - Z}{Z_{\mathrm{L}} + Z} = \frac{25 - 50}{25 + 50} = \frac{1}{3} \quad . \tag{A1.101}$$

Splitting an ethernet signal requires an active device to match the impedance.

A1.7 ANTENNAS

(7.1) *Find the electric field for an infinitesimal dipole radiator.*
   We want to evaluate

$$\vec{E} = \frac{1}{i\omega\mu_0\epsilon_0} \nabla(\nabla \cdot \vec{A}) - iw\vec{A} \tag{A1.102}$$

for

$$A_r = \mu_0 \frac{I_0 d e^{-ikr}}{4\pi r} \cos\theta \qquad A_\theta = -\mu_0 \frac{I_0 d e^{-ikr}}{4\pi r} \sin\theta \quad . \tag{A1.103}$$

The divergence is

$$\begin{aligned}
\nabla \cdot \vec{A} &= \frac{1}{r^2} \frac{\partial}{\partial r}\left(r^2 A_r\right) + \frac{1}{r\sin\theta}\frac{\partial}{\partial\theta}\left(\sin\theta A_\theta\right) \\
&= \frac{\mu_0 I_0 d}{4\pi}\left[\frac{1}{r^2}\frac{\partial}{\partial r}\left(re^{-ikr}\cos\theta\right) + \frac{1}{r\sin\theta}\frac{\partial}{\partial\theta}\left(-\frac{e^{-ikr}}{r}\sin^2\theta\right)\right] \\
&= -\frac{\mu_0 I_0 d}{4\pi}e^{-ikr}\left[\frac{1}{r^2} + \frac{ik}{r}\right]\cos\theta \quad .
\end{aligned} \tag{A1.104}$$

The gradient terms are then

$$\begin{aligned}
\nabla(\nabla \cdot \vec{A})_\theta &= \frac{1}{r}\frac{\partial}{\partial\theta}\left(\nabla \cdot \vec{A}\right) \\
&= \frac{\mu_0 I_0 d}{4\pi}e^{-ikr}\left[\frac{1}{r^3} + \frac{ik}{r^2}\right]\sin\theta \\
\nabla(\nabla \cdot \vec{A})_r &= \frac{\partial}{\partial r}\left(\nabla \cdot \vec{A}\right) \\
&= \frac{\mu_0 I_0 d}{4\pi}e^{-ikr}\left[\frac{2ik}{r^2} - \frac{k^2}{r} + \frac{2}{r^3}\right]\cos\theta \quad .
\end{aligned} \tag{A1.105}$$

Combining the components,

$$\begin{aligned}
E_\theta &= \frac{1}{i\omega\mu_0\epsilon_0}\frac{\mu_0 I_0 d}{4\pi}e^{-ikr}\left[\frac{1}{r^3} + \frac{ik}{r^2}\right]\sin\theta + i\omega\frac{\mu_0 I_0 d}{4\pi}e^{-ikr}\frac{1}{r}\sin\theta \\
&= \frac{I_0 d}{4\pi}e^{-ikr}\left[\frac{i\omega\mu_0}{r} + \frac{1}{r^2}\sqrt{\frac{\mu_0}{\epsilon_0}} + \frac{1}{i\omega\epsilon_0 r^3}\right]\sin\theta \\
E_r &= \frac{1}{i\omega\mu_0\epsilon_0}\frac{\mu_0 I_0 d}{4\pi}e^{-ikr}\left[\frac{2ik}{r^2} - \frac{k^2}{r} + \frac{2}{r^3}\right]\cos\theta - i\omega\frac{\mu_0 I_0 d}{4\pi}e^{-ikr}\frac{1}{r}\cos\theta \\
&= \frac{I_0 d}{4\pi}e^{-ikr}\left[\frac{2}{r^2}\sqrt{\frac{\mu_0}{\epsilon_0}} + \frac{2}{i\omega\epsilon_0 r^3}\right]\cos\theta \quad .
\end{aligned} \tag{A1.106}$$

(7.2) *What is the magnitude of the Poynting vector at a distance of 1 km from an antenna radiating 1 kW of power, assuming that it is an isotropic radiator with a wavelength much less than 1 km? What is the peak electric field strength at that distance?*

For a spherical wave

$$\langle P \rangle = \frac{10^3 \text{ W}}{4\pi \left(10^3 \text{ m}\right)^2} = 7.96 \times 10^{-5} \frac{\text{W}}{\text{m}^2} \quad . \tag{A1.107}$$

At this distance the radiation will appear to be a plane wave, so that

$$\vec{P} = \vec{E} \times \vec{H}$$

$$= \vec{E} \times \left(\sqrt{\frac{\epsilon_0}{\mu_0}} \hat{k} \times \vec{E} \right)$$

$$\langle P \rangle = \sqrt{\frac{\epsilon_0}{\mu_0}} \langle E^2 \rangle$$

$$= \frac{1}{2} \sqrt{\frac{\epsilon_0}{\mu_0}} E_{\text{peak}}^2$$

$$E_{\text{peak}} = \left[2\langle P \rangle \sqrt{\frac{\mu_0}{\epsilon_0}} \right]^{1/2}$$

$$= \left[2 \times 7.96 \times 10^{-5} \frac{\text{W}}{\text{m}^2} \times \sqrt{\frac{4\pi \times 10^{-7} \text{ H} \cdot \text{m}^{-1}}{8.85 \times 10^{-12} \text{ F} \cdot \text{m}^{-1}}} \right]^{1/2}$$

$$= 0.24 \frac{\text{V}}{\text{m}} \quad . \tag{A1.108}$$

(7.3) *For what value of R_{load} is the maximum power delivered to the load in Figure 7.3?*

$$I = \frac{V}{R_{\text{rad}} + R_{\text{load}}}$$

$$W_{\text{load}} = I^2 R_{\text{load}}$$

$$= \frac{V^2 R_{\text{load}}}{\left(R_{\text{rad}} + R_{\text{load}}\right)^2}$$

$$\frac{\partial W_{\text{load}}}{\partial R_{\text{load}}} = 0 = V^2 \left[\frac{1}{\left(R_{\text{rad}} + R_{\text{load}}\right)^2} - \frac{2R_{\text{load}}}{\left(R_{\text{rad}} + R_{\text{load}}\right)^3} \right]$$

$$1 = \frac{2R_{\text{load}}}{R_{\text{rad}} + R_{\text{load}}}$$

$$R_{\text{rad}} = R_{\text{load}} \quad . \tag{A1.109}$$

Therefore

$$W_{\text{load}} = \frac{V^2}{4R_{\text{rad}}} \tag{A1.110}$$

or for a periodic signal

$$\langle W_{\text{load}} \rangle = \frac{\langle V^2 \rangle}{4R_{\text{rad}}}$$

$$= \frac{V_{\text{peak}}^2}{8R_{\text{rad}}} \quad . \tag{A1.111}$$

(7.4) *For an infinitesimal dipole antenna, what are the gain and the area, and what
is their ratio?*

The gain is

$$G \equiv \max_{\theta,\varphi} \frac{P(r=1,\theta,\varphi)}{W/4\pi}$$

$$= \frac{I_0^2 k^2 d^2}{32\pi^2 r^2} \sqrt{\frac{\mu_0}{\epsilon_0}} \frac{3}{I_0^2 \pi} \sqrt{\epsilon_0 \mu_0} \left(\frac{\lambda}{d}\right)^2 4\pi$$

$$= \frac{3k^2\lambda^2}{8\pi^2}$$

$$= \frac{3}{2} \quad . \tag{A1.112}$$

Since the peak voltage is related to the peak electric field by $V_{\text{peak}} = E_{\text{peak}}d$, the
area is

$$\langle W_{\text{load}} \rangle = \frac{V_{\text{peak}}^2}{8R_{\text{rad}}}$$

$$= \frac{E_{\text{peak}}^2 d^2}{8R_{\text{rad}}}$$

$$= \frac{E_{\text{peak}}^2 d^2}{8} \frac{3}{2\pi} \sqrt{\frac{\epsilon_0}{\mu_0}} \left(\frac{\lambda}{d}\right)^2$$

$$= \langle P \rangle A$$

$$= \frac{1}{2} \sqrt{\frac{\epsilon_0}{\mu_0}} E_{\text{peak}}^2 A$$

$$\Rightarrow A = \frac{3}{8\pi}\lambda^2 \quad . \tag{A1.113}$$

Therefore

$$\frac{A}{G} = \frac{3}{8\pi}\lambda^2 \frac{2}{3} = \frac{\lambda^2}{4\pi} \quad . \tag{A1.114}$$

A1.8 OPTICS

(8.1) *Optics (as well as most of physics) can be derived from a global law as well as
a local one, in this case* Fermat's Principle: *a light ray chooses the path between
two points that minimizes the time to travel between them. Apply this to two
points on either side of a dielectric interface to derive Snell's Law.*

The total time to go between the endpoints in Figure A1.7 is

$$T = \frac{r_1}{v_1} + \frac{r_2}{v_2}$$

$$= \frac{n_1}{c}\sqrt{(x-x_1)^2 + z_1^2} + \frac{n_2}{c}\sqrt{(x_2-x)^2 + z_2^2} \quad , \tag{A1.115}$$

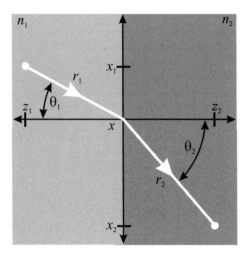

Figure A1.7. Interface for Snell's Law.

with a minimum at

$$\frac{dT}{dx} = \frac{n_1}{c}\frac{(x - x_1)}{r_1} - \frac{n_2}{c}\frac{(x_2 - x)}{r_2} = 0$$

$$n_1 \sin\theta_1 - n_2 \sin\theta_2 = 0$$

$$\frac{n_1}{n_2} = \frac{\sin\theta_2}{\sin\theta_1} \qquad . \tag{A1.116}$$

(8.2) (a) *Use Fresnel's equations and the Poynting vectors to find the reflectivity and transmisivity of a dielectric interface, defined by the ratios of incoming and outgoing energy.*

$$\vec{P} = \vec{E} \times \vec{H} = \vec{E} \times \sqrt{\frac{\epsilon}{\mu}}\hat{k} \times \vec{E} = \sqrt{\frac{\epsilon}{\mu}}E^2\hat{k} = \frac{n}{\mu c}E^2\hat{k} \qquad . \tag{A1.117}$$

Therefore, the ratio of the reflected to the incoming energy is

$$R = \frac{|E_1|^2}{|E_0|^2} \tag{A1.118}$$

and the ratio of the transmitted to the incoming energy is

$$T = \frac{n_2|E_2|^2}{n_0|E_0|^2} \qquad . \tag{A1.119}$$

There are two cases to consider:

• E perpendicular to plane of incidence

$$R = \frac{|E_1|^2}{|E_0|^2}$$

$$= \frac{\sin^2(\theta_2 - \theta_0)}{\sin^2(\theta_2 + \theta_0)} \tag{A1.120}$$

$$T = \frac{n_2 |E_2|^2}{n_0 |E_0|^2}$$

$$= \frac{4n_2 \cos^2 \theta_0 \sin^2 \theta_2}{n_0 \sin^2(\theta_2 + \theta_0)}$$

$$= \frac{4 \sin \theta_0 \cos^2 \theta_0 \sin^2 \theta_2}{\sin \theta_2 \sin^2(\theta_2 + \theta_0)} \quad . \tag{A1.121}$$

- E in the plane of incidence

$$R = \frac{\tan^2(\theta_0 - \theta_2)}{\tan^2(\theta_0 + \theta_2)} \tag{A1.122}$$

$$T = \frac{4 \sin \theta_0 \cos^2 \theta_0 \sin^2 \theta_2}{\sin \theta_2 \sin^2(\theta_0 + \theta_2) \cos^2(\theta_0 - \theta_2)} \quad . \tag{A1.123}$$

(b) *For a glass–air interface* $(n = 1.5)$ *what is the reflectivity at normal incidence?*

$$R = \frac{\sin^2(\theta_2 - \theta_0)}{\sin^2(\theta_2 + \theta_0)}$$

$$\lim_{\theta \to 0} R = \frac{\sin^2(\theta_2(1 - n_2/n_1))}{\sin^2(\theta_2(1 + n_2/n_1))}$$

$$= \frac{(1 - n_2/n_1)^2}{(1 + n_2/n_1)^2}$$

$$= \frac{(n_1 - n_2)^2}{(n_1 + n_2)^2}$$

$$= \frac{(1 - 1.5)^2}{(1 + 1.5)^2}$$

$$= 0.04 \tag{A1.124}$$

$$T = 1 - R = \frac{4n_1 n_2}{(n_1 + n_2)^2} = 0.96 \quad . \tag{A1.125}$$

(c) *What is the Brewster angle?*

$$\theta_B = \tan^{-1} \frac{n_2}{n_1} = 56.3° \text{ (air} \to \text{glass)}$$

$$= 33.7° \text{ (glass} \to \text{air)} \quad . \tag{A1.126}$$

The output windows in gas laser tubes are tilted at this angle to define the direction of polarization for the laser.

(d) *What is the critical angle?*

$$\theta_C = \sin^{-1} \frac{n_2}{n_1} = 41.8° \text{ (glass} \to \text{air)} \quad . \tag{A1.127}$$

Total internal reflection is possibly only when going from a higher to a lower index of refraction.

(8.3) *Consider a wave at normal incidence to a dielectric layer with index n_2 between layers with indices n_1 and n_3 (Figure 8.6).*

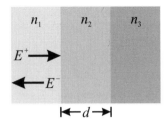

Figure A1.8. An antireflection coating.

(a) *What is the reflectivity? Think about matching the boundary conditions, or about the multiple reflections.*

At each interface, the ratio of the reflected and the incoming field strength is given by

$$r_{12} = \frac{n_1 - n_2}{n_1 + n_2} \tag{A1.128}$$

and the ratio of the transmitted and the incoming field strength is

$$t_{12} = \frac{2n_1}{n_1 + n_2} \quad . \tag{A1.129}$$

If the incoming wave has magnitude E^+, the total reflected wave E^- will be an infinite sum of reflections between the interfaces:

$$
\begin{aligned}
E^- &= E^+[r_{12} + t_{12}r_{23}e^{ik_2 2d}t_{21} + t_{12}r_{23}(r_{21}r_{23})e^{ik_2 4d}t_{21} \\
&\quad + t_{12}r_{23}(r_{21}r_{23})(r_{21}r_{23})e^{ik_2 6d}t_{21} + \ldots] \\
&= E^+[r_{12} + t_{12}r_{23}t_{21}e^{ik_2 2d}\sum_{n=0}^{\infty}(r_{21}r_{23}e^{ik_2 2d})^n] \\
&= E^+\left[r_{12} + t_{12}r_{23}t_{21}e^{ik_2 2d}\frac{1}{1 - r_{21}r_{23}e^{ik_2 2d}}\right] \\
&= E^+\left[\frac{r_{12}(1 - r_{21}r_{23}e^{ik_2 2d}) + t_{12}r_{23}t_{21}e^{ik_2 2d}}{1 - r_{21}r_{23}e^{ik_2 2d}}\right] \\
&= E^+\left[\frac{r_{12}(1 + r_{12}r_{23}e^{ik_2 2d}) + t_{12}r_{23}t_{21}e^{ik_2 2d}}{1 + r_{12}r_{23}e^{ik_2 2d}}\right] \\
&= E^+\left[\frac{r_{12} + r_{23}e^{ik_2 2d}(r_{12}^2 + t_{12}t_{21})}{1 + r_{12}r_{23}e^{ik_2 2d}}\right] \\
&= E^+\left[\frac{r_{12} + r_{23}e^{ik_2 2d}}{1 + r_{12}r_{23}e^{ik_2 2d}}\right] \quad . \tag{A1.130}
\end{aligned}
$$

Therefore, the total reflection coefficient is

$$
\begin{aligned}
R &= \frac{|E^-|^2}{|E^+|^2} \\
&= \left[\frac{r_{12} + r_{23}e^{ik_2 2d}}{1 + r_{12}r_{23}e^{ik_2 2d}}\right]^2 \quad . \tag{A1.131}
\end{aligned}
$$

This answer could also be found by solving the system of equations for matching the boundary conditions.

(b) *Can you find values for n_2 and d such that the reflection vanishes?*

Since r_{12} and r_{23} are real numbers, for the the numerator to vanish

$$e^{ik_2 2d} = -1 \Rightarrow k_2 2d = \pi$$

$$\frac{2\pi}{\lambda_2} 2d = \pi$$

$$d = \frac{\lambda_2}{4} \quad . \tag{A1.132}$$

This in turn implies that

$$R = \left[\frac{r_{12} - r_{23}}{1 - r_{12} r_{23}} \right]^2$$

$$= \left[\frac{\dfrac{n_1 - n_2}{n_1 + n_2} - \dfrac{n_2 - n_3}{n_2 + n_3}}{1 - \dfrac{n_1 - n_2}{n_1 + n_2} \dfrac{n_2 - n_3}{n_2 + n_3}} \right]^2$$

$$= \left[\frac{n_1 n_3 - n_2^2}{n_1 n_3 + n_2^2} \right]^2$$

$$R = 0 \Rightarrow n_2 = \sqrt{n_1 n_3} \quad . \tag{A1.133}$$

There will be no reflection if the intermediate layer has a thickness of 1/4 the wavelength in that layer, and if its index of refraction is the geometrical mean of that of the first and last layers. This is called ... can you guess? ... an *antireflection* coating, and is found on high-quality lenses and windows. With multiple dielectric layers it is also possible to design coatings that act as windows or mirrors for desired frequency bands and orientations [Fink *et al.*, 1998; Weber *et al.*, 2000].

(8.4) *Consider a ray starting with a height r_0 and some slope, a distance d_1 away from a thin lens with focal length f. Use ray matrices to find the image plane where all rays starting at this point rejoin, and discuss the magnification of the height r_0.*

The product of the matrices is

$$\begin{bmatrix} 1 & d_2 \\ 0 & 1 \end{bmatrix} \begin{bmatrix} 1 & 0 \\ -1/f & 1 \end{bmatrix} \begin{bmatrix} 1 & d_1 \\ 0 & 1 \end{bmatrix}$$

$$= \begin{bmatrix} 1 & d_2 \\ 0 & 1 \end{bmatrix} \begin{bmatrix} 1 & d_1 \\ -1/f & 1 - d_1/f \end{bmatrix}$$

$$= \begin{bmatrix} 1 - d_2/f & d_1 + d_2(1 - d_1/f) \\ -1/f & 1 - d_1/f \end{bmatrix} \quad . \tag{A1.134}$$

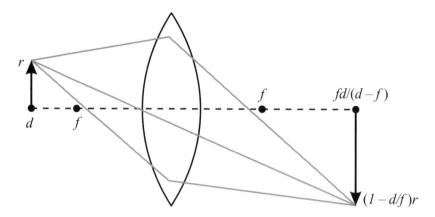

Figure A1.9. Focusing by a thin lens.

The position will be independent of the starting angle if

$$d_1 + d_2(1 - d_1/f) = 0$$

$$d_1 + d_2 = \frac{d_1 d_2}{f}$$

$$\frac{1}{d_1} + \frac{1}{d_2} = \frac{1}{f} \quad . \tag{A1.135}$$

At d_2, the height will be $(1 - d_2/f)r_0$, and so the magnification is $1 - d_2/f$ (Figure A1.9). This will change sign for d_2 greater than or less than f (the image inverts), and when $d_1 = f$ then $d_2 = \infty$.

(8.5) *Common CD players use an AlGaAs laser with a 790 nm wavelength.*

(a) *The pits that are read on a CD have a diameter of roughly 1 μm and the optics are diffraction-limited; what is the beam divergence angle?*

$$\theta = \frac{\lambda}{\pi n w_0}$$

$$= \frac{790 \times 10^{-9} \text{ m}}{\pi \times 1 \times 10^{-6} \text{ m}}$$

$$= 0.25 \text{ rad}$$

$$= 14° \quad . \tag{A1.136}$$

(b) *Assuming the same geometry, what wavelength laser would be needed to read 0.1 μm pits?*

$$0.25 = \frac{\lambda}{\pi \times 10^{-7} \text{ m}} \Rightarrow \lambda = 78 \text{ nm} \quad . \tag{A1.137}$$

(c) *How large must a telescope mirror be if it is to be able to read a car's license plate in visible light ($\lambda \sim 600$ nm) from a Low Earth Orbit (LEO) of 200 km?*

The required divergence angle is

$$\frac{600 \times 10^{-9} \text{ m}}{\pi \times 1 \times 10^{-2} \text{ m}} = 1.9 \times 10^{-5} \text{ rad} \quad . \tag{A1.138}$$

This corresponds to a mirror radius of

$$r = h \tan \theta$$
$$\approx h\theta$$
$$= 200{\times}10^3 \text{ m} \times 1.9{\times}10^{-5} \text{ rad}$$
$$= 3.8 \text{ m} \quad , \tag{A1.139}$$

which just happens to be the size of the planned next-generation space tele-scope. Because orbits this low decay very quickly, they're used primarily for reconnaissance and low-power communications.

A1.9 LENSLESS IMAGING AND INVERSE PROBLEMS

(9.1) *Is a thin spherical lens (like the ones we studied in the last chapter) a matched filter with a response like equation (9.8)?*

Yes. For a thin spherical lens with radius of curvature R and thickness on axis $2d_0$, the thickness $2d$ as a function of radius r from the axis is found from

$$[(R - d_0) + d]^2 + r^2 = R^2$$
$$d = d_0 - R + \sqrt{R^2 - r^2}$$
$$\approx d_0 - \frac{r^2}{2R} \quad . \tag{A1.140}$$

Therefore, the phase shift for an approximately horizontal paraxial ray passing through the lens is a quadratic function of the radius, and we have already seen that in the paraxial approximation the path length to a detector on the axis of the lens will be a quadratic function of the radius.

(9.2) *The resolving power of a lens can be defined in terms of the distance between the two points at which the point spread function has decreased from its maximum by 3 dB. Two objects can be resolved if they are separated by this distance. This is also sometimes defined by* Rayleigh's criterion: *the principal maximum of the point spread function from one object is at the first minimum of the point spread function of the other.*

(a) *For equation (9.9) what is the resolving power in terms of the wavelength, the lens aperture size, and the distance from the lens?*

$$20 \log_{10} \frac{a}{b} = -3 \Rightarrow \frac{a}{b} = 0.708$$

$$\frac{\sin(\frac{\pi x D}{z\lambda})}{\frac{\pi x D}{z\lambda}} = 0.708 \Rightarrow \frac{xD}{z\lambda} \approx 0.443$$

$$\Rightarrow x = \frac{0.443 z\lambda}{D}$$

$$\Rightarrow 2x = \frac{0.89 z\lambda}{D} \quad . \tag{A1.141}$$

The $\sin(x)/x = 0$ is a transcendental equation; its roots must be found by numerical root finding techniques.

(b) *For an ultrasonic signal (100 kHz) in air (~ 350 m/s), what size aperture is needed to resolve 1 cm at a distance of 1 m?*

$$\lambda = \frac{350 \text{ m} \cdot \text{s}^{-1}}{10^5 \text{ s}^{-1}} = 3.5 \text{ mm} \qquad (A1.142)$$

$$D = \frac{0.89 \times 1 \text{ m} \times 0.0035 \text{ m}}{0.01 \text{ m}} = 0.31 \text{ m} \quad . \qquad (A1.143)$$

(c) *For a radar satellite (10 GHz) in low Earth orbit (~ 200 km) what size aperture is needed to resolve 1 cm? What is the angle subtended by this aperture relative to the Earth's surface?*

$$\lambda = \frac{3 \times 10^8 \text{ m} \cdot \text{s}^{-1}}{10^{10} \text{ s}^{-1}} = 0.03 \text{ m} \qquad (A1.144)$$

$$D = \frac{0.89 \times 2 \times 10^5 \text{ m} \times 0.03 \text{ m}}{0.01 \text{ m}} = 5.34 \times 10^5 \text{ m} \qquad (A1.145)$$

$$\theta = 2 \tan^{-1} \frac{5.34 \times 10^5 / 2 \text{ m}}{2 \times 10^5 \text{ m}} = 106° \quad . \qquad (A1.146)$$

This huge aperture points to why synthetic apertures are needed for radar imaging from satellites.

(9.3) *Work out the delay function $g(x')$ to implement a matched filter in the plane for a paraxial synthetic aperture radar. Assume that the transceiver is moving along a straight line, at each point sending out a spherical wave and accumulating the return signal convolved by $g(x')$, and assume that the transceiver velocity is slow compared to the wave speed.*

The geometry is the same as in Figure 9.1, but now the signal starts at $(x', 0)$, travels to (x, z), and then returns to $(x' + vt, 0) \approx (x' + v2z/c, 0)$. Therefore, the phase shift over this path length becomes

$$f = e^{ik\{[z^2 + (x-x')^2]^{1/2} + [z^2 + (x-x'-vt)^2]^{1/2}\}}$$

$$\approx e^{ik2z} e^{ik[(x-x')^2 + (x-x'-2vz/c)^2]/2z}$$

$$\Rightarrow g(x') = f^*(-x')$$

$$= e^{-ik2z} e^{-ik[(x+x')^2 + (x+x'-2vz/c)^2]/2z} \quad . \qquad (A1.147)$$

(9.4) *Estimate the typical resonance frequency for a nuclear spin (NMR) and an electronic spin (ESR) in a 1 T field.*

Using the semiclassical relationship between field and frequency,

$$\begin{aligned}
\nu &= \frac{\omega}{2\pi} \\
&= \frac{\gamma B}{2\pi} \\
&\approx \frac{qB}{4\pi m} \\
&= \frac{1.602 \times 10^{-19} \text{ C} \times 1 \text{ T}}{4\pi \times 1.672 \times 10^{-27} \text{ kg}} \approx 7 \text{ MHz} \quad (\text{NMR}) \\
&= \frac{1.602 \times 10^{-19} \text{ C} \times 1 \text{ T}}{4\pi \times 9.109 \times 10^{-31} \text{ kg}} \approx 14 \text{ GHz} \quad (\text{ESR}) \quad . \qquad (A1.148)
\end{aligned}$$

(9.5) *Consider a point charge about an infinite conducting ground plane.*

(a) *Using the method of images, find the surface charge distribution on the plane. Using the geometry in Figure 5.7, if a charge q is located at (μ_x, μ_y, z), then at the surface*

$$E_z = \frac{qz}{2\pi \epsilon r^3}$$

$$= \frac{qz}{2\pi \epsilon [(x - \mu_x)^2 + (y - \mu_y)^2 + z^2]^{3/2}} \quad . \quad (A1.149)$$

Integrating Gauss' Law over an infinitesimal volume at the surface,

$$\int \rho \, dV = \int \nabla \cdot \vec{D} \, dV$$

$$Q = \int \epsilon \vec{E} \cdot d\vec{A}$$

$$= \epsilon E_z A$$

$$\frac{Q}{A} = \epsilon E_z$$

$$\sigma = \frac{qz}{2\pi [(x - \mu_x)^2 + (y - \mu_y)^2 + z^2]^{3/2}} \quad . \quad (A1.150)$$

(b) *Assume that the plane is divided up into a grid of square electrodes, and analytically integrate the charge density to find the measured charge at each electrode.*

$$\int \int \sigma(x, y) \, dx \, dy = \frac{q}{2\pi} \tan^{-1} \left[\frac{(x - \mu_x)(y - \mu_y)}{z\sqrt{(x - \mu_x)^2 + (y - \mu_y)^2 + z^2}} \right] \quad . \quad (A1.151)$$

(c) *Numerically evaluate the electrode charge distribution generated by the point charge.*

The solution is plotted in Figure A1.10.

(d) *Use these measurements to estimate the source charge distribution as a function of height above the center of the surface charge distribution. Take a least-squares likelihood function, and for a regularizer use the sum of squares of the source charges. Plot the charge distribution as the relative weight of the regularizer term is varied, showing the minimum amount of charge compatible with the measurements for a given total error.*

Define Q_m to be the charge measured in the mth electrode, q_n to be the source charge in the nth location, and G_{mn} to be the integral in equation (A1.151) evaluated between them. Then we want to make extremal

$$I = -\sum_m \left(Q_m - \sum_n G_{mn} q_n \right)^2 - \lambda \sum_n q_n^2 \quad . \quad (A1.152)$$

Exponentiating this (remember equation 9.32), the first term is a Gaussian around the measured charge, and the second is a Gaussian around the as-

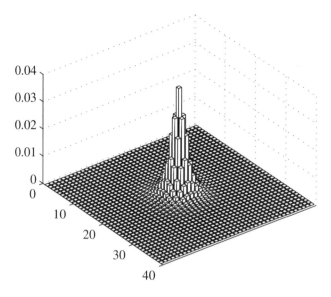

Figure A1.10. Surface charge distribution generated by a point charge above a ground plane.

sumption that $q_n = 0$, with a variance set by λ. Defining $\vec{q} = (q_1, \ldots, q_N)$, $\vec{Q} = (Q_1, \ldots, Q_M)$, $\mathbf{G}$ to be the matrix of the G_{mn}'s, and $\mathbf{I}$ to be the identity matrix, we need to solve

$$
\begin{aligned}
0 &= \frac{\partial I}{\partial q_i} \\[2mm]
&= \frac{\partial}{\partial q_i} \left[-\sum_m \left(Q_m - \sum_n G_{mn} q_n \right)^2 - \lambda \sum_n q_n^2 \right] \\[2mm]
&= -\sum_m 2 \left(Q_m - \sum_n G_{mn} q_n \right) (-G_{mi}) - \lambda 2 q_i \\[2mm]
&= \mathbf{G}^T \cdot (\vec{Q} - \mathbf{G} \cdot \vec{q}) - \lambda \vec{q} \\[2mm]
(\mathbf{G}^T \cdot \mathbf{G} + \lambda \mathbf{I}) \cdot \vec{q} &= \mathbf{G}^T \cdot \mathbf{Q} \\[2mm]
&\Rightarrow \vec{q} = (\mathbf{G}^T \cdot \mathbf{G} + \lambda \mathbf{I})^{-1} \cdot \mathbf{G}^T \cdot \mathbf{Q} \quad .
\end{aligned}
\tag{A1.153}
$$

The solution is shown in Figure A1.11, calculated by the following Matlab program. This does not match the original point charge because these distributions have a smaller error for a given amount of charge; recovering point charges requires doing a search in that basis [Smith, 1996].

```
%
% qreg.m
% electrostatic inverse problem
% Neil Gershenfeld  (c) 12/1/99
%
```

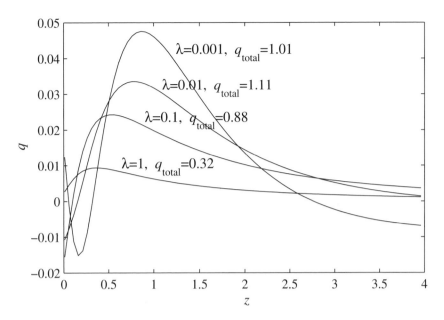

Figure A1.11. Regularized estimation for a unit charge located at $z = 1$.

```
clear all
npts = 40;
limit = 10;
delta = 2*limit/(npts-1);
q = 1;
muz = 1;
mux = 0;
muy = 0;
x = kron(ones(npts,1),(-limit:delta:limit));
y = kron((-limit:delta:limit)',ones(1,npts));
I = (q/(2*pi))*atan((x-mux).*(y-muy)./(muz*sqrt((x-mux).^2 ...
    + (y-muy).^2+muz^2)));
Qmeas = I(1:(npts-1),1:(npts-1)) + I(2:npts,2:npts) ...
        - I(1:(npts-1),2:npts) - I(2:npts,1:(npts-1));
bar3(Qmeas);
set(gca,'FontName','Times-Roman','FontSize',10)
print -deps qsurf.eps
z = .01:.05:4;
nz = length(z);
nmeas = (npts-1)*(npts-1);
Q = reshape(Qmeas,nmeas,1);
for i = 1:nz
    I = (1/(2*pi))*atan((x-mux).*(y-muy) ./ ...
        (z(i)*sqrt((x-mux).^2+(y-muy).^2+z(i)^2)));
    Iint = I(1:(npts-1),1:(npts-1)) + I(2:npts,2:npts) ...
        - I(1:(npts-1),2:npts) - I(2:npts,1:(npts-1));
```

```
    G(:,i) = reshape(Iint,nmeas,1);
    end
Ident = diag(ones(1,nz));
lambda = 1;
qpred = inv(G'*G+lambda*Ident)*G'*Q;
sum(qpred)
plot(z,qpred);
hold on
lambda = 0.1;
qpred = inv(G'*G+lambda*Ident)*G'*Q;
sum(qpred)
plot(z,qpred);
lambda = 0.01;
qpred = inv(G'*G+lambda*Ident)*G'*Q;
sum(qpred)
plot(z,qpred);
lambda = 0.001;
qpred = inv(G'*G+lambda*Ident)*G'*Q;
sum(qpred)
plot(z,qpred);
hold off
set(gca,'FontName','Times-Roman','FontSize',10)
set(gcf,'PaperPosition',[0.5 3 5 3.2]);
ylabel('{\it q}','FontSize',11)
xlabel('{\it z}','FontSize',11)
text(.64,.01,'\lambda=1, {\it q}_{total}=0.32')
text(.90,.023,'\lambda=0.1, {\it q}_{total}=0.88')
text(1.09,.032,'\lambda=0.01, {\it q}_{total}=1.11')
text(1.24,.042,'\lambda=0.001, {\it q}_{total}=1.01')
print -deps qreg.eps
```

....

A1.10 SEMICONDUCTOR MATERIALS AND DEVICES

(10.1) (a) *Derive equation (10.28) by taking the integral and limit of equation (10.27).*

Remembering that

$$
\begin{aligned}
\psi(x) &= Ae^{iqx} + Be^{-iqx} \\
&= e^{ikx} u(x) \\
&= e^{ikx}\left(Ae^{i(q-k)x} + Be^{-i(q+k)x} \right) \quad ,
\end{aligned}
\tag{A1.154}
$$

and that the limit must be taken within the interval between delta functions,

$$\underbrace{\lim_{\epsilon \to 0} \int_{-\epsilon}^{\epsilon} E\psi(x)\, dx}_{0} \quad + \quad \underbrace{\int_{-\epsilon}^{\epsilon} \frac{\hbar^2}{2m} \frac{d^2\psi(x)}{dx^2}\, dx}_{}$$

$$\underbrace{\frac{\hbar^2}{2m} \left(\frac{d\psi}{dx}\right)\Big|_{\epsilon}}_{\frac{\hbar^2}{2m} e^{ik\epsilon} \left(\frac{du}{dx}\right)\Big|_{\epsilon}} \quad - \quad \underbrace{\frac{\hbar^2}{2m} \left(\frac{d\psi}{dx}\right)\Big|_{-\epsilon}}_{\frac{\hbar^2}{2m} e^{-ik\epsilon} \left(\frac{du}{dx}\right)\Big|_{\alpha-\epsilon}}$$

$$= \underbrace{\int_{-\epsilon}^{\epsilon} \sum_{n=-\infty}^{\infty} V_0 \delta(x-n\alpha)\psi(x)}_{V_0 \psi(0)}$$

$$\frac{\hbar^2}{2m} iq\left(A - B - Ae^{i(q-k)\alpha} + Be^{-i(q+k)\alpha}\right) = V_0(A+B) \quad . \qquad (A1.155)$$

(b) *Show that equation (10.2) follows.*

$$A + B = Ae^{i(q-k)\alpha} + Be^{-i(q+k)\alpha}$$

$$\Rightarrow B = A\frac{1 - e^{i(q-k)\alpha}}{e^{-i(q+k)\alpha} - 1} \equiv A\frac{1-\beta}{\gamma-1} \quad . \qquad (A1.156)$$

Substitute into:

$$\frac{\hbar^2}{2m} iq(A - B - Ae^{i(q-k)\alpha} + Be^{-i(q+k)\alpha}) = V_0(A+B)$$

$$\frac{\hbar^2}{2m} iqA[(\gamma - 1) - (1 - \beta) - \beta(\gamma - 1) + \gamma(1 - \beta)] = V_0 A[(\gamma - 1) + (1 - \beta)]$$

$$iA(2\gamma + 2\beta - 2\beta\gamma - 2) = \frac{2mV_0}{\hbar^2 q} A(\gamma - \beta)$$

$$iA(\underbrace{e^{iq\alpha} + e^{-iq\alpha}}_{2\cos(q\alpha)}\,\underbrace{-e^{-ik\alpha} - e^{ik\alpha}}_{-2\cos(k\alpha)}) = \frac{mV_0}{\hbar^2 q} A(\underbrace{e^{-iq\alpha} - e^{iq\alpha}}_{i2\sin(q\alpha)})$$

$$\left[\cos(k\alpha) - \cos(q\alpha) - \frac{mV_0}{q\hbar^2}\sin(q\alpha)\right] A = 0 \quad . \qquad (A1.157)$$

(10.2) *What is the expected occupancy of a state at the conduction band edge for Ge, Si, and diamond at room temperature (300 K)?*

$$kT = 8.617 \times 10^{-5} \frac{\text{eV}}{\text{K}} \times 300 \text{ K} = 0.026 \text{ eV}$$

$$\frac{1}{1 + e^{(E-E_F)/kT}} = \frac{1}{1 + e^{(E_F + E_g/2 - E_F)/kT}} = \frac{1}{1 + e^{E_g/2kT}}$$

$$= \frac{1}{1 + e^{0.67 \text{ eV}/2 \cdot 0.026 \text{ eV}}} = 2.5 \times 10^{-6} \quad \text{(Ge)}$$

$$= \frac{1}{1 + e^{1.11 \text{ eV}/2 \cdot 0.026 \text{ eV}}} = 5.4 \times 10^{-10} \quad \text{(Si)}$$

$$= \frac{1}{1 + e^{0.67 \text{ eV}/2 \cdot 0.026 \text{ eV}}} = 1.7 \times 10^{-42} \quad \text{(diamond)} \quad . \quad \text{(A1.158)}$$

(10.3) *Consider Si doped with 10^{17} As atoms/cm^3.*

   (a) *What is the equilibrium hole concentration at 300 K?*
   For Si,

$$N_c = 2 \left(\frac{2\pi m_n^* kT}{h^2} \right)^{3/2}$$

$$= 2 \left(\frac{2\pi \times 1.1 \times 9.1095 \times 10^{-28} \text{ g} \times 1.3807 \times 10^{-16} \text{ erg} \cdot \text{K}^{-1} \times 300 \text{ K}}{(6.6262 \times 10^{-27} \text{ erg} \cdot \text{s})^2} \right)^{3/2}$$

$$= 2.895 \times 10^{19} \text{ cm}^{-3}$$

$$N_v = 2 \left(\frac{2\pi m_p^* kT}{h^2} \right)^{3/2}$$

$$= 2 \left(\frac{2\pi \times 0.56 \times 9.1095 \times 10^{-28} \text{ g} \times 1.3807 \times 10^{-16} \text{ erg} \cdot \text{K}^{-1} \times 300 \text{ K}}{(6.6262 \times 10^{-27} \text{ erg} \cdot \text{s})^2} \right)^{3/2}$$

$$= 1.052 \times 10^{19} \text{ cm}^{-3} \quad . \quad \text{(A1.159)}$$

Therefore

$$n_i^2 = N_c N_v e^{-E_g/kT}$$

$$= 2.895 \times 10^{19} \text{ cm}^{-3} \times 1.052 \times 10^{19} \text{ cm}^{-3} \cdot$$

$$e^{-1.11 \text{ eV}/(8.617 \times 10^{-5} \text{ eV} \cdot \text{K}^{-1} \times 300 \text{ K})}$$

$$= 6.851 \times 10^{19} \text{ cm}^{-6}$$

$$\Rightarrow n_i = 8.277 \times 10^9 \text{ cm}^{-3} \quad . \quad \text{(A1.160)}$$

Then the hole concentration is

$$np = n_i^2 \Rightarrow p = \frac{n_i^2}{n}$$

$$= \frac{6.851 \times 10^{19} \text{ cm}^{-6}}{10^{17} \text{ cm}^{-3}}$$

$$= 685 \text{ cm}^{-3} \quad . \quad \text{(A1.161)}$$

   (b) *How much does this move E_F relative to its intrinsic value?*

$$n = n_i e^{(E_F - E_i)/kT}$$

$$\Rightarrow E_F - E_i = kT \ln \frac{n}{n_i}$$

$$= 8.617 \times 10^{-5} \text{ eV} \cdot \text{K}^{-1} \times 300 \text{ K}$$

$$\times \ln \frac{10^{17} \text{ cm}^{-3}}{8.277 \times 10^9 \text{ cm}^{-6}}$$

$$= 0.422 \text{ eV} \quad . \quad \text{(A1.162)}$$

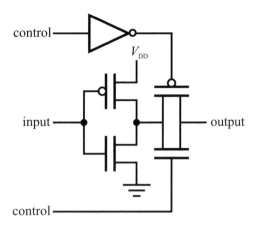

Figure A1.12. A tristate CMOS inverter.

(10.4) *Design a* tristate *CMOS inverter by adding a control input to a conventional inverter that can force the output to a high impedance (disconnected) state. These are useful for allowing multiple gates to share a single wire.*

The tristate output is produced by passing the inverter output through a *transmission gate* (Figure A1.12). This connects a PMOS and an NMOS inverter in parallel, with one side driven by the control and the other by its inverse. Therefore either both transistors will be off and the output will be in a high-impedance state, or they will both be on and hence equally able to pass a 0 or a 1 without any V_T drops. The transmission gate requires both the control signal and its complement; this is a simple example of *dual-rail logic*. Having every gate always produce and use the output and its complement can have a number of advantages for high-speed, low-power, or steady-load circuitry. This comes at the expense of real estate on the chip, but that is frequently not a severe constraint.

(10.5) *Let the output of a logic circuit be connected by a wire of resistance R to a load of capacitance C (i.e., the gate of the next FET). The load capacitor is initially discharged, then when the gate is turned on it is charged up to the supply voltage V. Assume that the output is turned on instantly, and take the supply voltage to be 5 V and the gate capacitance to be 10 fF.*

(a) *How much energy is stored in the capacitor?*

$$\text{energy stored} = \frac{1}{2}CV^2$$
$$= \frac{1}{2}10\times10^{-15} \text{ F} \times (5 \text{ V})^2$$
$$= 1.25\times10^{-13} \text{ J} \quad . \tag{A1.163}$$

(b) *How much energy was dissipated in the wire?*

$$I = \frac{V - V_C}{R} = C\frac{dV_C}{dt} \Rightarrow \frac{dV_C}{dt} = \frac{1}{RC}(V - V_C)$$
$$\Rightarrow V_C = V(1 - e^{-t/RC})$$
$$\Rightarrow I = \frac{V}{R}e^{-t/RC} \quad , \tag{A1.164}$$

$$\text{power dissipated} = \int_0^\infty I^2 R \, dt$$

$$= \int_0^\infty \frac{V^2}{R^2} e^{-2t/RC} R \, dt$$

$$= \frac{V^2}{R^2} \frac{RC}{2} R$$

$$= \frac{1}{2} C V^2 \quad . \tag{A1.165}$$

(c) *Approximately how much energy is dissipated in the wire if the supply voltage is linearly ramped from 0 to 5 V during a long time τ?*

During the time the voltage is ramping up, the differential equation is

$$\frac{dV_C}{dt} = \frac{1}{RC} \left(\frac{V}{\tau} t - V_C \right) \quad . \tag{A1.166}$$

Substituting an ansatz

$$V_C = \frac{V}{\tau} t + A \tag{A1.167}$$

gives

$$\frac{V}{\tau} = \frac{1}{RC} \left(\frac{V}{\tau} t - \frac{V}{\tau} t - A \right) \tag{A1.168}$$

or

$$A = -\frac{RC}{\tau} V \quad . \tag{A1.169}$$

The voltage lags by RC/τ, which can be made arbitrarily small by making τ large. This ignores the transients at the beginning and the end of the interval; they can be found by using Laplace transforms [Gershenfeld, 1999a]. The power dissipated in the resistor is then

$$\text{power dissipated} = \int_0^\tau I^2 R \, dt$$

$$= \int_0^\tau \frac{(V - V_C)^2}{R} \, dt$$

$$= \int_0^\tau \frac{RC^2 V^2}{\tau^2} \, dt$$

$$= \frac{RC}{\tau} C V^2 \quad . \tag{A1.170}$$

Increasing τ decreases the dissipation, and the relevant time scale is RC. This is the energy savings in adiabatic logic.

(d) *How often must the capacitor be charged and discharged for it to draw 1 W from the power supply?*

Charging (non-adiabatically) dissipates $CV^2/2$, and discharging eliminates the stored energy of $CV^2/2$, so the total energy cost is CV^2

$$10 \times 10^{-15} \text{ F} \times (5 \text{ V})^2 = 2.5 \times 10^{-13} \text{ J}$$

$$\frac{1 \text{ J} \cdot \text{s}^{-1}}{2.5 \times 10^{-13} \text{ J}} = 4 \times 10^{12} \text{ Hz} \quad . \tag{A1.171}$$

(*e*) *If a large IC has* 10^6 *transistors, each dissipating this energy once every cycle of a 100 MHz clock, how much power would be dissipated in this worst-case estimate?*

$$2.5 \times 10^{-13} \text{ J} \times 10^6 \text{ transistors} \times 100 \times 10^6 \text{ Hz} = 25 \text{ W} \quad . \tag{A1.172}$$

The most powerful CMOS microprocessors do in fact dissipate on the order of 10-100 W. Because the energy is quadratic in the supply voltage this is one of the reasons for the move to lower voltages. Supplies have dropped from 5 V to 3.3 V, and while chips have been operated down to about 1 V, they can't go much lower in CMOS because the supply voltage needs to be large compared to $kT \sim 0.02$ eV.

(*f*) *How many electrons are stored in the capacitor?*

$$Q = CV = 10 \times 10^{-15} \text{ F} \times 5 \text{ V} = 5 \times 10^{-14} \text{ C}$$

$$5 \times 10^{-14} \text{ C} \times \frac{1 \text{ electron}}{1.6 \times 10^{-19} \text{ C}} = 3.1 \times 10^5 \text{ electrons} \quad . \tag{A1.173}$$

As this number shrinks, statistical fluctuations become more significant.

A1.11 GENERATING, MODULATING, AND DETECTING LIGHT

(11.1) (*a*) *How many watts of power are contained in the light from a 1000 lumen video projector?*
Assuming that the light is white,

$$1000 \text{ lm} \times \frac{1 \text{ W}}{200 \text{ lm}} = 5 \text{ W} \quad . \tag{A1.174}$$

(*b*) *What spatial resolution is needed for the printing of a page in a book to match the eye's limit?*

$$
\begin{aligned}
r \, d\theta &= 10^{-2} \text{ m} \times \frac{1}{60}^\circ \times \frac{\pi \text{ rad}}{180^\circ} \\
&= 2.9 \times 10^{-6} \text{ m} \\
&= 8731 \text{ Dots Per Inch (DPI)} \quad . \tag{A1.175}
\end{aligned}
$$

While far beyond what's needed for most viewing conditions and image sources, the best scanners and typesetters do reach this limit.

(11.2) (*a*) *What is the peak wavelength for black-body radiation from a person?*

$$U(\nu) \, d\nu = \frac{8\pi h \nu^3}{c^3 (e^{h\nu/kT} - 1)} \, d\nu$$

$$\frac{dU}{d\nu} = 0 \quad \Rightarrow \quad 1 = \frac{h\nu_{\text{max}}}{3kT} + e^{-h\nu_{\text{max}}/kT}$$

$$\Rightarrow \quad \nu_{\text{max}} = 2.82 \frac{kT}{h} \quad . \tag{A1.176}$$

The linear relationship between T and ν_{max} is *Wien's Displacement Law*, which was first found experimentally.

$$\nu_{max} = 2.82\frac{1.38{\times}10^{-23}\ J\cdot K^{-1} \times (37 + 273.15)\ K}{6.63{\times}10^{-34}\ J\cdot s}$$

$$= 1.8{\times}10^{13}\ Hz$$

$$= 16\ \mu m\ \ .$$ (A1.177)

This radiation is detected by occupancy sensors that respond to the presence of people in a room.

From the cosmic background radiation at 2.74 K?

$$\nu_{max} = 2.82\frac{1.38{\times}10^{-23}\ J\cdot K^{-1} \times 2.74\ K}{6.63{\times}10^{-34}\ J\cdot s}$$

$$= 1.6{\times}10^{11}\ Hz$$

$$= 2\ mm\ \ .$$ (A1.178)

The distribution of this "echo" of the Big Bang provides important clues to the structure of the universe [Jungman *et al.*, 1996].

(b) *Approximately how hot is a material if it is "red-hot"?*

$$700\ nm = 4.3{\times}10^{14}\ Hz$$

$$\Rightarrow T = \frac{4.3{\times}10^{14}\ Hz \times 6.63{\times}10^{-34}\ J\cdot s}{2.82 \times 1.38{\times}10^{-23}\ J\cdot K^{-1}}$$

$$= 7325\ K$$

$$= 7052\ ^{\circ}C\ \ .$$ (A1.179)

(c) *Estimate the total power thermally radiated by a person.*

$$P = \sigma T^4 \times \text{ surface area}$$

$$= 5.67{\times}10^{-8}\ W\cdot m^{-2}\cdot K^{-4} \times (310\ K)^4 \times 1.8\ m \times 0.3\ m \times 2$$

$$= 565\ W\ \ .$$ (A1.180)

This overestimates the output, because a person in a room is not an ideal black-body radiator, but is the correct order of magnitude: a person radiates about the same amount of heat as a light bulb (i.e., wear your hat).

(11.3) (a) *Find a thickness and an orientation for a birefringent material that rotates a linearly polarized wave by* $90°$. *What is that thickness for calcite with visible light* $(\lambda \sim 600\ nm)$?

A linearly polarized wave has components whose ratio is a real number. Rotation by $90°$ corresponds to interchanging the components and flipping the sign of one of them, so we are looking for a solution to

$$\begin{pmatrix} -\alpha B \\ \alpha A \end{pmatrix} = \begin{pmatrix} e^{-i\gamma} & 0 \\ 0 & e^{i\gamma} \end{pmatrix} \begin{pmatrix} A \\ B \end{pmatrix}\ ,$$ (A1.181)

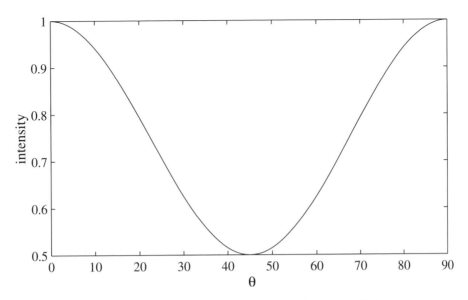

Figure A1.13. Rotating a quarter-wave plate between polarizers.

where α is an arbitrary phase factor. This will be solved if $A = B$ (the wave is oriented at $45°$ to the polarizer axes), and if

$$\gamma = (n_{\text{slow}} - n_{\text{fast}})\frac{\omega l}{2c} = \frac{\pi}{2}$$

$$\Rightarrow l = \frac{\pi}{2}\frac{2c}{\omega(n_{\text{slow}} - n_{\text{fast}})}$$

$$= \frac{\lambda}{2(n_{\text{slow}} - n_{\text{fast}})}$$

$$= \frac{600\times10^{-9} \text{ m}}{2(1.658 - 1.486)}$$

$$= 1.7\times10^{-6} \text{ m} \quad . \tag{A1.182}$$

This (or a thicker multiple) is called a *half-wave plate*. More generally, if the wave is aligned at an arbitrary angle θ to the axes then it will be rotated by 2θ.

(*b*) *Find a thickness and an orientation that converts linearly polarized light to circularly polarized light, and evaluate the thickness for calcite.*

For circularly polarized light the two components have equal magnitude but they are out of phase by i. This will happen if once again the components coming in have equal magnitudes, and if

$$\frac{e^{-i\gamma}}{e^{i\gamma}} = e^{-i2\gamma} = -i$$

$$\Rightarrow \gamma = \pi/4$$

$$\Rightarrow l = \frac{\lambda}{4(n_{\text{slow}} - n_{\text{fast}})} = 8.7 \times 10^{-7} \text{ m} \quad . \tag{A1.183}$$

This is called (can you guess?) a *quarter-wave plate*.

(11.4) *Consider two linear polarizers oriented along the same direction, and a bire-fringent material placed between them. What is the transmitted intensity as a function of the orientation of the birefringent material relative to the axis of the polarizers?*

The output of the second polarizer will be

$$\mathbf{L}\,\mathbf{R}(-\theta)\,\mathbf{B}(d)\,\mathbf{R}(\theta)\begin{pmatrix} E \\ 0 \end{pmatrix}$$

$$= \begin{pmatrix} 1 & 0 \\ 0 & 0 \end{pmatrix}\begin{pmatrix} \cos\theta & -\sin\theta \\ \sin\theta & \cos\theta \end{pmatrix}\begin{pmatrix} e^{-i\gamma} & 0 \\ 0 & e^{i\gamma} \end{pmatrix}\begin{pmatrix} \cos\theta & \sin\theta \\ -\sin\theta & \cos\theta \end{pmatrix}\begin{pmatrix} E \\ 0 \end{pmatrix}$$

$$= \begin{pmatrix} 1 & 0 \\ 0 & 0 \end{pmatrix}\begin{pmatrix} \cos\theta & -\sin\theta \\ \sin\theta & \cos\theta \end{pmatrix}\begin{pmatrix} e^{-i\gamma}\cos\theta & e^{-i\gamma}\sin\theta \\ -e^{i\gamma}\sin\theta & e^{i\gamma}\cos\theta \end{pmatrix}\begin{pmatrix} E \\ 0 \end{pmatrix}$$

$$= \begin{pmatrix} 1 & 0 \\ 0 & 0 \end{pmatrix}$$

$$\times \begin{pmatrix} e^{-i\gamma}\cos^2\theta + e^{i\gamma}\sin^2\theta & e^{-i\gamma}\sin\theta\cos\theta - e^{i\gamma}\sin\theta\cos\theta \\ e^{-i\gamma}\sin\theta\cos\theta - e^{i\gamma}\sin\theta\cos\theta & e^{-i\gamma}\sin^2\theta + e^{i\gamma}\cos^2\theta \end{pmatrix}\begin{pmatrix} E \\ 0 \end{pmatrix}$$

$$= \begin{pmatrix} e^{-i\gamma}\cos^2\theta + e^{i\gamma}\sin^2\theta & e^{-i\gamma}\sin\theta\cos\theta - e^{i\gamma}\sin\theta\cos\theta \\ 0 & 0 \end{pmatrix}\begin{pmatrix} E \\ 0 \end{pmatrix}$$

$$= \begin{pmatrix} E(e^{-i\gamma}\cos^2\theta + e^{i\gamma}\sin^2\theta) \\ 0 \end{pmatrix}\quad . \tag{A1.184}$$

Therefore the transmitted intensity will be

$$I = \left| E(e^{-i\gamma}\cos^2\theta + e^{i\gamma}\sin^2\theta) \right|^2$$

$$= E^2\left(\cos^4\theta + \sin^4\theta + \sin^2\theta\cos^2\theta(e^{i2\gamma} + e^{-i2\gamma})\right)$$

$$= E^2\left(\cos^4\theta + \sin^4\theta + \sin^2\theta\cos^2\theta(2\cos 2\gamma)\right)$$

$$= E^2\left(\cos^4\theta + \sin^4\theta + 2\sin^2\theta\cos^2\theta(2\cos^2\gamma - 1)\right)\quad . \tag{A1.185}$$

This function is shown in Figure A1.13 for $\gamma = \pi/4$. For $\theta = 45°$ it simplifies to

$$I = E^2\cos^2\gamma\quad , \tag{A1.186}$$

a sinusoidal function of the thickness.

A1.12 MAGNETIC STORAGE

(12.1) (a) *Estimate the diamagnetic susceptibility of a typical solid.*

$$\chi = -\mu_0\frac{q^2 Z r^2}{4mV}$$

$$= -4\pi \times 10^{-7}\frac{(1.602\times10^{-19}\text{ C})^2 \times (10^{-10}\text{ m})^2}{4 \times 9.11\times10^{-31}\text{ kg} \times (10^{-10}\text{ m})^3}$$

$$\approx -10^{-4}\quad . \tag{A1.187}$$

(b) *Using this, estimate the field strength needed to levitate a frog, assuming a gradient that drops to zero across the frog. Express your answer in teslas.*

$$F = -V\mu_0\chi_m H\frac{H}{d}$$

$$1\text{ kg} \times 9.8\ \frac{\text{m}}{\text{s}^2} = -10^{-3}\text{ m}^3 \times 4\pi\times10^{-7}\ \frac{\text{H}}{\text{m}} \times 10^{-4} \times \frac{H^2}{0.1\text{ m}}$$

$$\Rightarrow H \sim 3\times10^6\ \frac{\text{A}}{\text{m}}$$

$$\Rightarrow B \sim 3\text{ T} \quad . \tag{A1.188}$$

This is indeed possible [Berry & Geim, 1997].

(12.2) *Estimate the size of the direct magnetic interaction energy between two adjacent free electrons in a solid, and compare this to size of their electrostatic interaction energy. Remember that the field of a magnetic dipole $\vec{m}$ is*

$$\vec{B} = \frac{\mu_0}{4\pi}\left[\frac{3\hat{x}(\hat{x}\cdot\vec{m}) - \vec{m}}{|\vec{x}|^3}\right] \quad . \tag{A1.189}$$

For an electron, the dipole moment is μ_B, the interaction energy is $\mu_B B$, and the characteristic spacing is 1Å, therefore with their axes aligned

$$\begin{aligned}
U &= \mu_B\frac{\mu_0}{4\pi}\frac{\mu_B}{|x|^3} \\
&= \frac{(9.274\times10^{-24}\text{ J}\cdot\text{T}^{-1})^2 \times 4\pi\times10^{-7}\text{ H}\cdot\text{m}^{-1}}{4\pi \times (10^{-10}\text{ m})^3} \\
&= 8.6\times10^{-24}\text{ J} \times \frac{1\text{ eV}}{1.602\times10^{-19}\text{ J}} \\
&= 5\times10^{-5}\text{ eV} \quad . \tag{A1.190}
\end{aligned}$$

The electrostatic energy is

$$\begin{aligned}
U &= qV \\
&= q\frac{q}{4\pi\epsilon_0 r} \\
&= \frac{(1.602\times10^{-19}\text{ C})^2}{4\pi \times 8.854\times10^{-12}\text{ F}\cdot\text{m}^{-1} \times 10^{-10}\text{ m}} \\
&= 2.3\times10^{-18}\text{ J} \times \frac{1\text{ eV}}{1.602\times10^{-19}\text{ J}} \\
&= 14\text{ eV} \quad . \tag{A1.191}
\end{aligned}$$

Comparing these to kT, which is $\sim$0.026 eV at room temperature, shows why ordering from electrostatic forces is stable, and ordering from magnetic forces is irrelevant.

(12.3) *Using the equation for the energy in a magnetic field, describe why:*

(a) *A permanent magnet is attracted to an unmagnetized ferromagnet.*

The energy stored in a magnetic field is

$$U = \frac{1}{2}\int \vec{B}\cdot\vec{H}\ dV = \frac{1}{2\mu}\int B^2\ dV \quad . \tag{A1.192}$$

Since $\mu \gg 1$, the energy in the magnet's return flux will be much lower if it goes through the ferromagnet. The gradient of this energy reduction gives the attractive force.

(b) *The opposite poles of permanent magnets attract each other.*

The energy in one of the magnets is

$$
\begin{aligned}
E &= \frac{1}{2} \int \vec{B} \cdot \vec{H} \, d\vec{V} \\
&= \frac{1}{2} \int \vec{B} \cdot \left(\frac{1}{\mu_0} \vec{B} - \vec{M} \right) d\vec{V} \\
&= \frac{1}{2} \int \frac{1}{\mu_0} B^2 - \vec{B} \cdot \vec{M} \, d\vec{V} \quad .
\end{aligned}
\tag{A1.193}
$$

This will be minimized if the return flux from one magnet points in the direction of the remnant magnetization of the other.

Note that these macroscopic forces are unlike the quantum mechanical origin of ferromagnetic ordering. If a magnet is broken into pieces and then reassembled, the parts will adhere from this classical force, but they will remain separate because there will be negligible overlap of the wave functions between the pieces due to surface irregularity and contamination. However, if the surfaces are cleaned and pressed in an ultrahigh vacuum chamber, it can be possible to get them close enough together for the wave function overlap to be sufficiently strong for the parts to adhere back into a single piece.

(12.4) *Estimate the saturation magnetization for iron at 0 K.*

Assume that each atom contributes one free electron, each with a Bohr magneton:

$$
\frac{6.02 \times 10^{23} \text{ atoms}}{55.85 \text{ g}} \times \frac{7.86 \text{ g}}{1 \text{ cm}^3} \times \frac{(100 \text{ cm})^3}{1 \text{ m}^3} \times \frac{9.27 \times 10^{-24} \text{ J} \cdot \text{T}^{-1}}{\text{atom}}
$$

$$
= 7.8 \times 10^5 \, \frac{\text{A}}{\text{m}} \quad .
\tag{A1.194}
$$

This corresponds to a B field of $4\pi \times 10^{-7}$ H$\cdot$m^{-1} $\times\ 7.8 \times 10^5$ A$\cdot$m^{-1} = 0.98 T. The experimental value is 1.75×10^6 A$\cdot$m^{-1}, showing that there are roughly two free electrons per atom.

(12.5) (a) *Show that the area enclosed in a hysteresis loop in the (B,H) plane is equal to the energy dissipated in going around the loop.*

The B–H hysteresis loop is similar to the M–H loop, but sheared because when M saturates, B is proportional to H (Figure A1.14). From equation (5.116),

$$
\frac{\partial U}{\partial t} = \vec{E} \cdot \frac{\partial \vec{D}}{\partial t} + \vec{H} \cdot \frac{\partial \vec{B}}{\partial t} \quad .
\tag{A1.195}
$$

Therefore

$$
\begin{aligned}
\oint \frac{\partial U}{\partial t} \, dt &= \oint H \frac{\partial B}{\partial t} \, dt \\
&= \int_{\text{right curve}} H \, dB - \int_{\text{left curve}} H \, dB \quad .
\end{aligned}
\tag{A1.196}
$$

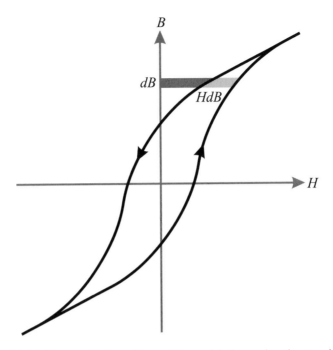

Figure A1.14. Energy dissipated in a differential change in a hysteresis loop.

The difference in the differential elements is the light shaded region shown; integrating this over the curve gives the area enclosed.

(b) *Estimate the power dissipated if 1 kg of iron is cycled through a hysteresis loop at 60 Hz; the coercivity of iron is 4×10^3 A/m.*

$$2 \times 0.98 \text{ T} \times 2 \times 4 \times 10^3 \, \frac{\text{A}}{\text{m}} = 15680 \, \frac{\text{J}}{\text{m}^3}$$

$$15680 \, \frac{\text{J}}{\text{m}^3} \times \frac{1 \text{ m}^3}{7860 \text{ kg}} \times 60 \text{ s}^{-1} = 120 \text{ W} \quad . \tag{A1.197}$$

(12.6) *Approximately what current would be required in a straight wire to be able to erase a γ-Fe_2O_3 recording at a distance of 1 cm?*

To erase the recording, a field on the order of the coercivity must be applied:

$$\oint \vec{H} \cdot d\vec{l} = \int \vec{J} \cdot d\vec{A} \tag{A1.198}$$

$$2\pi r H = I$$

$$H = \frac{I}{2\pi r}$$

$$I = 2\pi r H$$

$$= 2\pi \times 0.01 \text{ m} \times 300 \text{ Oe} \times \frac{1 \text{ A} \cdot \text{m}^{-1}}{4\pi / 10^3 \text{ Oe}}$$

$$= 1500 \text{ A} \quad .$$

Erasing a tape is done in practice with many more turns, and a much smaller spacing.

(12.7) *Assuming digital recording with a bit size equal to the shortest wavelength that is recorded in the medium, how long would a videotape need to be to store 1 Gbyte? 1 Tbyte?*

For a VHS-SP tape, the tape width is 0.5 in, the track pitch is 58 μm = 23 mils, and the track angle is 6°. Taking $\cos(6°) \approx 1$, there are 0.5 in/0.023 in $\approx$ 200 tracks across the tape. The length of the shortest wavelength is about 1 μm. Therefore,

$$\frac{1~\mu m}{bit/track} \times \frac{1}{200~tracks} \times 8 \times 10^9~bits \times 10^{-9}~\frac{km}{\mu m} = 0.04~km \quad , \quad (A1.199)$$

$$\frac{1~\mu m}{bit/track} \times \frac{1}{200~tracks} \times 8 \times 10^{12}~bits \times 10^{-9}~\frac{km}{\mu m} = 40~km \quad . \quad (A1.200)$$

Storing a gigabyte on a videotape is straightforward; much above that becomes prohibitive.

A1.13 MEASUREMENT AND CODING

(13.1) *(a) Show that the circuits in Figures 13.1 and 13.2 differentiate, integrate, sum, and difference.*

$$\frac{V_{in}}{R} + C\dot{V}_{out} = 0 \Rightarrow V_{out} = -\frac{1}{RC} \int^t V_{in}~dt$$

$$C\dot{V}_{in} + \frac{V_{out}}{R} = 0 \Rightarrow V_{out} = -RC~\dot{V}_{in}$$

$$\frac{V_1}{R_{in}} + \frac{V_2}{R_{in}} + \frac{V_{out}}{R_{out}} = 0 \Rightarrow V_{out} = -\frac{R_{out}}{R_{in}}(V_1 + V_2)$$

$$\frac{V_1 - V_2 \dfrac{R_{out}}{R_{in} + R_{out}}}{R_{in}} + \frac{V_{out} - V_2 \dfrac{R_{out}}{R_{in} + R_{out}}}{R_{out}} = 0$$

$$\Rightarrow V_{out} = \frac{R_{out}}{R_{in}}(V_2 - V_1) \quad . \quad (A1.201)$$

(b) Design a non-inverting op-amp amplifier. Why are they used less commonly than inverting ones?

The circuit is shown in Figure A1.15. It has a higher input impedance than an inverting amplifier because the feedback network does not draw any current from the input, but it's less flexible because the gain is restricted to be greater than 1 and inputs can't be added at a current summing junction, and precision performance is compromised by the need for common-mode rejection unlike the single-ended input in the inverting configuration.

(c) Design a transimpedance (voltage out proportional to current in) and a transconductance (current out proportional to voltage in) op-amp circuit.

These are shown in Figure A1.16.

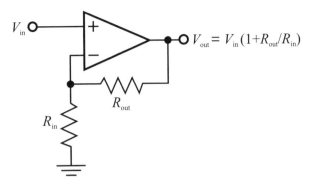

Figure A1.15. A non-inverting amplifier.

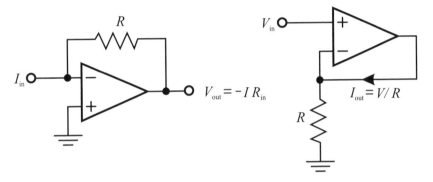

Figure A1.16. Transimpedance (left) and transconductance (right) amplifiers.

(*d*) *Derive equation (13.16).*

The sum of the input and output currents must vanish:

$$I_I + I_O = 0$$

$$I_I = \frac{V_{PD}}{R_I}$$

$$I_O = -I_I = -\frac{1}{R_I} V_{PD}$$

$$\frac{dI_O}{dt} = -\frac{1}{R_I} \frac{dV_{PD}}{dt}$$

$$I_O = C \left(\frac{dV_F}{dt} - R_O \frac{dI_O}{dt} \right)$$

$$-\frac{1}{R_I} V_{PD} - \frac{R_O C}{R_I} \frac{dV_{PD}}{dt} = C \frac{dV_F}{dt}$$

$$\frac{dV_F}{dt} = -\frac{R_O}{R_I} \frac{dV_{PD}}{dt} - \frac{1}{R_I C} V_{PD} \quad . \quad (A1.202)$$

(13.2) *If an op-amp with a gain–bandwidth product of 10 MHz and an open-loop DC gain of 100 dB is configured as an inverting amplifier, plot the magnitude and phase of the gain as a function of frequency as R_{out}/R_{in} is varied.*

The open-loop filter frequency is

$$\omega_{ol} = \frac{\omega_1}{G_{ol}}$$
$$= \frac{10^6 \text{ s}^{-1}}{10^{100/20}}$$
$$= 100 \text{ s}^{-1} \quad . \tag{A1.203}$$

To relate the gain to the feedback network, one equation comes from the op-amp's gain on its inverting input

$$V_{out} = -G(\omega) \, V_- \quad , \tag{A1.204}$$

and a second comes from recognizing that V_- is determined by the resistive divider between the output and the input

$$V_- = V_{in} + \frac{V_{out} - V_{in}}{R_{out} + R_{in}} R_{in} \quad . \tag{A1.205}$$

Eliminating V_- between these equations gives

$$\frac{V_{out}}{V_{in}} = -\frac{G(\omega)R_{out}/R_{in}}{G(\omega) + 1 + R_{out}/R_{in}} \quad . \tag{A1.206}$$

Substituting in

$$G(\omega) = \frac{G_{ol}}{1 + i\frac{\omega}{\omega_{ol}}} \tag{A1.207}$$

results in the plots shown in Figure A1.17. The high-frequency gain, and high-gain bandwidth, are much smaller than the open-loop gain and gain-bandwidth product.

(13.3) *A lock-in has an oscillator frequency of 100 kHz, a bandpass filter Q of 50 (remember that the Q or quality factor is the ratio of the center frequency to the width between the frequencies at which the power is reduced by a factor of 2), an input detector that has a flat response up to 1 MHz, and an output filter time constant of 1 s. For simplicity, assume that both filters are flat in their passbands and have sharp cutoffs. Estimate the amount of noise reduction at each stage for a signal corrupted by additive uncorrelated white noise.*

Since the noise spectrum is flat, the noise power is proportional to the bandwidth. The input filter width is

$$\frac{100 \text{ kHz}}{50} = 2 \text{ kHz} \quad . \tag{A1.208}$$

Therefore, the input filter reduces the noise by

$$10 \log \frac{2 \times 10^3 \text{ Hz}}{10^6 \text{ Hz}} = -27 \text{ dB} \quad . \tag{A1.209}$$

After demodulation the signal will extend up to 1 kHz, so a 1 s filter further reduces the noise bandwidth by

$$10 \log \frac{1 \text{ Hz}}{10^3 \text{ Hz}} = -30 \text{ dB} \quad . \tag{A1.210}$$

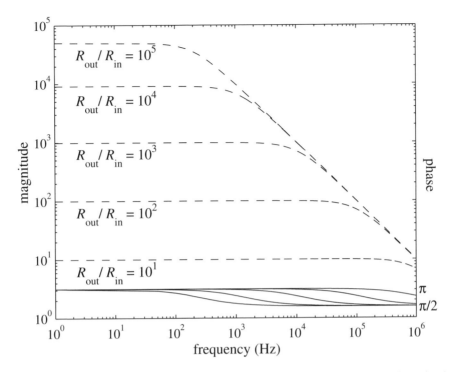

Figure A1.17. Op-amp gain magnitude $(- - -)$ and phase (——) for the indicated values of the feedback network.

The noise component that passes through both filters will be slightly smaller than this because the mean square expectation value of the noise must be averaged over the relative phase of the noise and the carrier:

$$\frac{\dfrac{1}{2\pi}\displaystyle\int_0^{2\pi}\dfrac{\omega}{2\pi}\int_0^{2\pi/\omega}\sin(\omega t + \varphi)\sin(\omega t)^2\,dt\,d\varphi}{\dfrac{\omega}{2\pi}\displaystyle\int_0^{2\pi/\omega}[\sin(\omega t)\sin(\omega t)]^2\,dt} = \frac{2}{3} \quad . \tag{A1.211}$$

(13.4) (a) *For an order 4 maximal LFSR work out the bit sequence.*
The recursion is

$$x_n = x_{n-1} + x_{n-4} \quad . \tag{A1.212}$$

This produces the output shown in Table A1.1.

(b) *If an LFSR has a chip rate of 1 GHz, how long must it be for the time between repeats to be the age of the universe?*
The age of the universe is $\sim 10^{10}$ years, therefore

$$(2^N - 1)\,10^{-9}\,\text{s} = 10^{10}\,\text{y} \times 10^7\,\frac{\text{s}}{\text{y}}$$

$$\Rightarrow N \approx 86 \quad . \tag{A1.213}$$

Table A1.1. *Order 4 maximal LFSR sequence.*

x_n	x_{n-1}	x_{n-2}	x_{n-3}	x_{n-4}
0	1	1	1	1
1	0	1	1	1
0	1	0	1	1
1	0	1	0	1
1	1	0	1	0
0	1	1	0	1
0	0	1	1	0
1	0	0	1	1
0	1	0	0	1
0	0	1	0	0
0	0	0	1	0
1	0	0	0	1
1	1	0	0	0
1	1	1	0	0
1	1	1	1	0
0	1	1	1	1

(c) *Assuming a flat noise power spectrum, what is the coding gain if the entire sequence is used to send one bit?*

$$10 \log_{10} 2^{86} = 259 \text{ dB} \tag{A1.214}$$

(as long as you're willing to wait that long to average over a complete cycle).

(13.5) *What is the SNR due to quantization noise in an 8-bit A/D? 16-bit?*

$$20 \log_{10} 2^{8} = 48 \text{ dB} \quad ,$$
$$20 \log_{10} 2^{16} = 96 \text{ dB} \quad . \tag{A1.215}$$

How much must the former be averaged to match the latter?

$$2^8 \sqrt{N} = 2^{16} \Rightarrow N = \left(\frac{2^{16}}{2^8} \right)^2 = 2^{16} = 65536 \quad . \tag{A1.216}$$

(13.6) *The message 00 10 01 11 00 (c_1, c_2) was received from a noisy channel. If it was sent by the convolutional encoder in Figure 13.20, what data were transmitted?*

The trellis in Figure A1.18 shows the errors into each node found in the forward pass, and then backtracks to find the path with the smallest final error. The error-corrected sequence 00 11 01 01 00 fixes two bit errors, to give the decoded data 0 1 1 0 1.

A1.14 TRANSDUCERS

(14.1) *Do a Taylor expansion of equation (14.6) around $V = 0$.*

This is a historically-important trick question. The expansion terms blow up because the function is not analytic at $V = 0$, which means that Cooper pairing

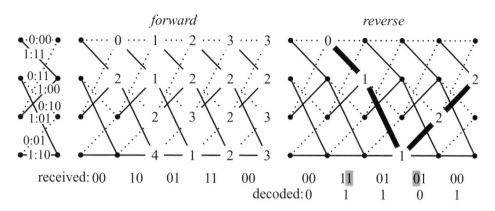

Figure A1.18. Decoding a Viterbi code.

cannot be found (and hence wasn't found) by perturbation theory around the free-particle solution for small V.

(14.2) *If a SQUID with an area of $A = 1$ cm^2 can detect 1 flux quantum, how far away can it sense the field from a wire carrying 1 A?*

$$H = \frac{I}{2\pi r}$$

$$\frac{\mu_0 I}{2\pi r} = \frac{\Phi_0}{A}$$

$$4\pi \times 10^{-7} \text{ H} \cdot \text{m}^{-1} \times \frac{1 \text{ A}}{2\pi r \text{ (m)}} \times \frac{10^4 \text{ G}}{1 \text{ T}} = \frac{2.07 \times 10^{-7} \text{ G} \cdot \text{cm}^2}{1 \text{ cm}^2}$$

$$r = 9.6 \times 10^3 \text{ m} \quad . \tag{A1.217}$$

Just the thing to communicate with your submarine [Davis *et al.*, 1977].

(14.3) *Typical parameters for a quartz resonator are $C_e = 5$ pF, $C_m = 20$ fF, $L_m = 3$ mH, $R_m = 6 \,\Omega$. Plot, and explain, the dependence of the reactance (imaginary part of the impedance), resistance (real part), and the phase angle of the impedance on the frequency.*

$$Z = \frac{\left[\dfrac{1}{i\omega C_m} + i\omega L_m + R_m \right] \dfrac{1}{i\omega C_e}}{\dfrac{1}{i\omega C_m} + i\omega L_m + R_m + \dfrac{1}{i\omega C_e}} \quad . \tag{A1.218}$$

This is plotted in Figure A1.19. In the low-frequency limit

$$\lim_{\omega \to 0} Z = \frac{\dfrac{1}{i\omega C_m} \dfrac{1}{i\omega C_e}}{\dfrac{1}{i\omega C_m} + \dfrac{1}{i\omega C_e}} = -i \frac{\dfrac{1}{\omega C_m} \dfrac{1}{\omega C_e}}{\dfrac{1}{\omega C_m} + \dfrac{1}{\omega C_e}} \tag{A1.219}$$

the phase is $-i$ ($-90°$). The same is true in the high-frequency limit

$$\lim_{\omega \to \infty} Z = \frac{i\omega L_m \dfrac{1}{i\omega C_e}}{i\omega L_m} = \frac{1}{i\omega C_e} = -i \frac{1}{\omega C_e} \quad . \tag{A1.220}$$

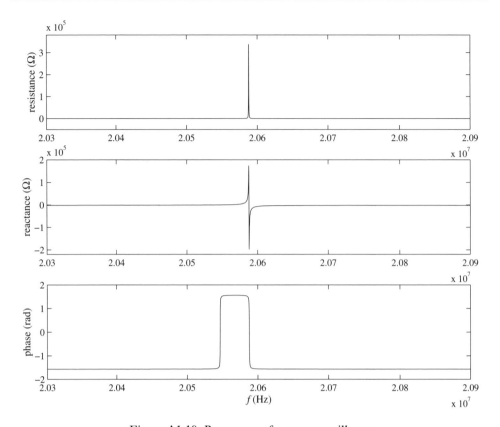

Figure A1.19. Resonance of a quartz oscillator.

In between there are two resonances. The first is a series one at $\omega = 1/\sqrt{L_m C_m}$

$$Z_{\text{series}} = \frac{R_m \dfrac{1}{i\omega C_e}}{R_m + \dfrac{1}{i\omega C_e}} \approx R_m \qquad (\text{A1.221})$$

because R_m is small. Just above that, expanding around $\omega - 1/\sqrt{L_m C_m} \equiv \delta\omega$ gives

$$Z \approx \frac{[2i\delta\omega L_m + R_m]\dfrac{1}{i\omega C_e}}{2i\delta\omega L_m + R_m + \dfrac{1}{i\omega C_e}} \approx 2i\delta\omega L_m \qquad . \qquad (\text{A1.222})$$

Now the phase is $+i$. Finally there is a parallel resonance at

$$\omega = 1/\sqrt{L_m C_m C_e/(C_m + C_e))} \qquad . \qquad (\text{A1.223})$$

This is very close because $C_m \ll C_e$, so

$$Z_{\text{parallel}} \approx \frac{R_m \dfrac{1}{i\omega C_e}}{R_m} \qquad (\text{A1.224})$$

and the phase comes back to $-i$. The region between these is where the clock frequency can be tuned (or drift).

(14.4) *If a ship traveling on the equator uses one of John Harrison's chronometers to navigate, what is the error in its position after one month? What if it uses a cesium beam atomic clock?*

$$\frac{60 \text{ sec}}{1 \text{ min}} \times \frac{60 \text{ min}}{1 \text{ hour}} \times \frac{24 \text{ hours}}{1 \text{ day}} \times \frac{30 \text{ days}}{1 \text{ month}} \times \frac{2\pi \times 6.38 \times 10^6 \text{ m}}{60 \times 60 \times 24 \text{ sec}} \times 10^{-5} = 11 \text{ km}$$

$$\frac{60 \text{ sec}}{1 \text{ min}} \times \frac{60 \text{ min}}{1 \text{ hour}} \times \frac{24 \text{ hours}}{1 \text{ day}} \times \frac{30 \text{ days}}{1 \text{ month}} \times \frac{2\pi \times 6.38 \times 10^6 \text{ m}}{60 \times 60 \times 24 \text{ sec}} \times 10^{-12} = 1 \text{ mm}$$

(14.5) *GPS satellites orbit at an altitude of 20 180 km.*

(a) *How fast do they travel?*

$$\frac{mv^2}{r} = \frac{GMm}{r^2}$$

$$v = \left(\frac{GM}{r}\right)^{1/2}$$

$$= \left(\frac{6.67 \times 10^{-11} \text{ m}^3 \cdot \text{kg}^{-1} \cdot \text{s}^{-2} \times 5.98 \times 10^{24} \text{ kg}}{2.02 \times 10^7 \text{ m} + 6.38 \times 10^6 \text{ m}}\right)^{1/2}$$

$$= 3.87 \times 10^3 \, \frac{\text{m}}{\text{s}} \quad . \tag{A1.225}$$

(b) *What is their orbital period?*

$$\frac{2\pi r}{v} = \frac{2\pi \times 2.66 \times 10^7 \text{ m}}{3.87 \times 10^3 \text{ m} \cdot \text{s}^{-1}}$$

$$= 4.31 \times 10^4 \text{ s}$$

$$= 12 \text{ hr} \quad . \tag{A1.226}$$

(c) *Estimate the special-relativistic correction over one orbit between a clock on a GPS satellite and one on the Earth. Which clock goes slower?*

Ignoring the slower surface velocity of the Earth,

$$t_{\text{Earth}} - t_{\text{satellite}} = t_{\text{Earth}} \left(1 - \sqrt{1 - v^2/c^2}\right)$$

$$= 4.31 \times 10^4 \text{ s}$$

$$\times \left(1 - \sqrt{1 - (3.87 \times 10^3 \text{ m}/3.00 \times 10^8 \text{ m} \cdot \text{s}^{-1})^2}\right)$$

$$= 3.6 \times 10^{-6} \text{ s} \quad . \tag{A1.227}$$

The satellite clock runs slower.

(d) *What is the general-relativistic correction over one orbit? Which clock goes slower?*

$$
t_{\text{Earth}} - t_{\text{satellite}} = t_{\text{Earth}} \left(1 - \frac{1 - \dfrac{GM}{r'c^2}}{1 - \dfrac{GM}{rc^2}} \right)
$$

$$
= 4.31 \times 10^4 \text{ (s)} \times
$$

$$
\times \left(1 - \frac{1 - \dfrac{6.67\times10^{-11} \text{ m}^3 \cdot \text{kg}^{-1} \cdot \text{s}^{-2} \times 5.98\times10^{24} \text{ kg}}{2.66\times10^7 \text{ m} \times \left(3.00\times10^8 \text{ m/s}\right)^2}}{1 - \dfrac{6.67\times10^{-11} \text{ m}^3 \cdot \text{kg}^{-1} \cdot \text{s}^{-2} \times 5.98\times10^{24} \text{ kg}}{6.38\times10^6 \text{ m} \times \left(3.00\times10^8 \text{ m/s}\right)^2}} \right)
$$

$$
= -22.8\times10^{-6} \text{ s} \quad . \tag{A1.228}
$$

The satellite runs faster.

These numbers are in good agreement with what is observed, and with a calculation of all of the relativistic effects [Schwarze *et al.*, 1993].

A1.15 QUANTUM COMPUTING AND COMMUNICATIONS

(15.1) *How is Ehrenfest's theorem changed if the observable has an explicit time dependence?*

$$
\frac{d}{dt}\langle \hat{A}(t)\rangle = \frac{d}{dt}\langle \psi(t)|\hat{A}(t)|\psi(t)\rangle
$$

$$
= \langle \psi(t)|\hat{A}\frac{d|\psi(t)\rangle}{dt} + \frac{d\langle\psi(t)|}{dt}\hat{A}|\psi(t)\rangle + \langle\psi(t)|\frac{d\hat{A}}{dt}|\psi(t)\rangle
$$

$$
= \frac{1}{i\hbar}\langle\psi(t)|\hat{A}\hat{\mathcal{H}}|\psi(t)\rangle - \frac{1}{i\hbar}\langle\psi(t)|\hat{\mathcal{H}}\hat{A}|\psi(t)\rangle + \langle\psi(t)|\frac{d\hat{A}}{dt}|\psi(t)\rangle
$$

$$
= \frac{1}{i\hbar}\langle\psi(t)|[\hat{A},\hat{\mathcal{H}}]|\psi(t)\rangle + \langle\psi(t)|\frac{d\hat{A}}{dt}|\psi(t)\rangle
$$

$$
= \frac{1}{i\hbar}\langle[\hat{A},\hat{\mathcal{H}}]\rangle + \left\langle\frac{d\hat{A}}{dt}\right\rangle \quad . \tag{A1.229}
$$

(15.2) *Show that that* $\text{Tr}(\hat{\rho}^2) \leq 1$.

$$
\text{Tr}(\hat{\rho}^2) = \sum_n \langle n|\hat{\rho}\hat{\rho}|n\rangle
$$

$$
= \sum_n \langle n|\sum_{n'} p_{n'}|n'\rangle\langle n'|\sum_{n''} p_{n''}|n''\rangle\langle n''|n\rangle
$$

$$
= \sum_n p_n \langle n|p_n|n\rangle
$$

$$
= \sum_n p_n^2
$$

$$
\leq 1 \quad . \tag{A1.230}
$$

(15.3) (*a*) *Using the Pauli matrices in the z eigenbasis of equations (15.71), find the eigenvectors for* $\hat{\sigma}_x$ *and* $\hat{\sigma}_y$.

$$\begin{pmatrix} 0 & 1 \\ 1 & 0 \end{pmatrix} \begin{pmatrix} 1 \\ 1 \end{pmatrix} = \begin{pmatrix} 1 \\ 1 \end{pmatrix}, \quad \begin{pmatrix} 0 & 1 \\ 1 & 0 \end{pmatrix} \begin{pmatrix} -1 \\ 1 \end{pmatrix} = \begin{pmatrix} 1 \\ -1 \end{pmatrix} \quad \text{(A1.231)}$$

The state $(| \uparrow \rangle + | \downarrow \rangle) / \sqrt{2}$ has $m_x = 1/2$, and the state $(-| \uparrow \rangle + | \downarrow \rangle) / \sqrt{2}$ has $m_x = -1/2$. Similarly,

$$\begin{pmatrix} 0 & -i \\ i & 0 \end{pmatrix} \begin{pmatrix} 1 \\ i \end{pmatrix} = \begin{pmatrix} 1 \\ i \end{pmatrix}, \quad \begin{pmatrix} 0 & -i \\ i & 0 \end{pmatrix} \begin{pmatrix} -1 \\ i \end{pmatrix} = \begin{pmatrix} 1 \\ -i \end{pmatrix} \quad \text{(A1.232)}$$

the state $(| \uparrow \rangle + i| \downarrow \rangle) / \sqrt{2}$ has $m_y = 1/2$, and the state $(-| \uparrow \rangle + i| \downarrow \rangle) / \sqrt{2}$ has $m_y = -1/2$.

(*b*) *Using the result of the previous problem, apply to the singlet state* $|\psi\rangle = (| \uparrow \rangle_1| \downarrow \rangle_2 - | \downarrow \rangle_1| \uparrow \rangle_2) / \sqrt{2}$ *a projector* $|m_x = 1/2\rangle_1 \langle m_x = 1/2|_1$ *onto the* $m_{1x} = 1/2$ *state of the first spin. What state is the second spin left in? Repeat this in the y direction.*

$$\begin{aligned} &|m_x = 1/2\rangle_1 \langle m_x = 1/2|_1 |\psi\rangle \\ =\; &(| \uparrow \rangle_1 + | \downarrow \rangle_1)\,((\langle \uparrow |_1 + \langle \downarrow |_1)\, |\psi\rangle \\ =\; &(| \uparrow \rangle_1 \langle \uparrow |_1 + | \downarrow \rangle_1 \langle \uparrow |_1 + | \uparrow \rangle_1 \langle \downarrow |_1 + | \downarrow \rangle_1 \langle \downarrow |_1)\, |\psi\rangle \\ =\; &(| \uparrow \rangle_1 \langle \uparrow |_1 + | \downarrow \rangle_1 \langle \uparrow |_1 + | \uparrow \rangle_1 \langle \downarrow |_1 + | \downarrow \rangle_1 \langle \downarrow |_1)\,(| \uparrow \rangle_1| \downarrow \rangle_2 - | \downarrow \rangle_1| \uparrow \rangle_2) \\ =\; &| \uparrow \rangle_1| \downarrow \rangle_2 + | \downarrow \rangle_1| \downarrow \rangle_2 - | \uparrow \rangle_1| \uparrow \rangle_2 - | \downarrow \rangle_1| \uparrow \rangle_2 \\ =\; &(| \uparrow \rangle_1 + | \downarrow \rangle_1)\,(-| \uparrow \rangle_2 + | \downarrow \rangle_2) \\ =\; &|m_x = 1/2\rangle_1 |m_x = -1/2\rangle_2 \quad \text{(A1.233)} \end{aligned}$$

(ignoring normalization). The second spin points in the opposite direction to the first, so that $m_{1x} = -m_{2x}$. A similar calculation shows that $m_{1y} = m_{2y}$.

(*c*) *How are the products of eigenvalues* $m_{1x}m_{2y}$ *and* $m_{1y}m_{2x}$ *related?*

Multiplying these,

$$m_{1x}m_{2y} = m_{1y}m_{2x} \quad . \quad \text{(A1.234)}$$

(*d*) *Recognizing that* $\hat{\vec{\sigma}}_1$ *commutes with* $\hat{\vec{\sigma}}_2$, *what does the last result imply for the relationship between* $\langle\psi|\hat{\sigma}_{1x}\hat{\sigma}_{2y}|\psi\rangle$ *and* $\langle\psi|\hat{\sigma}_{1y}\hat{\sigma}_{2x}|\psi\rangle$?

Since they commute their values can be determined independently. Going from the eigenvalues to the operators, equation (A1.234) would *appear* to suggest that

$$\langle\psi|\hat{\sigma}_{1x}\hat{\sigma}_{2y}|\psi\rangle = \langle\psi|\hat{\sigma}_{1y}\hat{\sigma}_{2x}|\psi\rangle \quad . \quad \text{(A1.235)}$$

(*e*) *In the z eigenbasis, work out the tensor products to evaluate* $\langle\psi|\hat{\sigma}_{1x}\hat{\sigma}_{2y} + \hat{\sigma}_{1y}\hat{\sigma}_{2x}|\psi\rangle$.

The state vector is

$$|\psi\rangle = \frac{1}{\sqrt{2}} \left(|\uparrow\rangle_1 |\downarrow\rangle_2 - |\downarrow\rangle_1 |\uparrow\rangle_2\right)$$

$$= \frac{1}{\sqrt{2}} \left(\begin{pmatrix} 1 \\ 0 \end{pmatrix} \otimes \begin{pmatrix} 0 \\ 1 \end{pmatrix} - \begin{pmatrix} 0 \\ 1 \end{pmatrix} \otimes \begin{pmatrix} 1 \\ 0 \end{pmatrix} \right)$$

$$= \frac{1}{\sqrt{2}} \left(\begin{pmatrix} 0 \\ 1 \\ 0 \\ 0 \end{pmatrix} - \begin{pmatrix} 0 \\ 0 \\ 1 \\ 0 \end{pmatrix} \right)$$

$$= \frac{1}{\sqrt{2}} \begin{pmatrix} 0 \\ 1 \\ -1 \\ 0 \end{pmatrix} \qquad\qquad (A1.236)$$

and the operators are

$$\hat{\sigma}_{1x}\hat{\sigma}_{2y} + \hat{\sigma}_{1y}\hat{\sigma}_{2x} = \begin{pmatrix} 0 & 1 \\ 1 & 0 \end{pmatrix} \otimes \begin{pmatrix} 0 & -i \\ i & 0 \end{pmatrix} + \begin{pmatrix} 0 & -i \\ i & 0 \end{pmatrix} \otimes \begin{pmatrix} 0 & 1 \\ 1 & 0 \end{pmatrix}$$

$$= \begin{pmatrix} 0 & 0 & 0 & -2i \\ 0 & 0 & 0 & 0 \\ 0 & 0 & 0 & 0 \\ 2i & 0 & 0 & 0 \end{pmatrix} . \qquad\qquad (A1.237)$$

Therefore

$$\langle \psi | \hat{\sigma}_{1x}\hat{\sigma}_{2y} + \hat{\sigma}_{1y}\hat{\sigma}_{2x} | \psi \rangle = 0 \quad . \qquad\qquad (A1.238)$$

(f) *Compare the results of Problems 15.3 (d) and 15.3 (e). What happened?*
These disagree. The mistake was made in going from the eigenvalues in equation (A1.234) to the operators in equation (A1.235). If m_{1x} is measured it determines m_{2x}, and if m_{1y} is measured it determines m_{2y}, but because $\hat{\sigma}_{1x}$ does not commute with $\hat{\sigma}_{2x}$ these measurements cannot be performed independently. Equation (A1.234) combines the results of two incompatible experiments into one algebraic expression. This is only possible if the system has internal information that lets it "remember" what to do in the succeeding measurements rather than having the results be determined at the time of the measurement. All of the available theoretical and experimental evidence points to the latter, so that the apparent randomness in quantum mechanics is fundamental rather than a consequence of incomplete information [Peres, 1990].

(15.4) *Using the spectral representation, work out the rotation operators associated with the Pauli spin matrices.*
Using the eigenvalues and eigenvectors found in the last problem,

$$\hat{\sigma}_x = 1 \times \frac{1}{2} \begin{pmatrix} 1 & 1 \\ 1 & 1 \end{pmatrix} + (-1) \times \frac{1}{2} \begin{pmatrix} 1 & -1 \\ -1 & 1 \end{pmatrix}$$

$$\hat{R}_x(\theta) = e^{-i\theta \hat{L}_x / \hbar}$$

$$= e^{-i\hat{\sigma}_x \theta/2}$$

$$= e^{-i\theta/2} \cdot \frac{1}{2} \begin{pmatrix} 1 & 1 \\ 1 & 1 \end{pmatrix} + e^{i\theta/2} \cdot \frac{1}{2} \begin{pmatrix} 1 & -1 \\ -1 & 1 \end{pmatrix}$$

$$= \cos\frac{\theta}{2} \begin{pmatrix} 1 & 0 \\ 0 & 1 \end{pmatrix} - i\sin\frac{\theta}{2} \begin{pmatrix} 0 & 1 \\ 1 & 0 \end{pmatrix}$$

$$= \cos\frac{\theta}{2}\hat{I} - i\sin\frac{\theta}{2}\hat{\sigma}_x$$

$$= \begin{pmatrix} \cos\frac{\theta}{2} & -i\sin\frac{\theta}{2} \\ -i\sin\frac{\theta}{2} & \cos\frac{\theta}{2} \end{pmatrix}$$

$$\hat{R}_y(\theta) = \cos\frac{\theta}{2}\hat{I} - i\sin\frac{\theta}{2}\hat{\sigma}_y$$

$$= \begin{pmatrix} \cos\frac{\theta}{2} & -\sin\frac{\theta}{2} \\ \sin\frac{\theta}{2} & \cos\frac{\theta}{2} \end{pmatrix}$$

$$\hat{R}_z(\theta) = \cos\frac{\theta}{2}\hat{I} - i\sin\frac{\theta}{2}\hat{\sigma}_z$$

$$= \begin{pmatrix} e^{-i\theta/2} & 0 \\ 0 & e^{i\theta/2} \end{pmatrix} \quad . \tag{A1.239}$$

(15.5) *Find the matrix representation of the $\sqrt{\text{NOT}}$ gate that when applied twice gives a* NOT *gate (don't worry about getting the signs of the final state right).*

$$\frac{1}{\sqrt{2}}\begin{pmatrix} 1 & -1 \\ 1 & 1 \end{pmatrix} \frac{1}{\sqrt{2}}\begin{pmatrix} 1 & -1 \\ 1 & 1 \end{pmatrix} = \begin{pmatrix} 0 & -1 \\ 1 & 0 \end{pmatrix} \quad . \tag{A1.240}$$

A single application creates a superposition; performing it twice swaps the state amplitudes.

(15.6) *Show that applying Hadamard transformations individually to N qubits each in the $|0\rangle$ state puts them into an equal superposition of the 2^N possible logical states.*

This can be seen by induction on an example:

$$\begin{aligned} |0\rangle|0\rangle|0\rangle &\rightarrow (|0\rangle + |1\rangle)(|0\rangle + |1\rangle)(|0\rangle + |1\rangle) \\ &= (|0\rangle + |1\rangle)(|00\rangle + |01\rangle + |10\rangle + |11\rangle) \\ &= |000\rangle + |001\rangle + |010\rangle + |011\rangle \\ &\quad + |100\rangle + |101\rangle + |110\rangle + |111\rangle \quad . \end{aligned} \tag{A1.241}$$

(15.7) *What does this circuit do?*

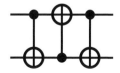

$$
\begin{bmatrix} 1 & 0 & 0 & 0 \\ 0 & 1 & 0 & 0 \\ 0 & 0 & 0 & 1 \\ 0 & 0 & 1 & 0 \end{bmatrix} \begin{bmatrix} 1 & 0 & 0 & 0 \\ 0 & 0 & 0 & 1 \\ 0 & 0 & 1 & 0 \\ 0 & 1 & 0 & 0 \end{bmatrix} \begin{bmatrix} 1 & 0 & 0 & 0 \\ 0 & 1 & 0 & 0 \\ 0 & 0 & 0 & 1 \\ 0 & 0 & 1 & 0 \end{bmatrix}
$$

$$
= \begin{bmatrix} 1 & 0 & 0 & 0 \\ 0 & 1 & 0 & 0 \\ 0 & 0 & 0 & 1 \\ 0 & 0 & 1 & 0 \end{bmatrix} \begin{bmatrix} 1 & 0 & 0 & 0 \\ 0 & 0 & 1 & 0 \\ 0 & 0 & 0 & 1 \\ 0 & 1 & 0 & 0 \end{bmatrix}
$$

$$
= \begin{bmatrix} 1 & 0 & 0 & 0 \\ 0 & 0 & 1 & 0 \\ 0 & 1 & 0 & 0 \\ 0 & 0 & 0 & 1 \end{bmatrix} . \tag{A1.242}
$$

This is a SWAP gate. It reverses the indices, interchanging the qubits.

Bibliography

[Abrams & Lloyd, 1998] Abrams, D.S., & Lloyd, S. (1998). Nonlinear Quantum Mechanics Implies Polynomial-Time Solution for NP-Complete and P Problems. *Physical Review Letters*, **81**, 3992–5.

[Adleman, 1994] Adleman, L.M. (1994). Molecular Computation of Solutions to Combinatorial Problems. *Science*, **5187**, 1021–3.

[Aharonov & Bohm, 1961] Aharonov, Y., & Bohm, D. (1961). Time in the Quantum Theory and the Uncertainty Relation for Time and Energy. *Physical Review*, **122**, 1649–1658.

[Alt & Pleshko, 1974] Alt, P.M., & Pleshko, P. (1974). Scanning Limitations of Liquid-Crystal Displays. *IEEE Transactions on Electron Devices*, **ED-21**, 146–55.

[Alvelda & Lewis, 1998] Alvelda, P., & Lewis, N.D. (1998). New Ultra-Portable Display Technology and Applications. *Proceedings of SPIE*, **3362**, 322–5.

[Anderson, 1989] Anderson, H. (ed). (1989). *A Physicist's Desk Reference*. New York: Institute of Physics.

[Anderson, 1992] Anderson, J.L. (1992). Why We Use Retarded Potentials. *American Journal of Physics*, **60**, 465–7.

[Anderson *et al.*, 1995] Anderson, M.H., Ensher, J.R., Matthews, M.R., Wieman, C.E., & Cornell, E.A. (1995). Observation of Bose–Einstein Condensation in a Dilute Atomic Vapor. *Science*, **269**, 198–201.

[Ash, 1990] Ash, Robert B. (1990). *Information Theory*. New York: Dover Publications.

[Ashcroft & Mermin, 1976] Ashcroft, N., & Mermin, N.D. (1976). *Solid State Physics*. New York: Holt, Rinehart and Winston.

[Aspect *et al.*, 1981] Aspect, A., Grangier, P., & Roger, G. (1981). Experimental Tests of Realistic Local Theories via Bell's Theorem. *Physical Review Letters*, **47**, 460–3.

[Athas *et al.*, 1994] Athas, W.C., Svensson, L.J., Koller, J.G., Tzartzanis, N., & Chou, E. Ying-Chin. (1994). Low-Power Digital Systems Based on Adiabatic-Switching Principles. *IEEE Transactions on VLSI Systems*, **2**, 398–407.

[Baibich *et al.*, 1988] Baibich, M.N., Broto, J.M, Fert, A., Dau, F. Nguyen Van, Petroff, F., Etienne, P., Creuzet, G., Friederich, A., & Chazelas, J. (1988). Giant Magnetoresistance of (001)Fe/(001)Cr Magnetic Superlattices. *Physical Review Letters*, **61**, 2472–5.

[Balanis, 1997] Balanis, Constantine. (1997). *Antenna Theory: Analysis and Design*. 2nd edn. New York: Wiley.

[Balian, 1991] Balian, Roger. (1991). *From Microphysics to Macrophysics: Methods and Applications of Statistical Physics*. New York: Springer–Verlag. Translated by D. ter Haar and J.F. Gregg, 2 volumes.

[Bardeen *et al.*, 1957] Bardeen, J., Cooper, L.N., & Schrieffer, J.R. (1957). Theory of Superconductivity. *Physical Review*, **108**, 1175–1204.

[Barenco *et al.*, 1995] Barenco, Adriano, Bennett, Charles H., Cleve, Richard, DiVincenzo,

David P., Margolus, Norman, Shor, Peter, Sleator, Tycho, Smolin, John, & Weinfurter, Harald. (1995). Elementary Gates for Quantum Computation. *Phys. Rev. A*, **52**, 3457–67.

[Baym, 1973] Baym, Gordon. (1973). *Lectures on Quantum Mechanics*. Reading: W.A. Benjamin.

[Beckman *et al.*, 1996] Beckman, D., Chari, A.N., Devabhaktuni, S., & Preskill, J. (1996). Efficient Networks for Quantum Factoring. *Physical Review A*, **54**, 1034–63.

[Bell, 1964] Bell, J. (1964). On The Einstein Podolsky Rosen Paradox. *Physics*, **1**, 195–200.

[Benioff, 1980] Benioff, P. (1980). The Computer as a Physical System: A Microscopic Quantum Mechanical Hamiltonian Model of Computers as Represented by Turing Machines. *Journal of Statistical Physics*, **22**, 563–91.

[Bennett, 1973] Bennett, C.H. (1973). Logical Reversibility of Computation. *IBM Journal of Research and Development*, **17**, 525.

[Bennett, 1988] Bennett, C.H. (1988). Notes on the History of Reversible Computation. *IBM Journal of Research and Development*, **32**, 16–23.

[Bennett & Brassard, 1984] Bennett, C.H., & Brassard, G. (1984). Quantum Cryptography: Public Key Distribution and Coin Tossing. Pages 175–9 of: *Proceedings of IEEE International Conference on Computers, Systems, and Signal Processing*. New York: IEEE.

[Bennett & Brassard, 1989] Bennett, C.H., & Brassard, G. (1989). The Dawn of a New Era for Quantum Cryptography: the Experimental Prototype is Working. *Sigact News*, **20**, 78–82.

[Bennett *et al.*, 1993] Bennett, C.H., Brassard, G., Crepeau, C., Jozsa, R., & Wootters, A. Peres W.K. (1993). Teleporting an Unknown Quantum State via Dual Classical and Einstein–Podolsky–Rosen channels. *Physical Review Letters*, **70**, 1895–9.

[Benton, 1969] Benton, S.A. (1969). Hologram Reconstructions With Extended Incoherent Sources. *Journal of the Optical Society of America*, **59**, 1545–6.

[Berry & Geim, 1997] Berry, M.V., & Geim, A.K. (1997). Of Flying Frogs and Levitrons Magnetic Levitation. *European Journal of Physics*, **18**, 307–13.

[Bertram *et al.*, 1998] Bertram, H.N., Zhou, H., & Gustafson, R. (1998). Signal to Noise Ratio Scaling and Density Limit Estimates in Longitudinal Magnetic Recording. *IEEE Transactions on Magnetics*, **34**, 1845–7.

[Besag *et al.*, 1995] Besag, J., Green, P.J., Higdon, D., & Mengersen, K. (1995). Bayesian Computation and Stochastic Systems. *Statistical Science*, **10**, 3–66.

[Bhaskar *et al.*, 1996] Bhaskar, N.D., White, J., Mallette, L.A., McClelland, T.A., & Hardy, J. (1996). A Historical Review of Atomic Frequency Standards used in Space Systems. Pages 24–32 of: *Proceedings of the 1996 IEEE International Frequency Control Symposium*. New York: IEEE.

[Binnig *et al.*, 1986] Binnig, G., Quate, C.F., & Gerber, C. (1986). Atomic Force Microscope. *Physical Review Letters*, **56**, 930–3.

[BIPM, 1998] BIPM. (1998). *The International System of Units (SI)*. Organisation Intergouvernementale de la Convention du Métre.

[Birtwistle & Davis, 1995] Birtwistle, G., & Davis, A. (eds). (1995). *Asynchronous Digital Circuit Design*. New York: Springer-Verlag.

[Bitzer, 1999] Bitzer, D.L. (1999). Inventing the AC Plasma Panel. *Information Display*, **15**, 22–7.

[Black, 1934] Black, H.S. (1934). Stabilized Feedback Amplifiers. *Bell System Technical Journal*, **13**, 1–18.

[Blahut, 1988] Blahut, Richard E. (1988). *Principles and Practice of Information Theory*. Reading: Addison-Wesley.

[Boguna & Corral, 1997] Boguna, M., & Corral, A. (1997). Long-Tailed Trapping Times and Levy Flights in a Self-Organized Critical Granular System. *Physical Review Letters*, **78**, 4950–3.

[Born & Wolf, 1999] Born, Max, & Wolf, Emil. (1999). *Principles of Optics:*

Electromagnetic Theory of Propagation, Interference and Diffraction of Light. 7th edn. New York: Cambridge University Press.

[Bortoletto *et al.*, 1999] Bortoletto, F., Bonoli, C., Fantinel, D., Gardio, D., & Pernechele, C. (1999). An Active Telescope Secondary Mirror Control System. *Review of Scientific Instruments*, **70**, 2856–60.

[Bove, 1998] Bove, V.M. (1998). Object-Based Media and Stream-Based Computing. *Proceedings of SPIE*, **3311**, 24–9.

[Boyer *et al.*, 1998] Boyer, M., Brassard, G., Hoyer, P., & Tapp, A. (1998). Tight Bounds on Quantum Searching. *Progress of Physics*, **46**, 493–505.

[Boyle & Smith, 1971] Boyle, W.S., & Smith, G.E. (1971). Charge-Coupled Devices – A New Approach to MIS Device Structures. *IEEE Spectrum*, **8**, 18–27.

[Brodie & Muray, 1982] Brodie, Ivor, & Muray, Julius J. (1982). *The Physics of Microfabrication*. New York: Plenum Press.

[Brody, 1996] Brody, T.P. (1996). The Birth and Early Childhood of Active Matrix – A Personal Memoir. *Journal of the Society for Information Display*, **4**, 113–27.

[Brunel *et al.*, 1999] Brunel, C., Lounis, B., Tamarat, P., & Orrit, M. (1999). Triggered Source of Single photons based on Controlled Single Molecule Fluorescence. *Physical Review Letters*, **83**, 2722–5.

[Brush, 1976] Brush, Stephen G. (1976). *The Kind of Motion We Call Heat: A History of the Kinetic Theory of Gases in the 19th Century*. New York: North-Holland. 2 volumes.

[Buschow, 1991] Buschow, K.H.J. (1991). New Developments in Hard Magnetic Materials. *Reports on Progress in Physics*, **54**, 1123–213.

[Calderbank & Shor, 1996] Calderbank, A.R., & Shor, P.W. (1996). Good Quantum Error-Correcting Codes Exist. *Physical Review A*, **54**, 1098–105.

[Callen, 1985] Callen, Herbert B. (1985). *Thermodynamics and an Introduction to Thermostatistics*. 2nd edn. New York: Wiley.

[Campbell & Green, 1966] Campbell, F.W., & Green, D. (1966). Optical and Retinal Factors Affecting Visual Resolution. *Journal of Physiology*, **181**, 576–93.

[Chandrasekhar, 1992] Chandrasekhar, S. (1992). *Liquid Crystals*. 2nd edn. New York: Cambridge University Press.

[Chapin *et al.*, 1954] Chapin, D.M., Fuller, C.S., & Pearson, G.L. (1954). A New Silicon p–n Junction Photocell for Converting Solar Radiation into Electrical Power. *Journal of Applied Physics*, **25**, 676.

[Chow *et al.*, 1985] Chow, W.W., Gea-Banacloche, J., Pedrotti, L.M., Sanders, V.E., Schleich, W., & Scully, M.O. (1985). The Ring Laser Gyro. *Reviews of Modern Physics*, **57**, 61–104.

[Chuang *et al.*, 1998a] Chuang, I.L., Gershenfeld, N., Kubinec, M.G., & Leung, D.W. (1998a). Bulk Quantum Computation with Nuclear Magnetic Resonance: Theory and Experiment. *Proceedings of the Royal Society of London Series A*, **454**, 447–67.

[Chuang *et al.*, 1998b] Chuang, I.L., Gershenfeld, N., & Kubinec, M. (1998b). Experimental Implementation of Fast Quantum Searching. *Physical Review Letters*, **80**, 3408–11.

[Cirac & Zoller, 1995] Cirac, J.I., & Zoller, P. (1995). Quantum Computations with Cold Trapped Ions. *Physical Review Letters*, **74**, 4091–4.

[Clarke, 1999] Clarke, R.J. (1999). Image and Video Compression: A Survey. *International Journal of Imaging Systems & Technology*, **10**, 20–32.

[Comiskey *et al.*, 1998] Comiskey, B., Albert, J.D., Yoshizawa, H., & Jacobson, J. (1998). An Electrophoretic Ink for All-Printed Reflective Electronic Displays. *Nature*, **394**, 253–5.

[Conway, 1991] Conway, B.E. (1991). Transition from Supercapacitor to Battery Behavior in Electrochemical Energy-Storage. *Journal of the Electrochemical Society*, **138**, 1539–48.

[Conway & Sloane, 1993] Conway, J.H., & Sloane, N.J.A. (1993). *Sphere Packings, Lattices, and Groups*. 2nd edn. New York: Springer-Verlag.

[Cook, 1971] Cook, S.A. (1971). The Complexity of Theorem-Proving Procedures. Pages 151–8 of: *Proceedings of the 3rd Annual ACM Symposium on the Theory of Computing*. New York: Association for Computing Machinery.

[Cooper et al., 1999] Cooper, E.B., Manalis, S.R., Fang, H., Dai, H., Matsumoto, K., Minne, S.C., Hunt, T., & Quate, C.F. (1999). Terabit-per-Square-Inch Data Storage with the Atomic Force Microscope. *Applied Physics Letters*, **75**, 3566–8.

[Cooper, 1956] Cooper, L.N. (1956). *Physical Review*, **104**, 1189.

[Corney, 1978] Corney, Alan. (1978). *Atomic and Laser Spectroscopy*. Oxford: Clarendon Press.

[Cory et al., 1997] Cory, D.G., Fahmy, A.F., & Havel, T.F. (1997). Ensemble Quantum Computing by NMR Spectroscopy. *Proceedings of the National Academy of Science*, **94**, 1634–9.

[Cory et al., 1998] Cory, D.G., Price, M.D., Maas, W., Knill, E., Laflamme, R., Zurek, W.H., Havel, T.F., & Somaroo, S.S. (1998). Experimental Quantum Error Correction. *Physical Review Letters*, **81**, 2152–5.

[Cover & Thomas, 1991] Cover, Thomas M., & Thomas, Joy A. (1991). *Elements of Information Theory*. New York: Wiley.

[Cowper, 1998] Cowper, R. (1998). A View of Next Generation Optical Communication Systems – Possible Future High-Capacity Transport Implementations. *Proceedings of SPIE*, **3491**, 575–80.

[Crisanti et al., 1993] Crisanti, A., Jensen, M.H., Vulpiani, A., & Paladin, G. (1993). Intermittency and Predictability in Turbulence. *Physical Review Letters*, **70**, 166–9.

[Crommie et al., 1993] Crommie, M.F., Lutz, C.P., & Eigler, D.M. (1993). Confinement of Electrons to Quantum Corrals on a Metal Surface. *Science*, **262**, 218–20.

[Danzer, 1999] Danzer, Paul (ed). (1999). *The ARRL Handbook for Radio Amateurs*. 76th edn. Newington: American Radio Relay League.

[Davis et al., 1977] Davis, J.R., Dinger, R.J., & Goldstein, J.A. (1977). Development of a Superconducting ELF Receiving Antenna. *IEEE Transactions on Antennas & Propagation*, **AP-25**, 223–31.

[de Groot & Mazur, 1984] de Groot, S.R., & Mazur, P. (1984). *Non-Equilibrium Thermodynamics*. Mineola: Dover Publications, Inc.

[Delavaux & Nagel, 1995] Delavaux, J.-M.P., & Nagel, J.A. (1995). Multi-Stage Erbium-Doped Fiber Amplifier Designs. *Journal of Lightwave Technology*, **13**, 703–20.

[Denk et al., 1990] Denk, W., Strickler, J.H., & Webb, W.W. (1990). Two-Photon Laser Scanning Fluorescence Microscopy. *Science*, **248**, 73–6.

[Dennard, 1968] Dennard, R.H. (1968). *Field-Effect Transistor Memory*. US Patent No. 3 387 286.

[Denyer et al., 1995] Denyer, P.B., Renshaw, D., & Smith, S.G. (1995). Intelligent CMOS Imaging. *Proceedings of SPIE*, **2415**, 285–91.

[Deutsch, 1985] Deutsch, D. (1985). Quantum Theory, the Church–Turing Principle and the Universal Quantum Computer. *Proceedings of the Royal Society of London Series A*, **A400**, 97–117.

[Dickinson & Denker, 1995] Dickinson, A.G., & Denker, J.S. (1995). Adiabatic Dynamic Logic. *IEEE Journal of Solid-State Circuits*, **30**, 311–5.

[Diffie & Hellman, 1976] Diffie, W., & Hellman, M. (1976). New Directions in Cryptography. *IEEE Transactions on Information Theory*, **IT-22**, 644–54.

[DiVincenzo & Steinhardt, 1991] DiVincenzo, David P., & Steinhardt, Paul J. (eds). (1991). *Quasicrystals: The State of the Art*. Singapore: World Scientific.

[Dixon, 1984] Dixon, R.C. (1984). *Spread Spectrum Systems*. New York: John Wiley & Sons.

[Drexler, 1992] Drexler, K. Eric. (1992). *Nanosystems: Molecular Machinery, Manufacturing, and Computation*. New York: John Wiley & Sons.

[Durrani *et al.*, 1999] Durrani, Z.A.K., Irvine, A.C., Ahmed, H., & Nakazato, K. (1999). A Memory Cell with Single-Electron and Metal-Oxide-Semiconductor Transistor Integration. *Applied Physics Letters*, **74**, 1293–5.

[Dutta & Horn, 1981] Dutta, P., & Horn, P.M. (1981). Low-Frequency Fluctuations in Solids: $1/f$ Noise. *Reviews of Modern Physics*, **53**, 497–516.

[*Economist*, 1993] *Economist*. (1993). **326**, 49 (January 30th).

[Edelstein *et al.*, 1997] Edelstein, D., Heidenreich, J., Goldblatt, R., Cote, W., Uzoh, C., Lustig, N., Roper, P., McDevitt, T., Motsiff, W., Simon, A., Dukovic, J., Wachnik, R., Rathore, H., Schulz, R., Su, L., Luce, S., & Slattery, J. (1997). Full Copper Wiring in a Sub-0.25 μm CMOS ULSI Technology. Pages 773–776 of: *Proceedings of the IEEE International Electron Devices Meeting*. New York: IEEE.

[Einstein, 1905] Einstein, A. (1905). Zur Electrodynamik bewegter Körper. *Annalen der Physik*, **17**, 891–921.

[Einstein, 1916] Einstein, A. (1916). Grundlagen der allgemeinen Relativitätstheorie. *Annalen der Physik*, **49**, 769–822.

[Einstein *et al.*, 1935] Einstein, A., Podolsky, B., & Rosen, N. (1935). Can Quantum-Mechanical Description of Physical Reality be Considered Complete? *Physical Review*, **47**, 777–80.

[Ekert & Jozsa, 1996] Ekert, Artur, & Jozsa, Richard. (1996). Quantum Computation and Shor's Factoring Algorithm. *Reviews of Modern Physics*, **68**(3), 733–53.

[Ernst *et al.*, 1994] Ernst, R.R., Bodenhausen, G., & Wokaun, A. (1994). *Principles of Nuclear Magnetic Resonance in One and Two Dimensions*. Oxford: Oxford University Press.

[Everett, 1957] Everett, Hugh. (1957). Relative State Formulation of Quantum Mechanics. *Reviews of Modern Physics*, **29**, 454–62.

[Farhi *et al.*, 1998] Farhi, E., Goldstone, J., Gutmann, S., & Sipser, M. (1998). Limit on the Speed of Quantum Computation in Determining Parity. *Physical Review Letters*, **81**, 5442–4.

[Fauchet, 1998] Fauchet, P.M. (1998). Progress Toward Nanoscale Silicon Light Emitters. *IEEE Journal of Selected Topics in Quantum Electronics*, **4**, 1020–8.

[Feller, 1968] Feller, William. (1968). *An Introduction to Probability Theory and Its Applications*. 3rd edn. New York: Wiley.

[Feller, 1974] Feller, William. (1974). *An Introduction to Probability Theory and Its Applications*. 2nd edn. Vol. II. New York: Wiley.

[Fergason, 1985] Fergason, J.L. (1985). Polymer Encapsulated Nematic Liquid Crystals for Display and Light Control Applications. Pages 68–70 of: *1985 SID International Symposium*. New York: Pallisades Institute for Research Services.

[Feynman, 1982] Feynman, R.P. (1982). Simulating Physics with Computers. *International Journal of Theoretical Physics*, **21**, 467–88.

[Feynman, 1992] Feynman, R.P. (1992). There's Plenty of Room at the Bottom (Data Storage). *Journal of Microelectromechanical Systems*, **1**, 60–6.

[Fink *et al.*, 1998] Fink, Y., Winn, J.N., Shanhui, Fan, Chiping, Chen, Michel, J., Joannopoulos, J.D., & Thomas, E.L. (1998). A Dielectric Omnidirectional Reflector. *Science*, **282**, 1679–82.

[Fischer *et al.*, 1972] Fischer, A.G., Brody, T.P., & Escott, W.S. (1972). Design of a Liquid Crystal Color TV Panel. Pages 64–6 of: *Conference on Display Devices*. IEEE Conference Record of 1972. Piscataway: IEEE.

[Fitch, 1988] Fitch, J. Patrick. (1988). *Synthetic Aperture Radar*. New York: Springer-Verlag.

[Fletcher *et al.*, 1997] Fletcher, R., Levitan, J.A., Rosenberg, J., & Gershenfeld, N. (1997). Application of Smart Materials to Wireless ID Tags and Remote Sensors. George, E.P., Gotthardt, R., Otsuka, K., Trolier-McKinstry, S., & Wun-Fogle, M. (eds), *Materials for Smart Systems II*. Pittsburgh: Materials Research Society.

[Fletcher *et al.*, 1993] Fletcher, R.M., Kuo, K. Chihping, Osentowski, T.D., Jiann, G.Y., & Robbins, V.M. (1993). High-Efficiency Aluminum Indium Gallium Phosphide Light-Emitting Diodes. *Hewlett-Packard Journal*, **44**, 6–14.

[Fowler & Nordheim, 1928] Fowler, R.H., & Nordheim, L. (1928). Electron Emission in Intense Electric Fields. *Proceedings of the Royal Society of London*, **119**, 173–81.

[Fraden, 1993] Fraden, Jacob. (1993). *AIP Handbook of Modern Sensors: Physics, Designs and Applications*. New York: American Institute of Physics.

[Friend *et al.*, 1999] Friend, R.H., Gymer, R.W., Holmes, A.B., Burroughes, J.H., Marks, R.N., Taliani, C., Bradley, D.D.C, Santos, D.A. Dos, Bredas, J.L., Logdlund, M., & Salaneck, W.R. (1999). Electroluminescence in Conjugated Polymers. *Nature*, **397**, 121–8.

[Fukada & Yasuda, 1957] Fukada, E., & Yasuda, L. (1957). On the Piezoelectric Effect of Bone. *Journal of the Physical Society of Japan*, **12**, 1158.

[Fukuda, 1998] Fukuda, Y. (1998). Evidence for Oscillation of Atmospheric Neutrinos. *Physical Review Letters*, **81**, 1562–7.

[Fukushima & Roeder, 1981] Fukushima, Eiichi, & Roeder, Stephen B.W. (1981). *Experimental Pulse NMR: A Nuts and Bolts Approach*. Reading: Addison-Wesley.

[Furusawa *et al.*, 1998] Furusawa, A., Sorensen, J.L., Braunstein, S.L., Fuchs, C.A., Kimble, H.J., & Polzik, E.S. (1998). Unconditional Quantum Teleportation. *Science*, **282**, 706–9.

[Gabor, 1948] Gabor, D. (1948). A New Microscopic Principle. *Nature*, **161**, 777–8.

[Gabor, 1966] Gabor, D. (1966). Holography of the "Whole Picture". *New Scientist*, **29**, 74–8.

[Galtarossa *et al.*, 1994] Galtarossa, A., Someda, C.G., Matera, F., & Schiano, M. (1994). Polarization Mode Dispersion in Long Single-Mode-Fiber Links: A Review. *Fiber & Integrated Optics*, **13**, 215–29.

[Garey & Johnson, 1979] Garey, Michael R., & Johnson, David S. (1979). *Computers and Intractability: A Guide to the Theory of NP-completeness*. San Francisco: W.H. Freeman.

[Gershenfeld, 1993] Gershenfeld, N.A. (1993). Information in Dynamics. Pages 276–80 of: Matzke, Doug (ed), *Proceedings of the Workshop on Physics of Computation*. Piscataway: IEEE Press.

[Gershenfeld, 1996] Gershenfeld, N.A. (1996). Signal Entropy and the Thermodynamics of Computation. *IBM Systems Journal*, **35**, 577–86.

[Gershenfeld, 1999a] Gershenfeld, N.A. (1999a). *The Nature of Mathematical Modeling*. New York: Cambridge University Press.

[Gershenfeld, 1999b] Gershenfeld, N.A. (1999b). *When Things Start To Think*. New York: Henry Holt and Company.

[Gershenfeld & Chuang, 1997] Gershenfeld, N.A., & Chuang, I.L. (1997). Bulk Spin Resonance Quantum Computation. *Science*, **275**, 350–6.

[Gershenfeld & Grinstein, 1995] Gershenfeld, N.A., & Grinstein, G. (1995). Entrainment and Communication with Dissipative Pseudorandom Dynamics. *Physical Review Letters*, **74**, 5024–7.

[Ghrayeb *et al.*, 1997] Ghrayeb, J., Jackson, T.W., Daniels, R., & Hopper, D.G. (1997). Review of Field Emission Display Potential as a Future (Leap-Frog) Flat Panel Technology. *Proceedings of SPIE*, **3057**, 237–48.

[Gibble & Chu, 1993] Gibble, K., & Chu, S. (1993). Laser-Cooled Cs Frequency Standard and a Measurement of the Frequency Shift due to Ultracold Collisions. *Physical Review Letters*, **70**, 1771–4.

[Gilbert, 1975] Gilbert, B. (1975). A New Technique for Analog Multiplication. *IEEE Journal of Solid-State Circuits*, **SC-10**, 437–47.

[Ginzburg & Landau, 1950] Ginzburg, V.L., & Landau, L.D. (1950). Concerning the Theory of Superconductivity. *Soviet Physics JETP*, **20**, 1064–82.

[Girard, 1994] Girard, G. (1994). The Third Periodic Verification of National Prototypes of the Kilogram (1988–1992). *Metrologia*, **31**, 317–36.

[Giveon & Kutasov, 1999] Giveon, A., & Kutasov, D. (1999). Brane Dynamics and Gauge Theory. *Reviews of Modern Physics*, **71**, 983–1084.

[Goldstein, 1980] Goldstein, Herbert. (1980). *Classical Mechanics*. 2nd edn. Reading: Addison-Wesley.

[Grabert & Devoret, 1992] Grabert, Hermann, & Devoret, Michel H. (eds). (1992). *Single Charge Tunneling: Coulomb Blockade Phenomena in Nanostructures*. New York: Plenum Press.

[Greenberger *et al.*, 1990] Greenberger, D.M., Horne, M.A., Shimony, A., & Zeilinger, A. (1990). Bell's Theorem Without Inequalities. *American Journal of Physics*, **58**, 1131–43.

[Grochowski *et al.*, 1993] Grochowski, E.G., Hoyt, R.F., & Heath, J.S. (1993). Magnetic Hard Disk Drive Form Factor Evolution. *IEEE Transactions on Magnetics*, **29**, 4065–7. Part 2.

[Grover, 1997] Grover, L.K. (1997). Quantum Mechanics Helps in Searching for a Needle in a Haystack. *Physical Review Letters*, **79**, 325–8.

[Grover, 1998] Grover, L.K. (1998). Quantum Computers Can Search Rapidly by Using Almost Any Transformation. *Physical Review Letters*, **80**, 4329–32.

[Gruetter *et al.*, 1995] Gruetter, P., Mamin, H.J., & Rugar, D. (1995). Magnetic Force Microscopy (MFM). Pages 151–207 of: *Scanning Tunneling Microscopy II*. Berlin: Springer-Verlag.

[Gundlach *et al.*, 1996] Gundlach, J.H., Adelberger, E.G., Heckel, B.R., & Swanson, H.E. (1996). New Technique for Measuring Newton's Constant G. *Physical Review D*, **54**, 1256.

[Hagen, 1996] Hagen, Jon B. (1996). *Radio-Frequency Electronics: Circuits and Applications*. New York: Cambridge University Press.

[Hardy & Wright, 1998] Hardy, G.H., & Wright, E.M. (1998). *An Introduction to the Theory of Numbers*. 5th edn. New York: Oxford University Press.

[Hastings *et al.*, 1994] Hastings, M.B., Stone, A.D., & Baranger, H.U. (1994). Inequivalence of Weak Localization and Coherent Backscattering. *Physical Review B*, **50**, 8230–44.

[Hawking, 1993] Hawking, S.W. (1993). *Hawking on the Big Bang and Black Holes*. Singapore: World Scientific.

[Heald & Marion, 1995] Heald, Mark A., & Marion, Jerry B. (1995). *Classical Electromagnetic Radiation*. 3rd edn. Fort Worth: Saunders.

[Heath *et al.*, 1998] Heath, J.R., Kuekes, P.J., Snider, G.S., & Williams, R.S. (1998). A Defect-Tolerant Computer Architecture: Opportunities for Nanotechnology. *Science*, **280**, 1716–21.

[Herring, 1999] Herring, T.A. (1999). Geodetic Applications of GPS. *Proceedings of the IEEE*, **87**, 92–110.

[Herzig & Dandliker, 1987] Herzig, H.P., & Dandliker, R. (1987). Holographic Optical Scanning Elements: Analytical Method for Determining the Phase Function. *Journal of the Optical Society of America A*, **4**, 1063–70.

[Hill & Peterson, 1993] Hill, Fredrick J., & Peterson, Gerald R. (1993). *Computer Aided Logical Design with Emphasis on VLSI*. 4th edn. New York: Wiley.

[Hoffmann, 1988] Hoffmann, Roald. (1988). *Solids and Surfaces: A Chemist's View of Bonding in Extended Structures*. New York: VCH Publishers.

[Hollister, 1987] Hollister, D.D. (1987). Overview of Advances in Light Sources. *Proceedings of SPIE*, **692**, 170–7.

[Hornbeck, 1998] Hornbeck, Larry J. (1998). From Cathode Rays to Digital Micromirrors: A History of Electronic Projection Display Technology. *Texas Instruments Technical Journal*, **15**.

[Horowitz & Hill, 1993] Horowitz, Paul, & Hill, Winfield. (1993). *The Art of Electronics*. 2nd edn. New York: Cambridge University Press.

[Huang *et al.*, 1991] Huang, D., Swanson, E.A., Lin, C.P., Schuman, J.S., Stinson, W.G., Chang, W., Hee, M.R., Flotte, T., Gregory, K., Puliafito, C.A., & Fujimoto, J.G. (1991). Optical Coherence Tomography. *Science*, **254**, 1178–81.

[Hughes *et al.*, 2000] Hughes, R.J., Buttler, W.T., Kwiat, P.G., Lamoreaux, S.K., Morgan, G.L., Nordholt, J.E., & Peterson, C.G. (2000). Free-Space Quantum Key Distribution in Daylight. *Journal of Modern Optics*, **47**, 549–62.

[Hummel, 1993] Hummel, Rolf E. (1993). *Electronic Properties of Materials*. 2nd edn. Berlin: Springer-Verlag.

[Hunt & Fisher, 1990] Hunt, G.R., & Fisher, W.G. (1990). EMP Ship Trial, Planning, Execution and Result. Pages 308–17 of: *Seventh International Conference on Electromagnetic Compatibility*. IEE, London.

[Jackman *et al.*, 1998] Jackman, R.J., Brittain, S.T., Adams, A., Prentiss, M.G., & Whitesides, G.M. (1998). Design and Fabrication of Topologically Complex, Three-Dimensional Microstructures. *Science*, **280**, 2089–91.

[Jackson, 1999] Jackson, John David. (1999). *Classical Electrodynamics*. 3rd edn. New York: Wiley.

[Johnson & Rahmat-Samii, 1997] Johnson, J.M., & Rahmat-Samii, V. (1997). Genetic Algorithms in Engineering Electromagnetics. *IEEE Antennas & Propagation Magazine*, **39**, 7–21.

[Johnson & Jajodia, 1998] Johnson, N.F., & Jajodia, S. (1998). Steganography: Seeing the Unseen. *IEEE Computer*, **31**, 26–34.

[Jones & Mosca, 1998] Jones, J.A., & Mosca, M. (1998). Implementation of a Quantum Algorithm on a Nuclear Magnetic Resonance Quantum Computer. *Journal of Chemical Physics*, **109**, 1648–53.

[Josephson, 1962] Josephson, B.D. (1962). Possible New Effects in Superconductive Tunnelling. *Physics Letters*, **1**, 251.

[Jozsa & Schumacher, 1994] Jozsa, R., & Schumacher, B. (1994). A New Proof of the Quantum Noiseless Coding Theorem. *Journal of Modern Optics*, **41**, 2343–9.

[Jungman *et al.*, 1996] Jungman, G., Kamionkowski, M., Kosowsky, A., & Spergel, D.N. (1996). Weighing the Universe with the Cosmic Microwave Background. *Physical Review Letters*, **76**, 1007–10.

[Kak & Slaney, 1988] Kak, A.C., & Slaney, M. (1988). *Principles of Computerized Tomographic Imaging*. New York: IEEE Press.

[Kane, 1998] Kane, B.E. (1998). A Silicon-Based Nuclear Spin Quantum Computer. *Nature*, **393**, 133–7.

[Kawai, 1969] Kawai, H. (1969). The Piezoelectricity of Poly(vinylidene Fluoride). *Japanese Journal of Applied Physics*, **8**, 975–6.

[Keller *et al.*, 1999] Keller, Mark W., Eichenberger, Ali L., Martinis, John M., & Zimmerman, Neil M. (1999). A Capacitance Standard Based on Counting Electrons. *Science*, **285**, 1706–9.

[Keyes, 1987] Keyes, R.W. (1987). *The Physics of VLSI Systems*. Reading: Addison-Wesley.

[Kino, 1987] Kino, Gordon S. (1987). *Acoustic Waves: Devices, Imaging, and Analog Signal Processing*. Englewood Cliffs: Prentice-Hall.

[Knill *et al.*, 1998a] Knill, E., I., & Laflamme, R. (1998a). Effective Pure States for Bulk Quantum Computation. *Physical Review A*, **57**, 3348–63.

[Knill *et al.*, 1998b] Knill, E., Laflamme, R., & Zurek, W.H. (1998b). Resilient Quantum
    Computation. *Science*, **279**, 342–5.
[Koblitz, 1994] Koblitz, N. (1994). *A Course in Number Theory and Cryptography*. New
    York: Springer-Verlag.
[Koenen, 1999] Koenen, R. (1999). MPEG-4: Multimedia For Our Time. *IEEE
    Spectrum*, **36**, 26–33.
[Kusch, 1949] Kusch, P. (1949). Some Design Considerations of an Atomic Clock using
    Atomic Beam Techniques. *Physical Review*, **76**, 161.
[Kwong, 1995] Kwong, K.K. (1995). Functional Magnetic Resonance Imaging with Echo
    Planar Imaging. *Magnetic Resonance Quarterly*, **11**, 1–20.
[Land, 1951] Land, E.H. (1951). Some Aspects of the Development of Sheet Polarizers.
    Journal of the Optical Society of America, **41**, 956–63.
[Landauer, 1961] Landauer, Rolf. (1961). Irreversibility and Heat Generation in the
    Computing Process. *IBM Journal of Research and Development*, **5**, 183–91.
[Landman & Russo, 1971] Landman, E.S., & Russo, R.L. (1971). On a Pin Versus Block
    Relationships for Partitions of Logic Graphs. *IEEE Transactions on
    Computers*, **C20**, 1469–79.
[Larsen, 1973] Larsen, K.J. (1973). Short Convolutional Codes with Maximal Free
    Distance for Rates 1/2, 1/3, and 1/4. *IEEE Transactions on Information
    Theory*, **IT-19**, 371–2.
[Lauterbur, 1973] Lauterbur, P.C. (1973). Image Formation by Induced Local Interactions:
    Examples Employing Nuclear Magnetic Resonance. *Nature*, **242**, 190–1.
[Leff & Rex, 1990] Leff, Harvey S. & Rex, Andrew F. (eds). (1990). *Maxwell's Demon:
    Entropy, Information, Computing*. Princeton: Princeton University Press.
[Lehrman & Tully, 1993] Lehrman, Paul D., & Tully, Tim. (1993). *MIDI For The
    Professional*. New York: Amsco Publications.
[Lenstra & Lenstra, Jr., 1993] Lenstra, A.K., & Lenstra, Jr., H.W. (eds). (1993). *The
    Development of the Number Field Sieve*. Lecture Notes in Math, 1554. New
    York: Springer-Verlag.
[Lerner & Trigg, 1991] Lerner, Rita G., & Trigg, George L. (1991). *Encyclopedia of
    Physics*. 2nd edn. New York: VCH.
[Lichtman, 1994] Lichtman, J.W. (1994). Confocal Microscopy. *Scientific American*, **271**,
    30–5.
[Likharev, 1999] Likharev, K. (1999). Superconductor Devices for Ultrafast Computing.
    Weinstock, H. (ed), *Applications of Superconductivity*. Dordrecht: Kluwer.
[Likharev & Claeson, 1992] Likharev, K.K., & Claeson, T. (1992). Single Electronics.
    Scientific American, **266**, 50–5.
[Lind & Marcus, 1995] Lind, Douglas, & Marcus, Brian. (1995). *An Introduction to
    Symbolic Dynamics and Coding*. New York: Cambridge University Press.
[Linden *et al.*, 1999] Linden, N., Barjat, H., Kupce, E., & Freeman, R. (1999). How to
    Exchange Information Between Two Coupled Nuclear Spins: the Universal
    SWAP Operation. *Chemical Physics Letters*, **307**, 198–204.
[Lloyd, 1993] Lloyd, S. (1993). A Potentially Realizable Quantum Computer. *Science*, **261**,
    1569–71.
[Lloyd, 1996] Lloyd, S. (1996). Universal Quantum Simulators. *Science*, **273**, 1073–8.
[Lloyd, 1997] Lloyd, S. (1997). Capacity of the Noisy Quantum Channel. *Physical Review
    A*, **55**, 1613–22.
[Lloyd, 2000] Lloyd, S. (2000). Ultimate Physical Limits to Computation. *Nature*. To
    appear.
[Lo & Chau, 1999] Lo, H.K., & Chau, H.F. (1999). Unconditional Security of Quantum
    Key Distribution over Arbitrarily Long Distances. *Science*, **283**, 2050–6.
[Lott *et al.*, 1993] Lott, J.A., Schneider, R.P., Choquette, K.D., Kilcoyne, S.P., & Figiel,
    J.J. (1993). Room Temperature Continuous Wave Operation of Red Vertical
    Cavity Surface Emitting Laser Diodes. *Electronics Letters*, **29**, 1693–4.

[Lucente, 1997] Lucente, Mark. (1997). Interactive Three-Dimensional Holographic Displays: Seeing the Future in Depth. *Computer Graphics*, **31**, 63–7.

[Major, 1998] Major, Fouad G. (1998). *The Quantum Beat: The Physical Principles of Atomic Clocks.* New York: Springer.

[Mallinson, 1993] Mallinson, J.C. (1993). *The Foundations of Magnetic Recording.* 2nd edn. Boston: Academic Press.

[Mallinson, 1996] Mallinson, J.C. (1996). Scaling in Magnetic Recording. *IEEE Transactions on Magnetics*, **32**, 599–600.

[Mandelbrot, 1983] Mandelbrot, Benoit B. (1983). *The Fractal Geometry of Nature.* New York: W.H. Freeman.

[Mattis, 1988] Mattis, Daniel C. (1988). *The Theory of Magnetism I: Statics and Dynamics.* New York: Springer-Verlag.

[Maxwell, 1998] Maxwell, James Clerk. (1998). *A Treatise on Electricity and Magnetism.* 3rd edn. Oxford: Oxford University Press. First published in 1873.

[McCluskey, 1956] McCluskey, E.J. (1956). Minimization of Boolean Functions. *Bell System Technical Journal*, **35**, 1417–44.

[McKittrick *et al.*, 1999] McKittrick, J., Kassner, M.E., & Shea, L.E. (1999). Materials Issues in Flat Panel Displays: Phosphor Selection and Optimization. *Proceedings of SPIE*, **3582**, 565–70.

[Mee & Daniel, 1996] Mee, C. Denis, & Daniel, Eric D. (eds). (1996). *Magnetic Storage Handbook.* 2nd edn. New York: McGraw-Hill.

[Merkle, 1978] Merkle, R. (1978). Secure Communications over Insecure Channels. *Communications of the ACM*, **21**, 294–9.

[Merkle, 1993] Merkle, R.C. (1993). Reversible electronic logic using switches. *Nanotechnology*, **4**, 21–40.

[Merkle, 1998] Merkle, R.C. (1998). Making Smaller, Faster, Cheaper Computers. *Proceedings of the IEEE*, **86**, 2384–6.

[Mermin, 1985] Mermin, N.D. (1985). Is the moon there when nobody looks? Reality and the quantum theory. *Physics Today*, **38**, 38–47.

[Mermin, 1993] Mermin, N.D. (1993). Hidden Variables and the Two Theorems of John Bell. *Reviews of Modern Physics*, **65**, 803–15.

[Merzbacher *et al.*, 1996] Merzbacher, C.I., Kersey, A.D., & Friebele, E.J. (1996). Fiber Optic Sensors in Concrete Structures: A Review. *Smart Materials & Structures*, **5**, 196–208.

[Millman & Grabel, 1987] Millman, Jacob, & Grabel, Arvin. (1987). *Microelectronics.* 2nd edn. New York: McGraw-Hill.

[Minsky, 1957] Minsky, Marvin. (1957). *Microscopy Apparatus.* US Patent No. 3 013 467.

[Misner *et al.*, 1973] Misner, C.W., Wheeler, J.A., & Thorne, K.S. (1973). *Gravitation.* New York: W.H. Freeman & Co.

[Mitchell & George, 1998] Mitchell, S., & George, R. (1998). EMP Protection. *Electrotechnology*, **9**, 33–5.

[Miya *et al.*, 1979] Miya, T., Terunuma, Y., Hosaka, T., & Miyashita, T. (1979). Ultimate Low-Loss Single-Mode Fibre at 1.55 μm. *Electronics Letters*, **15**, 106–8.

[Mollenauer *et al.*, 1996] Mollenauer, L.F., Mamyshev, P.V., & Neubelt, M.J. (1996). Demonstration of Soliton WDM Transmission at 6 and 7*10 Gbit/s, Error Free Over Transoceanic Distances. *Electronics Letters*, **32**, 471–3.

[Montroll & Lebowitz, 1987] Montroll, E.W., & Lebowitz, J.L. (eds). (1987). *Fluctuation Phenomena.* New York: North-Holland.

[Mooij *et al.*, 1999] Mooij, J.E., Orlando, T.P., Levitov, L., Tian, L., van der Wal, C.H., & Lloyd, S. (1999). Josephson Persistent-Current Qubit. *Science*, **285**, 1036–9.

[Moore, 1979] Moore, G. (1979). VLSI: Some Fundamental Challenges. *IEEE Spectrum*, **16**, 30.

[Morrison & Morrison, 1982] Morrison, Philip, & Morrison, Phylis. (1982). *Powers Of*

Ten: A Book About the Relative Size of Things. Redding: Scientific American Library.

[Mukai *et al.*, 1999] Mukai, T., Yamada, M., & Nakamura, S. (1999). Characteristic of InGaN-based UV/Blue/Green/Amber/Red Light-Emitting Diodes. *Japanese Journal of Applied Physics*, **38**, 3976–81.

[Muller *et al.*, 1996] Muller, A., Zbinden, H., & Gisin, N. (1996). Quantum Cryptography over 23 km in Installed Under-Lake Telecom Fibre. *Europhysics Letters*, **33**, 335–9.

[Nachtmann, 1990] Nachtmann, Otto. (1990). *Elementary Particle Physics: Concepts and Phenomena*. New York: Springer-Verlag. Translated by A. Lahee and W. Wetzel.

[Nakashima, 1998] Nakashima, H. (1998). Present Status of Progress in MAGLEV Development. *Japanese Railway Engineering*, **37**, 6–8.

[Nakazawa *et al.*, 1993] Nakazawa, M., Kimura, Y., & Suzuki, K. (1993). Nonlinear Optics in Optical Fibers and Future Prospects for Optical Soliton Communications Technologies. *NTT R&D*, **42**, 1317–26.

[Nielsen & Chuang, 2000] Nielsen, M.A., & Chuang, I.L. (2000). *Quantum Computation and Quantum Information*. New York: Cambridge University Press.

[Ogawa *et al.*, 1990] Ogawa, S., Lee, T.M., Kay, A.R., & Tank, D.W. (1990). Brain Magnetic Resonance Imaging with Contrast Dependent on Blood Oxygenation. *Proceedings of the National Academy of Sciences*, **87**, 9868–72.

[O'Mara, 1993] O'Mara, William C. (1993). *Liquid Crystal Flat Panel Displays: Manufacturing Science & Technology*. New York: Van Nostrand Reinhold.

[Ono & Yano, 1998] Ono, T., & Yano, Y. (1998). Key Technologies for Terabit/Second WDM Systems with High Spectral Efficiency of over 1 bit/s/Hz. *IEEE Journal of Quantum Electronics*, **34**, 2080–8.

[Onsager, 1931] Onsager, L. (1931). *Physical Review*, **38**, 2265.

[Pai & Springett, 1993] Pai, D.M., & Springett, B.E. (1993). Physics of Electrophotography. *Reviews of Modern Physics*, **65**, 163–211.

[Parkin, 1994] Parkin, S.S.P. (1994). Materials Update: Giant Magnetoresistance in Magnetic Multilayers and Granular Alloys. *Materials Letters*, **20**, 1–4.

[Pavlidis, 1999] Pavlidis, D. (1999). HBT vs. PHEMT vs. MESFET: What's Best and Why. *Compound Semiconductor*, **5**, 56–9.

[Peres, 1990] Peres, A. (1990). Incompatible Results of Quantum Measurements. *Physics Letters A*, **151**, 107–8.

[Peres, 1993] Peres, Asher. (1993). *Quantum Theory: Concepts and Methods*. Boston: Kluwer Academic.

[Peters *et al.*, 1999] Peters, A., Chung, K.Y., & Chu, S. (1999). Measurement of Gravitational Acceleration by Dropping Atoms. *Nature*, **400**, 849–52.

[Phillips *et al.*, 1998] Phillips, P.M., Spindt, C.A., Holland, C.E., Schwoebel, P.R., & Brodie, I. (1998). Development of Spindt Cathodes for High Frequency Devices and Flat Panel Display Applications. *Proceedings of SPIE*, **3465**, 90–7.

[Posner & Stevens, 1984] Posner, E.C., & Stevens, R. (1984). Deep Space Communication – Past, Present, and Future. *IEEE Communications Magazine*, **22**, 8–21.

[Press *et al.*, 1992] Press, William H., Teukolsky, Saul A., Vetterling, William T., & Flannery, Brian P. (1992). *Numerical Recipes in C: The Art of Scientific Computing*. 2nd edn. New York: Cambridge University Press.

[Pritchard & Gibson, 1980] Pritchard, D., & Gibson, J. (1980). Worldwide Color Television Standards. *J. Soc. Motion. Pict. Telev. Eng.*, **89**, 111–120.

[Quine, 1952] Quine, W.V. (1952). The Problem of Simplifying Truth Functions. *American Mathematical Monthly*, **59**, 521–31.

[Radon, 1917] Radon, J. (1917). On The Determination Of Functions From Their Integrals Along Certain Manifolds. *Berichte Saechsische Akademie der Wissenschaften*, **29**, 262–77.

[Rallison, 1984] Rallison, R. (1984). Applications of Holographic Optical Elements. *Lasers & Applications*, **3**, 61–8.

[Ramirez *et al.*, 1997] Ramirez, A.P., Cheong, S-W, & Schiffer, P. (1997). Colossal Magnetoresistance and Charge Order in $La_{1-x}Ca_xMnO_3$. *Journal of Applied Physics*, **81**, 5337–42.

[Ramo *et al.*, 1994] Ramo, Simon, Whinnery, John R., & Duzer, Theodore Van. (1994). *Fields and Waves in Communication Electronics*. 3rd edn. New York: Wiley.

[Ramsey, 1972] Ramsey, N.F. (1972). History of Atomic and Molecular Standards of Frequency and Time. *IEEE Trans. Instrumentation and Measurement*, **IM 21**, 90–9.

[Ramsey, 1980] Ramsey, N.F. (1980). The Method of Successive Oscillatory Fields. *Physics Today*, **33**, 25–30.

[Rauf & Kushner, 1999] Rauf, S., & Kushner, M.J. (1999). Dynamics of a Coplanar-Electrode Plasma Display Panel Cell. I. Basic Operation. *Journal of Applied Physics*, **85**, 3460–9.

[Reichl, 1998] Reichl, L.E. (1998). *A Modern Course in Statistical Physics*. 2nd edn. New York: Wiley.

[Reif, 1965] Reif, F. (1965). *Fundamentals of Statistical and Thermal Physics*. New York: McGraw-Hill.

[Ridley *et al.*, 1999] Ridley, B., Nivi, B., & Jacobson, J. (1999). All-Inorganic Field Effect Transistors Fabricated by Printing. *Science*, **286**, 746–9.

[Rivest *et al.*, 1978] Rivest, R.L., Shamir, A., & Adleman, L.M. (1978). A Method of Obtaining Digital Signatures and Public-Key Cryptosystems. *Communications of the ACM*, **21**, 120–6.

[Rodgers *et al.*, 1997] Rodgers, M.S., Sniegowski, J.J., Miller, S.L., Barron, C., & PJ, P.J. McWhorter. (1997). Advanced Micromechanisms in a Multi-Level Polysilicon Technology. *Proceedings of SPIE*, **3224**, 120–30.

[Rogers & Buhrman, 1984] Rogers, C.T., & Buhrman, R.A. (1984). Composition of $1/f$ Noise in Metal–Insulator–Metal Tunnel Junctions. *Physical Review Letters*, **53**, 1272–5.

[Rüeger, 1990] Rüeger, J.M. (1990). *Electronic Distance Measurement*. 3rd edn. New York: Springer-Verlag.

[Sakurai, 1967] Sakurai, J.J. (1967). *Advanced Quantum Mechanics*. Reading: Addison-Wesley.

[Schack & Caves, 1999] Schack, R., & Caves, C.M. (1999). Classical Model for Bulk-Ensemble NMR Quantum Computation. *Physical Review A*, **60**, 4354–62.

[Schadt & Helfrich, 1971] Schadt, M., & Helfrich, W. (1971). Voltage-Dependent Optical Activity of a Twisted Nematic Liquid Crystal. *Applied Physics Letters*, **18**, 127–8.

[Schroeder, 1990] Schroeder, M.R. (1990). *Number Theory in Science and Communication*. 2nd edn. New York: Springer-Verlag.

[Schroeder *et al.*, 1979] Schroeder, M.R., Atal, B.S., & Hall, J.L. (1979). Optimizing digital speech coders by exploiting masking properties of the human ear. *Journal of the Acoustical Society of America*, **66**, 1647–52.

[Schwarze *et al.*, 1993] Schwarze, V.S., Hartmann, T., Leins, M., & Soffel, M.H. (1993). Relativistic Effects in Satellite Positioning. *Manuscripta Geodaetica*, **18**, 306–16.

[Scott, 1998] Scott, J.F. (1998). Status Report on Ferroelectric Memory Materials. *Integrated Ferroelectrics*, **20**, 15–23.

[Sheats *et al.*, 1996] Sheats, J.R., Antoniadis, H., Hueschen, M., Leonard, W., Miller, J., Moon, R., Roitman, D., & Stocking, A. (1996). Organic Electroluminescent Devices. *Science*, **273**, 884–8.

[Shepherd, 1990] Shepherd, G. (1990). *The Synaptic Organization of the Brain*. 3rd edn. New York: Oxford University Press.

[Shor, 1995] Shor, P.W. (1995). Scheme for Reducing Decoherence in Quantum Computer Memory. *Physical Review A*, **52**, 2493–6.

[Shor, 1996] Shor, P.W. (1996). Fault-Tolerant Quantum Computation. Pages 56–65 of: *Proceedings of the 37th Annual Symposium Foundations of Computer Science*. Los Alamitos: IEEE Computer Society Press.

[Shor, 1997] Shor, P.W. (1997). Polynomial-Time Algorithms for Prime Factorization and Discrete Logarithms on a Quantum Computer. *SIAM Journal on Computing*, **26**, 1484–509.

[Shung *et al.*, 1992] Shung, K. Kirk, Smith, Michael B., & Tsui, Benjamin. (1992). *Principles of Medical Imaging*. San Diego: Academic Press.

[Sikora, 1997] Sikora, T. (1997). MPEG Digital Video-Coding Standards. *IEEE Signal Processing Magazine*, **14**, 82–100.

[Simmons, 1992] Simmons, G.J. (ed). (1992). *Contemporary Cryptology: The Science of Information Integrity*. Piscataway: IEEE Press.

[Simon *et al.*, 1994] Simon, M.K., Omura, J.K., Scholtz, R.A., & Levitt, B.K. (1994). *Spread Spectrum Communications Handbook*. New York: McGraw-Hill.

[Sklar, 1988] Sklar, Bernard. (1988). *Digital Communications: Fundamentals and Applications*. Englewood Cliffs: Prentice Hall.

[Skolnik, 1990] Skolnik, Merrill I. (ed). (1990). *Radar Handbook*. 2nd edn. New York: McGraw-Hill.

[Slepian, 1974] Slepian, David (ed). (1974). *Key Papers in the Development of Information Theory*. New York: IEEE Press.

[Slichter, 1992] Slichter, Charles P. (1992). *Principles of Magnetic Resonance*. 3rd edn. New York: Springer-Verlag.

[Smith, 1996] Smith, J.R. (1996). Field Mice: Extracting Hand Geometry From Electric Field Measurements. *IBM Systems Journal*, **35**, 587–608.

[Smith, 1999] Smith, J.R. (1999). *Electric Field Imaging*. Ph.D. thesis, MIT.

[Sobel, 1996] Sobel, Dava. (1996). *Longitude: The True Story of a Lone Genius Who Solved the Greatest Scientific Problem of His Time*. New York: McGraw-Hill.

[Somaroo *et al.*, 1999] Somaroo, S., Tseng, C.H., Havel, T.F, Laflamme, R., & Cory, D.G. (1999). Quantum Simulations on a Quantum Computer. *Physical Review Letters*, **82**, 5381–4.

[Someya *et al.*, 1999] Someya, T., Werner, R., Forchel, A., Catalano, M., Cingolani, R., & Arakawa, Y. (1999). Room Temperature Lasing at Blue Wavelengths in Gallium Nitride Microcavities. *Science*, **285**, 1905–6.

[Song *et al.*, 1999] Song, Y.Q., Goodson, B.M., & Pines, A. (1999). NMR and MRI using Laser-Polarized Xenon. *Spectroscopy*, **14**, 26–33.

[Sourlas, 1989] Sourlas, N. (1989). Spin-Glass Models as Error-Correcting Codes. *Nature*, **339**, 693–5.

[Spuhler, 1983] Spuhler, H. (1983). Where Fluidics Still Makes Sense. *Machine Design*, **55**, 92–4.

[Starkweather, 1980] Starkweather, G.K. (1980). High-Speed Laser Printing Systems. Pages 125–89 of: *Laser Applications*, vol. 4. New York: Academic Press.

[Steane, 1996] Steane, A.M. (1996). Error Correcting Codes in Quantum Theory. *Physical Review Letters*, **77**, 793–7.

[Stehling *et al.*, 1991] Stehling, M.K., Turner, R., & Mansfield, P. (1991). Echo-planar Imaging: Magnetic Resonance Imaging In A Fraction Of A Second. *Science*, **254**, 43–50.

[Stern, 1996] Stern, M.B. (1996). Binary Optics: a VLSI-based microoptics Technology. *Microelectronic Engineering*, **32**, 369–88.

[Stofan *et al.*, 1995] Stofan, E.R., Evans, D.L., Schmullius, C., Holt, B., Plaut, J.J., van Zyl, J., Wall, S.D., & Way, J. (1995). Overview of Results of Spaceborne Imaging Radar-C, X-Band Synthetic Aperture Radar (SIR-C/X-SAR). *IEEE Transactions on Geoscience & Remote Sensing*, **33**, 817–28.

[Strang, 1988] Strang, Gilbert. (1988). *Linear Algebra and its Applications*. 3rd edn. San Diego: Harcourt, Brace, Jovanovich.

[Streetman, 1990] Streetman, B. (1990). *Solid State Electronic Devices*. Englewood Cliffs: Prentice-Hall.

[Stroscio & Eigler, 1991] Stroscio, J.A., & Eigler, D.M. (1991). Atomic and Molecular Manipulation with the Scanning Tunneling Microscope. *Science*, **254**, 319–26.

[Surguy, 1993] Surguy, P.W.H. (1993). The Development of Ferroelectric LCDs for Display Applications. *Journal of the Society for Information Display*, **1**, 247–54.

[Sweatt, 1979] Sweatt, W.C. (1979). Mathematical Equivalence Between a Holographic Optical Element and an Ultra-High Index Lens. *Journal of the Optical Society of America*, **69**, 486–7.

[Sze, 1981] Sze, S.M. (1981). *Physics of Semiconductor Devices*. 2nd edn. New York: Wiley-Interscience.

[Sze, 1998] Sze, S.M. (ed). (1998). *Modern Semiconductor Device Physics*. New York: Wiley-Interscience.

[Takahashi, 1993] Takahashi, S. (1993). Fibers for Optical Communications. *Advanced Materials*, **5**, 187–91.

[Taylor & Wheeler, 1992] Taylor, Edwin F., & Wheeler, John Archibald. (1992). *Spacetime Physics: Introduction to Special Relativity*. 2nd edn. New York: W.H. Freeman.

[Thompson & Best, 2000] Thompson, D.A., & Best, J.S. (2000). The Future of Magnetic Data Storage Technology. *IBM Journal of Research and Development*, **44**, 311–22.

[Tinkham, 1995] Tinkham, Michael. (1995). *Introduction to Superconductivity*. 2nd edn. New York: McGraw-Hill.

[Tittel *et al.*, 1998] Tittel, W., Brendel, J., Zbinden, H., & Gisin, N. (1998). Violation of Bell Inequalities by Photons More than 10 km Apart. *Physical Review Letters*, **81**, 3563–6.

[Todorovic *et al.*, 1999] Todorovic, M., Schultz, S., Wong, J., & Scherer, A. (1999). Writing and Reading of Single Magnetic Domain Per Bit Perpendicular Patterned Media. *Applied Physics Letters*, **74**, 2516–18.

[Turing, 1936] Turing, A.M. (1936). On Computable Numbers, with an Application to the *Entscheidungsproblem*. *Proc. London Math. Soc.*, **42**, 230–65.

[Turing, 1950] Turing, A.M. (1950). Computing Machinery and Intelligence. *Mind*, **59**, 433–560.

[Underkoffler *et al.*, 1999] Underkoffler, J., Ullmer, B., & Ishii, H. (1999). Emancipated Pixels: Real-World Graphics in the Luminous Room. Pages 385–92 of: *Proceedings of SIGGRAPH '99*. New York: ACM Press.

[Unruh, 1995] Unruh, W.G. (1995). Maintaining Coherence in Quantum Computers. *Physical Review A*, **51**, 992–7.

[van Kessel *et al.*, 1998] van Kessel, P.F., Hornbeck, L.J., RE, R.E. Meier, & Douglass, M.R. (1998). A MEMS-Based Projection Display. *Proceedings of the IEEE*, **86**, 1687–704.

[Vilkelis, 1982] Vilkelis, W.V. (1982). Lead Reduction among Combinatorial Logic Circuits. *IBM Journal of Research and Development*, **26**, 342–348.

[Viterbi & Omura, 1979] Viterbi, Andrew J., & Omura, Jim K. (1979). *Principles of Digital Communication and Coding*. New York: McGraw-Hill.

[von Neumann, 1956] von Neumann, J. (1956). Probabilistic Logics and the Synthesis of Reliable Organisms from Unreliable Components. Pages 43–98 of: Shannon, C., & McCarthy, J. (eds), *Automata Studies*. Princeton: Princeton University Press.

[Walls & Vig, 1995] Walls, F.L., & Vig, J.R. (1995). Fundamental Limits on the Frequency Stabilities of Crystal Oscillators. *IEEE Transactions on Ultrasonics Ferroelectrics & Frequency Control*, **42**, 576–89.

[Wang, 1989] Wang, Shyh. (1989). *Fundamentals of Semiconductor Theory and Device Physics*. Englewood Cliffs: Prentice-Hall.

[Weber *et al.*, 2000] Weber, M.F., Stover, C.A., Gilbert, L.R., Nevitt, T.J., & Ouderkirk, A.J. (2000). Giant Birefringent Optics in Multilayer Polymer Mirrors. *Science*, **287**, 2451–2455.

[Weinacht *et al.*, 1999] Weinacht, T.C., Ahn, J., & Bucksbaum, P.H. (1999). Controlling the Shape of a Quantum Wavefunction. *Nature*, **397**, 233–5.

[Weinberg, 1989] Weinberg, S. (1989). Testing Quantum Mechanics. *Annals of Physics*, **194**, 336–86.

[Welch, 1984] Welch, Terry A. (1984). A Technique for High Performance Data Compression. *IEEE Computer*, **17**, 8–19.

[Wieman *et al.*, 1999] Wieman, C.E., Pritchard, D.E., & Wineland, D.J. (1999). Atom Cooling, Trapping, and Quantum Manipulation. *Reviews of Modern Physics*, **71**, S253–62.

[Wiesner, 1983] Wiesner, S. (1983). Conjugate Coding. *Sigact News*, **15**, 78–88.

[Williams, 1993] Williams, Edgar M. (1993). *The Physics and Technology of Xerographic Processes*. Malabar: Krieger.

[Winograd & Cowan, 1963] Winograd, S., & Cowan, J.D. (1963). *Reliable Computation in the Presence of Noise*. Cambridge: MIT Press.

[Wolaver, 1991] Wolaver, Dan H. (1991). *Phase-Locked Loop Circuit Design*. Englewood Cliffs: Prentice Hall.

[Wooters & Zurek, 1982] Wooters, W.K., & Zurek, W.H. (1982). A Single Quantum Cannot Be Cloned. *Nature*, **299**, 802–3.

[Wright, 1998] Wright, H. (1998). Observe Digital Modualtion Through Diagrams. *Test and Measurement World*, 61–64.

[Yariv, 1987] Yariv, A. (1987). Operator Algebra for Propagation Problems involving Phase Conjugation and Nonreciprocal Elements. *Applied Optics*, **26**, 4538–40.

[Yariv, 1991] Yariv, A. (1991). *Optical Electronics*. 4th edn. Philadelphia: Saunders College Publishing.

[Yariv & Pepper, 1977] Yariv, A., & Pepper, D.M. (1977). Amplified Reflection, Phase Conjugation, and Oscillation in Degenerate Four-Wave Mixing. *Optics Letters*, **1**, 16–18.

[Ye *et al.*, 1999] Ye, J., Vernooy, D.W., & Kimble, H.J. (1999). Trapping of Single Atoms in Cavity QED. *Physical Review Letters*, **83**, 4987–90.

[Yoo *et al.*, 1989] Yoo, K.M., Takiguchi, Y., & Alfano, R.R. (1989). Dynamic Effect of Weak Localization on the Light Scattering from Random Media using Ultrafat Laser Technology. *Applied Optics*, **28**, 2343–9.

[Younis & Knight, 1993] Younis, S., & Knight, T. (1993). Practical Implementation of Charge Recovering Asymptotically Zero Power CMOS. Pages 234–50 of: *Proceeding of the 1993 Symposium on Integrated Systems*. Cambridge: MIT Press.

[Yourgrau *et al.*, 1982] Yourgrau, Wolfgang, van der Merwe, Alwyn, & Raw, Gough. (1982). *Treatise on Irreversible and Statistical Thermophysics: An Introduction to Nonclassical Thermodynamics*. New York: Dover.

[Zabusky, 1981] Zabusky, N.J. (1981). Computational Synergetics and Mathematical Innovation. *Journal of Computational Physics*, **43**, 195–249.

[Zimmerman, 1998] Zimmerman, NM. (1998). A Primer on Electrical Units in the Systeme International. *American Journal of Physics*, **66**, 324–31.

[Zurek, 1998] Zurek, W.H. (1998). Decoherence, Einselection and the Existential Interpretation (The Rough Guide). *Philosophical Transactions of the Royal Society of London Series A*, **356**, 1793–821.

Index

RETURN TO: PHYSICS-ASTRONOMY LIBRARY
351 LeConte Hall

| LOAN PERIOD 1 | 2 | 3 |
1-MONTH		
4	5	6

ALL BOOKS MAY BE RECALLED AFTER 7 DAYS
Books may be renewed by calling 510-642-3122

DUE AS STAMPED BELOW

FORM NO. DD 22
1M 7-11

UNIVERSITY OF CALIFORNIA, BERKELEY
Berkeley, California 94720–6000